SMOOTH HOMOGENEOUS STRUCTURES IN OPERATOR THEORY

CHAPMAN & HALL/CRC
Monographs and Surveys in Pure and Applied Mathematics

SMOOTH HOMOGENEOUS STRUCTURES IN OPERATOR THEORY

Daniel Beltiţă

CRC Press is an imprint of the
Taylor & Francis Group, an **informa** business
A CHAPMAN & HALL BOOK

CRC Press
Taylor & Francis Group
6000 Broken Sound Parkway NW, Suite 300
Boca Raton, FL 33487-2742

First issued in paperback 2019

ISBN-13: 978-1-58488-617-4 (hbk)
ISBN-13: 978-0-367-39189-8 (pbk)

Library of Congress Card Number 2005051891

Library of Congress Cataloging-in-Publication Data

Beltita, Daniel, 1971-
Smooth homogeneous structures in operator theory / by Daniel Beltita.
p. cm. -- (Monographs and surveys in pure and applied mathematics ; no. 137)
Includes bibliographical references and index.
ISBN 1-58488-617-X (alk. paper)
1. Operator theory. 2. Homogeneous spaces. I. Title. II. Chapman & Hall/CRC monographs and surveys in pure and applied mathematics ; 137.

QA329.B45 2005
515'.724--dc22 2005051891

Visit the Taylor & Francis Web site at
http://www.taylorandfrancis.com

and the CRC Press Web site at
http://www.crcpress.com

Contents

Preface

Ideas of functional analysis and differential geometry often stick together. It thus happens in various fields like complex analysis, the theory of partial differential equations, or in representation theory of Lie groups. In the present book we focus on the influence which certain techniques and ideas of differential geometry have upon operator theory and upon the theory of operator algebras. The literature devoted to this subject is quite extensive, as a look at the reference list of the present book will convince anyone. However, with a few notable exceptions, the corresponding literature consists only of journal articles. One of the exceptions to the rule is the exciting book by Pierre de la Harpe, *Classical Banach-Lie algebras and Banach-Lie groups of operators in Hilbert space* (Lecture Notes in Mathematics, vol. 285, Springer-Verlag, 1972). Another exception is the impressive monograph by Harald Upmeier, *Symmetric Banach manifolds and Jordan C^*-algebras* (North-Holland Mathematics Studies, 104. Notas de Matemática, 96, North-Holland Publishing Co., 1985).

In this new book on methods of differential geometry in operator theory we shall report on several recent developments in order to draw the reader's attention to this fruitful field of research. At the same time, we raise the problem of a systematic investigation of operator ideals from the point of view of Lie theory. And we suggest that this problem could be approached by means of an appropriate version of abstract reproducing kernels, which we call equivariant monotone operators.

We plan to introduce the reader to a circle of ideas where the infinite-dimensional Lie groups play a central role. With this objective in mind, we designed Part I of the book as an initiation to Lie theory in infinite dimensions, which could well be used (and we did use it) as a one-semester course for third year graduate students. Only some familiarity with functional analysis and calculus is assumed. Part II of the book concerns the geometry of homogeneous spaces, for instance unitary orbits of operators. In Part III we investigate orbits of certain unitary groups that are closely related to admissible pairs of operator ideals. We endow such unitary orbits with structures of homogeneous Kähler manifolds and formulate a number of questions that might lead to further developments along this line of research. The material included in the third part of the book claims its origins from the construction of the restricted Grassmann manifold in the theory of loop groups.

We hope this book will be useful to advanced graduate students who want to work on interesting open problems in functional analysis and operator theory. The book is also intended for professional functional analysts and operator theorists who are interested in symmetry groups of various structures and would like to get acquainted with the basic principles of Lie theory. Last but not least, the present text could prove useful to people interested in differential geometry, by showing the applicability of their methods in operator theory.

I am indebted to many people whose encouragement, questions, suggestions, or kindly sent reprints greatly influenced me while I was writing this book: I. Beltiţă, L. Beznea, J.E. Galé, A. Gheondea, H. Glöckner, L. Lempert, M. Martin, K.-H. Neeb, V.G. Pestov, B. Prunaru, T.S. Ratiu, M. Şabac, H. Upmeier; and finally G. Weiss. I would like to thank the people from CRC Press/Taylor & Francis for their help and patience, and particularly P. Board, C. Brannigan, E. Meany, J. Vakili, K. Craig, S. Nair, and the series editor R.G. Douglas.

I gratefully acknowledge the partial support coming from grant CNCSIS no. 1620/22.11.2002.

The author

Bucharest, May 2005

Introduction

Aim

The first aim of the present book is to provide an introduction to the theory of certain infinite-dimensional homogeneous spaces that show up in several of the classical settings of operator theory. Homogeneous spaces are sets transitively acted on by groups. That is, given a set X and a group G, we say that X is a *homogeneous space of* G if we have a group homomorphism $g \mapsto \alpha_g$ from G into the group of all bijections $X \to X$ such that the following transitivity condition is satisfied: for every two points $x, x' \in X$ we have $\alpha_g(x) = x'$ for some $g \in G$. In other words, homogeneous spaces are nothing more than orbits of group actions. It is under this disguise that the homogeneous spaces usually show up in operator theory. Thus unitary orbits of operators, functionals, representations, etc., are homogeneous spaces of the unitary groups under consideration. And what is most interesting is that these unitary orbits are often smooth manifolds (sometimes complex manifolds or symplectic manifolds or both) and carry differential geometric structures that encode a lot of operator theoretic information.

Many papers have previously been published which describe one of these manifolds or another (see the reference list at the end of the book). Besides providing an introduction to the ideas of differential geometry and Lie theory that constitute the background of these papers, we construct invariant complex structures on certain new classes of infinite-dimensional symplectic homogeneous manifolds, turning the latter ones into Kähler homogeneous manifolds in a suitable sense. The main illustration of this construction is provided by homogeneous spaces of Banach-Lie groups associated with certain admissible pairs of ideals of compact operators on Hilbert spaces.

The main tool of the aforementioned construction is the notion of equivariant monotone operator. Given a real Banach-Lie algebra $\mathfrak{g}$, an *equivariant monotone operator* is simply a bounded linear operator

$$\iota\colon \mathfrak{g}^{\#} \to \mathfrak{g}$$

defined on the topological dual $\mathfrak{g}^{\#}$ of $\mathfrak{g}$, which intertwines the coadjoint representation of $\mathfrak{g}$ and the adjoint one (equivariance), equals its own adjoint, and is monotone in the sense that

$$(\forall f \in \mathfrak{g}^{\#}) \qquad \langle f, \iota(f) \rangle \geq 0,$$

where $\langle \cdot, \cdot \rangle \colon \mathfrak{g}^{\#} \times \mathfrak{g} \to \mathbb{R}$ is the duality pairing. From the point of view of Lie theory, it is one of the main themes of the present book that, in some respects, the equivariant monotone operators play a crucial role in the theory of certain classes of infinite-dimensional Lie groups and their Lie algebras.

Motivation

There are several fields where the existence of invariant differential geometric structures on infinite-dimensional homogeneous spaces satisfying various additional conditions is a basic problem. We now mention briefly some of them, which motivated the present work, and where the results/methods developed in the present book are applicable.

1. *Representation Theory*: There is a recent growth of interest in extensions of the Bott-Borel-Weil Theorem to various classes of infinite-dimensional Lie groups (see e.g., [NRW01], [Ne04], [Ne01b]). And the geometric realization of Lie group representations involves complex homogeneous spaces of the group under consideration. From this point of view, the representation theory of the groups whose homogeneous spaces are studied here in Chapter 9 becomes interesting.

2. *Operator Theory*: A crucial role in the theory of Cowen-Douglas operators is played by (invariant) complex structures on flag manifolds in C^*-algebras (see [CD78], [MS97], [MS98] and the references therein). Since the methods of the present paper allow us to construct invariant complex structures on flag manifolds in arbitrary associative Banach algebras (in Chapter 6), the foundation is laid for an investigation of Cowen-Douglas classes of operators on Banach spaces.

3. *Theory of Operator Algebras*: Several sets naturally associated with an operator algebra have structures of smooth homogeneous spaces: the unitary orbits of conditional expectations, unitary orbits of representations and so on (see Chapter 4 in the present book for more examples). And one can exploit the differential geometric properties of the corresponding homogeneous spaces in order to distinguish various classes of operator algebras. Several interesting results were already obtained in that promising direction of investigation; see e.g., the papers [ACS95a] and [ArS01].

4. *Complex Geometry*: There is a rich literature on various classification problems for complex homogeneous spaces with invariant pseudo-Kähler structure (see e.g., [DN88], [DG92], [Gu02]). It seems a challenging task to extend some of the corresponding classification theorems to infinite dimensions. To this end, it might be worth taking into consideration the new classes of symplectic (respectively Kähler) manifolds constructed here in Chapters 6 and 9.

Some particular tools

Since the constructions carried out in Part III of this book might seem overly complicated, we note that this is actually due to the key role which the equivariant monotone operators play in the whole story. What we mean is that to handle the powerful instrument which those operators provide, one needs ideas and techniques belonging to various fields and it appears appropriate to single them out. We briefly comment on the main topics below.

(a) *Operator ranges*: We develop the needed machinery in Section 7.3. It combines ideas from the classical theory of operator ranges in complex Hilbert spaces (see e.g., [FW71]) and the abstract theory of reproducing kernel Hilbert spaces (see [Scw64], and also [GG04]). Roughly speaking, there is a tight connection between equivariant monotone operators and a special type of ideals ("L^*-ideals") of Banach-Lie algebras (Definition 7.12). These ideals are usually non-closed, but they are ranges of homomorphisms from L^*-algebras (whence the name of these ideals) into the Banach-Lie algebra we are concerned with. The basic observation is that one can have such ideals, and they play an important role, even in the case of topologically simple Banach-Lie algebras.

(b) *H^*-algebras* (cf. e.g., [CMMR94]): These are very close to the topic of operator ranges, since the equivariant monotone operators allow one to employ the L^*-algebras (which are nothing else than Lie H^*-algebras) in the investigation of more general Banach-Lie algebras. A place where their role is quite prominent is in Chapter 8, where they lead to an enlargibility criterion in terms of equivariant monotone operators (see Corollary 8.36). It is noteworthy that, to a remarkable extent, the properties we need for L^*-algebras actually hold for general H^*-algebras (see Sections 7.4 and 7.5).

(c) *Weyl functional calculus* (cf. [An69], [An70]): This is a rather technical topic, and we collect in Chapter 5 the facts we need. This chapter is fairly self-contained, although it is not meant as a systematic treatment of the topic of Weyl functional calculus and its generalizations. We just collected the results necessary in the present book, stating them appropriately. Our need for such techniques stems from the fact that in Chapters 6 and 9 we have to deal with tuples of hermitian operators, and need the concepts like joint spectrum, spectral subspace, and functional calculus for such tuples. For example, we study vector-valued symplectic forms in Chapter 9 and one can think of them as families of scalar-valued ones. In order to construct a complex polarization which works simultaneously for all symplectic forms in such a family, we have to consider a "non-positive spectral subspace" for a certain family of hermitian operators. That is why we develop here tools which replace the classical local spectral theory used in the case of single operators:

the latter theory suffices when one deals with scalar-valued symplectic forms (see [Be05a]), but it no longer suffices when one is concerned with the vector-valued ones.

More detailed description of contents

Part I as a whole provides an elementary introduction to the theory of infinite-dimensional Lie groups. No previous knowledge of Lie theory is assumed. The few notions we need from differential geometry are reviewed in Appendix A.

Chapter 1 exposes the basic ideas and notions on topological Lie algebras. We especially emphasize the Banach-Lie algebras as the most important class for the applications we have in mind. A fairly complete treatment of the Baker-Campbell-Hausdorff series is included by using the method of [Dj75].

In Chapter 2 we introduce the notions of Lie group and exponential map. In the first part of the chapter we deal with infinite-dimensional Lie groups modeled on locally convex spaces. We strove to explain in detail in Section 2.2 the fact that each Lie group is associated with a certain Lie algebra, which is critical in order to understand the structure of the Lie group under consideration. In the last part of the chapter we prove a number of basic properties of Banach-Lie groups (i.e., Lie groups modeled on Banach spaces). Among these properties we mention the fact that the exponential map provides a local chart, and the fact that Banach-Lie groups are real analytic (Theorem 2.42). In the proof of these properties we heavily lean on the results of Appendix B.

Chapter 3 provides an elementary introduction to the phenomena related to enlargibility of Lie algebras. That is, the question of when a Lie algebra comes from a Lie group. In the first section of this chapter we prove the basic theorem on integrating Lie algebra homomorphisms to Lie group homomorphisms (Theorem 3.5). In Section 3.2 we give a proof of the part we need from the celebrated theorem of N. Kuiper (see [Ku65]) that the group of unitary operators on an infinite-dimensional separable Hilbert space is contractible. In Section 3.3 we use the corresponding result to construct the Douady-Lazard example of a non-enlargible Banach-Lie algebra. The reason we include this discussion of enlargibility is that it is closely related to one of the basic notions introduced in this book (the equivariant monotone operators; see Corollary 8.36 for the precise statement in this connection).

Part II describes the differential geometric setting that turns out to be related to a number of quite interesting problems in operator theory. That setting is provided by the idea of homogeneous space, which is nothing more than a space that is transitively acted on by a certain group. In order to make sure that this structure can be described by differential geometric techniques, we naturally need to have a *smooth* homogeneous space acted on by a *Lie* group.

In the first part of Chapter 4, we introduce the notion of Banach-Lie subgroup and prove a number of basic results in this connection such as a theorem on algebraic groups of invertible elements of Banach algebras (cf. [HK77]). We then define the basic differential geometric structures on a homogeneous space that are investigated later on, e.g., various types of symplectic manifolds. They slightly extend usually encountered concepts (cf. e.g., [Ne04]) in the sense that we are dealing here with vector-valued symplectic forms. In the second part of this chapter we describe a number of quite interesting smooth homogeneous spaces that were discovered recently by various authors working in operator theory and operator algebras.

Chapter 5 is a technical one and develops several auxiliary facts used later. It is centered on the notion of quasimultiplicative map, which is an extension of the well-known Weyl functional calculus for finite families of self-adjoint operators (developed by R.F.V. Anderson in [An69] and [An70]). Essentially, we are working with operator-valued distributions on a finite-dimensional real vector space. The first part of the chapter is devoted to the study of a kind of spectral subspace (see Definition 5.5 and Corollary 5.11). One of the main results of Chapter 5 is Theorem 5.14 concerning separate parts of the support of a quasimultiplicative operator-valued distribution. We then prove other results to be used, notably, the connection between the supports of two quasimultiplicative maps which are intertwined in a natural way (see Proposition 5.18 and Proposition 5.22). We conclude Chapter 5 by important examples provided by hermitian maps which contain the Weyl functional calculus (Example 5.25) with its special situation of maps constructed by means of spectral measures (Example 5.26).

Our aim in Chapter 6 is to provide a general method to endow homogeneous spaces with invariant complex structures. It turns out that the quasimultiplicative maps studied in Chapter 6 constitute a very effective tool for exploiting an already known sufficient condition for the existence of complex structures (see Theorem 6.1 below). We record the corresponding results in Proposition 6.2, Corollary 6.3 and Proposition 6.4 and use them in the last part of Chapter 6, in order to construct certain suitable complex structures on flag manifolds in associative Banach algebras (see Theorem 6.18). We thus extend some results of M. Martin and N. Salinas (see [MS97] and [MS98]) from C^*-algebras to arbitrary associative Banach algebras. In Chapter 6 we also introduce the various types of pseudo-Kähler manifolds, which describe the interaction between the symplectic and complex structures on a manifold.

In Part III of the book we introduce the notion of equivariant monotone operators and explore the way this notion is involved in the local and global theory of large classes of Lie groups and their homogeneous spaces.

One of the most important points of Chapter 7 is Definition 7.1, where the equivariant monotone operators are introduced. In Remarks 7.5 and 7.6 we then discuss the existence of these operators in some simple special cases. In Section 7.2 we collect some necessary notions involving H^*- and L^*-algebras. Section 7.3 concerns a special kind of theorem of factorization through Hilbert spaces, based on a familiar procedure of construction of square roots of positive operators on Hilbert spaces. More precisely, if $\mathfrak{Y}$ is a real Banach space with the topological dual $\mathfrak{Y}^\#$ and $\iota\colon \mathfrak{Y}^\# \to \mathfrak{Y}$ is a monotone operator, then $\iota = \varphi \circ \varphi^\#$ for some operator $\varphi\colon \mathfrak{X} \to \mathfrak{Y}$, where $\mathfrak{X}$ is a real Hilbert space and $\operatorname{Ran}\varphi$ depends only upon ι (Proposition 7.23). We then describe some basic properties of the map $\iota \mapsto \operatorname{Ran}\varphi$ from monotone operators to subspaces of $\mathfrak{Y}$ which are ranges of operators from Hilbert spaces into $\mathfrak{Y}$. The results of Section 7.3 are more or less well-known (e.g., see the paper [Scw64]). We emphasize that the key steps and constructions carried out in the proofs of Section 7.3 are needed later on, particularly in Chapter 8. It is also worth mentioning that many results included in Section 7.3 claim their origins from the classical theory of operator ranges (see [FW71]) and from the abstract theory of reproducing kernels (see [Scw64]). In Section 7.4 we describe the H^*-ideals of real H^*-algebras (see Theorem 7.34). The motivation of this section is the following: it is one of the leading ideas of the present work that the equivariant monotone operators allow one to use the L^*-algebras in order to understand the structure of more general Banach-Lie algebras and this is done by means of their L^*-ideals (see Chapter 8). In this context, it is natural to wonder what happens when the object of the investigation is already an L^*-algebra. In other words, what about the L^*-ideals of an L^*-algebra? We answer this question in Section 7.4 in the more general/natural case of H^*-algebras. Section 7.5 is a very short one and concerns some basic properties of the set of H^*-ideals of an arbitrary involutive Banach algebra (not necessarily associative!). Namely, that set is closed under set intersection (Proposition 7.35) and it contains at most one element in the case of topologically simple algebras (Theorem 7.38).

Chapter 8 concerns the crucial correspondence

$$\text{equivariant monotone operators} \longleftrightarrow L^*\text{-ideals}$$

for a real involutive Banach-Lie algebra. The easy part of that correspondence is settled in Proposition 8.1, concerning the direction from ideals to operators. The reverse direction proves to be a much more difficult matter and our main result in this connection is contained in Theorem 8.8. It reads as follows:

> *Let $\mathfrak{g}$ be a real involutive Banach-Lie algebra and $\iota\colon \mathfrak{g}^\# \to \mathfrak{g}$ an equivariant monotone operator. Then there exists a real L^*-algebra $\mathfrak{X}$ and a continuous injective $*$-homomorphism $\varphi\colon \mathfrak{X} \to \mathfrak{g}$ such that $\iota = \varphi \circ \varphi^\#$ and $\operatorname{Ran}\varphi$ is an L^*-ideal of $\mathfrak{g}$ which depends only on ι and is denoted $\Phi(\iota)$.*

We then develop in Section 8.3 the basic properties of the correspondence $\iota \mapsto \Phi(\iota)$ between equivariant monotone operators and L^*-ideals of a real involutive Banach-Lie algebra. We call simply "adequate the Banach-Lie algebras which arise by means of that correspondence, i.e., the algebras for which that correspondence is surjective. It turns out that several important classes of algebras are adequate: the topologically simple ones, the L^*-ones, and the elliptic centerless ones (see Example 8.17, Example 8.18, and Corollary 8.27, respectively). The L^*-ideals of an adequate algebra constitute a lattice with respect to vector addition and set intersection (Proposition 8.19). Another interesting consequence of the aforementioned correspondence $\iota \mapsto \Phi(\iota)$ is that if a topologically simple involutive Banach-Lie algebra possesses injective equivariant monotone operators, then its automorphism group has a natural uniformly continuous irreducible representation on a real Hilbert space (Theorem 8.29). We conclude Chapter 8 by an enlargibility criterion for Banach-Lie algebras. It asserts that if $\mathfrak{g}$ is a real involutive Banach-Lie algebra possessing injective equivariant monotone operators, then $\mathfrak{g}$ is enlargible (i.e., $\mathfrak{g}$ is associated to some Banach-Lie group).

In Chapter 9 we construct complex polarizations for a class of real Banach-Lie algebras which we call pseudo-restricted algebras (see Definition 9.4), this terminology being inspired by the restricted groups and algebras occurring in the theory of loop groups (e.g., see Section 6.2 in [PS90]). It turns out that, under favorable circumstances, a pseudo-restricted algebra is naturally endowed with a vector-valued continuous 2-cocycle (Theorem 9.10). Our main aim in Chapter 9 is to construct complex polarizations in these cocycles (see Theorem 9.1) using the results of Chapter 8. We obtain in Section 9.3 a result concerning positive complex polarizations (Theorem 9.15), extending Theorem 2.10 in [Be05a]. Section 9.5 provides a quick review of some ideas we need from the theory of operator ideals. Moreover we introduce the notion of an admissible pair of ideals of $\mathcal{B}(\mathfrak{H})$ (Definition 9.22). In Section 9.7, we explore the global consequences of the results of Section 9.3 in a framework provided by certain ideals of compact operators on a Hilbert space. More precisely, we construct some new examples of infinite-dimensional symplectic homogeneous spaces (Theorem 9.29) and Kähler homogeneous spaces (Theorem 9.33 and Corollary 9.34).

Appendix A is a collection of miscellaneous auxiliary facts and fundamental definitions. In connection with the latter ones, a word of caution is in order. We decided to include definitions of certain very basic and well-known notions such as Banach algebras, inasmuch as we do *not* use them with the most widely used significance. For instance, our Banach algebras will be most often *non-associative*, unlike the usual significance of that term in operator theory. (See also Remark 7.9(a) for an analogous situation.) Appendix B includes a development of the differential equations that lead to the conclusion that in every Banach-Lie group the multiplication is locally expressed by the Baker-Campbell-Hausdorff series, which eventually leads to the fact that all Banach-Lie groups are real analytic. We collected in Appendix C a few basic results from the theory of topological groups that are needed in Part I of the book.

Lie theory

Chapter 1

Topological Lie Algebras

Abstract. Lie algebras play a critical role in the investigations on Lie groups. In the first part of this chapter we define the notion of topological Lie algebra and present some basic examples: associative algebras, the skew-adjoint operators on a Hilbert space, and the smooth vector fields on a finite-dimensional manifold. We then discuss in more detail the relationship between Lie algebras and associative algebras. In order to do this, we introduce the universal enveloping algebras and state some of their basic properties. The universal enveloping algebras also provide the algebraic foundation of the idea of exponential map. That is, the formal series of elements in certain enveloping algebras provide an algebraic setting where exponentials and logarithms can be defined. The central result of the corresponding construction is that, roughly speaking, if x and y are noncommuting variables, then $\log(\exp x \cdot \exp y) = x + y + \frac{1}{2}[x,y] + \frac{1}{12}[x,[x,y]] + \frac{1}{12}[y,[y,x]] + \cdots$, which is a formal series whose terms are linear combinations of iterated commutators of x and y. That series is called the Baker-Campbell-Hausdorff series. When x and y are replaced by elements of a Banach-Lie algebra, the series is convergent provided $\|x\|$ and $\|y\|$ are small enough.

1.1 Fundamentals

The goal of this section is to introduce the topological Lie algebras and to illustrate this notion by a number of examples. To begin with, we explain what a Lie algebra is.

DEFINITION 1.1 A *Lie algebra* over $\mathbb{K} \in \{\mathbb{R}, \mathbb{C}\}$ is a vector space $\mathfrak{g}$ over $\mathbb{K}$ equipped with a bilinear mapping

$$\mathfrak{g} \times \mathfrak{g} \to \mathfrak{g}, \quad (x,y) \mapsto [x,y],$$

such that for all $x, y, z \in \mathfrak{g}$ we have

$$[x, y] = -[y, x] \qquad \text{(anti-symmetry)}$$

and

$$[[x, y], z] + [[y, z], x] + [[z, x], y] = 0 \qquad \text{(Jacobi identity).}$$

In this case, the bilinear mapping $[\cdot, \cdot]$ is called a *Lie bracket* and we define the *adjoint representation* of $\mathfrak{g}$ by

$$\mathrm{ad}_{\mathfrak{g}} : \mathfrak{g} \to \mathrm{End}\,(\mathfrak{g}), \quad \mathrm{ad}_{\mathfrak{g}} x := [x, \cdot].$$

Note that the properties of the Lie bracket imply that

$$(\forall x, y \in \mathfrak{g}) \qquad \mathrm{ad}_{\mathfrak{g}}[x, y] = (\mathrm{ad}_{\mathfrak{g}} x)(\mathrm{ad}_{\mathfrak{g}} y) - (\mathrm{ad}_{\mathfrak{g}} y)(\mathrm{ad}_{\mathfrak{g}} x).$$

If $\mathfrak{h}$ is another Lie algebra over $\mathbb{K}$, then a *homomorphism of Lie algebras* from $\mathfrak{g}$ into $\mathfrak{h}$ is a $\mathbb{K}$-linear mapping $\psi \colon \mathfrak{g} \to \mathfrak{h}$ such that $\psi([x, y]) = [\psi(x), \psi(y)]$ whenever $x, y \in \mathfrak{g}$.

DEFINITION 1.2 A *topological Lie algebra* over $\mathbb{K} \in \{\mathbb{R}, \mathbb{C}\}$ is a topological vector space $\mathfrak{g}$ over $\mathbb{K}$ equipped with a continuous Lie bracket.

If the topological vector space $\mathfrak{g}$ is locally convex, Fréchet, Banach, or Hilbert, then we say that $\mathfrak{g}$ is a *locally convex*, *Fréchet-*, *Banach-*, or *Hilbert-Lie algebra*, respectively.

If $\mathfrak{g}$ and $\mathfrak{h}$ are two topological Lie algebras over $\mathbb{K}$, then a *homomorphism of topological Lie algebras* from $\mathfrak{g}$ into $\mathfrak{h}$ is a continuous mapping $\psi \colon \mathfrak{g} \to \mathfrak{h}$ which is moreover a homomorphism of Lie algebras over $\mathbb{K}$.

EXAMPLE 1.3 If $\mathcal{A}$ is an associative algebra over $\mathbb{K} \in \{\mathbb{R}, \mathbb{C}\}$, then we always can make $\mathcal{A}$ into a Lie algebra with the Lie bracket defined by

$$(\forall a, b \in \mathcal{A}) \qquad [a, b] := ab - ba.$$

Any subalgebra of that Lie algebra will be called a *Lie subalgebra of $\mathcal{A}$*.

If $\mathcal{A}$ is moreover a topological algebra, in the sense that it has a structure of topological vector space such that the multiplication mapping

$$\mathcal{A} \times \mathcal{A} \to \mathcal{A}, \quad (a, b) \mapsto ab,$$

is continuous, then the above defined Lie bracket makes $\mathcal{A}$ into a topological Lie algebra.

EXAMPLE 1.4 If $\mathcal{H}$ is a Hilbert space over $\mathbb{K} \in \{\mathbb{R}, \mathbb{C}\}$, then

$$\mathfrak{g} := \{T \in \mathcal{B}(\mathcal{H}) \mid T^* = -T\}$$

is a real Banach-Lie algebra with the norm topology and the Lie bracket defined by $[T, S] = TS - ST$ whenever $T, S \in \mathfrak{g}$. The only thing to be noted here is that for all $T, S \in \mathfrak{g}$ we have $[T, S] \in \mathfrak{g}$. Then the fact that $[\cdot, \cdot]$ is indeed a Lie bracket follows according to Example 1.3.

Topological Lie algebras of vector fields

Let M be a smooth manifold modeled on $\mathbb{R}^n$, with a family of local coordinate systems $\{\varphi_\alpha \colon V_\alpha \to M_\alpha\}_{\alpha\in A}$ as in Definition A3.3. For all $\alpha \in A$, K a compact subset of M_α and k a positive integer, define

$$p_{\alpha,K,k} \colon \mathfrak{V}(M) \to \mathbb{R}_+, \quad p_{\alpha,K,k}(v) := \sup\{\|(\widetilde{v}_\alpha)^{(j)}(x)\| \mid x \in K, 0 \le j \le k\},$$

where, for each $v \in \mathfrak{V}(M)$, we denote by $\widetilde{v}_\alpha \colon V_\alpha \to \mathbb{R}^n$ the smooth mapping defined by $(T(\varphi_\alpha^{-1}) \circ v \circ \varphi_\alpha)(\cdot) = (\cdot, \widetilde{v}_\alpha(\cdot))$.

Then each $p_{\alpha,K,k}$ is a seminorm on $\mathfrak{V}(M)$, and $0 \in \mathfrak{V}(M)$ is the only vector field $v \in \mathfrak{V}(M)$ with $p_{\alpha,K,k}(v) = 0$ for all α, K and k. Thus, the family of seminorms

$$\{p_{\alpha,K,k}\}_{\alpha\in A, k\ge 0, K \text{ compact } \subseteq U_\alpha}$$

defines a locally convex (Hausdorff) topology on the vector space $\mathfrak{V}(M)$. From now on, we always think of $\mathfrak{V}(M)$ equipped with this topology.

THEOREM 1.5 *The topology of $\mathfrak{V}(M)$ is complete and makes the Lie algebra $\mathfrak{V}(M)$ into a locally convex Lie algebra. If the topology of the manifold M is second countable, then $\mathfrak{V}(M)$ is in fact a Fréchet-Lie algebra.*

PROOF We use the notation preceding the statement. If for each open subset V of $\mathbb{R}^n$ we endow $\mathcal{C}^\infty(V, \mathbb{R}^n)$ with its natural Fréchet topology, then the definition of the topology of $\mathcal{V}(M)$ shows that we have an exact sequence of topological vector spaces and continuous linear mappings

$$0 \to \mathcal{V}(M) \xrightarrow{\iota} \prod_{\alpha\in A} \mathcal{C}^\infty(V_\alpha, \mathbb{R}^n) \xrightarrow{\varphi} \prod_{\alpha,\beta\in A} \mathcal{C}^\infty(V_\alpha \cap V_\beta, \mathbb{R}^n),$$

where

$$(\forall v \in \mathcal{V}(M)) \qquad \iota(v) = (\widetilde{v}_\alpha)_{\alpha\in A}$$

and

$$\varphi\big((g_\alpha)_{\alpha\in A}\big) = (g_\alpha|_{V_\alpha\cap V_\beta} - g_\beta|_{V_\alpha\cap V_\beta})_{\alpha,\beta\in A}$$

whenever $(g_\alpha)_{\alpha\in A} \in \prod\limits_{\alpha\in A} \mathcal{C}^\infty(V_\alpha, \mathbb{R}^n)$.

Moreover, ι is actually a topological embedding, hence $\mathcal{V}(M)$ is isomorphic (as a topological vector space) to $\operatorname{Ker}\varphi$. But $\operatorname{Ker}\varphi$ is complete as a closed subspace of a complete topological vector space. Thus $\mathcal{V}(M)$ is a complete topological vector space.

To see that the bracket of $\mathcal{V}(M)$ is continuous, note that for all $v, w \in \mathcal{V}(M)$ we have, according to formulas (A.2) and (A.3) in the proof of Lemma A.66 that

$$\iota([v,w]) = \big(d\widetilde{v}_\alpha(\cdot)\widetilde{w}_\alpha(\cdot) - d\widetilde{w}_\alpha(\cdot)\widetilde{v}_\alpha(\cdot)\big)_{\alpha\in A}.$$

Since we already noted that $v \mapsto \iota(v)$ is a topological embedding, this formula clearly shows that the bracket $[\cdot,\cdot]\colon \mathcal{V}(M) \times \mathcal{V}(M) \to \mathcal{V}(M)$ is continuous.

In the case when M is second countable, the index set A can be chosen countable. Since each $\mathcal{C}^\infty(V_\alpha, \mathbb{R}^n)$ is a Fréchet space, it then follows that $\mathcal{V}(M) \simeq \operatorname{Ker}\varphi$ is a closed subspace of the product of countably many Fréchet spaces, hence is in turn a Fréchet space. ∎

1.2 Universal enveloping algebras

We have seen in Example 1.3 that each associative algebra can be viewed as a Lie algebra. In the present section we shall have a closer look at the relationship between Lie algebras and associative algebras. To this end we need the tensor algebras introduced in the following definition.

DEFINITION 1.6 Let $\mathcal{V}$ be a vector space over $\mathbb{K} \in \{\mathbb{R}, \mathbb{C}\}$. Then the infinite direct sum

$$\begin{aligned}\mathcal{T}(\mathcal{V}) :=& \mathbb{K}\mathbf{1} \dotplus \mathcal{V} \dotplus (\mathcal{V}\otimes\mathcal{V}) \dotplus (\mathcal{V}\otimes\mathcal{V}\otimes\mathcal{V}) \dotplus \cdots \\ =& \Big\{\sum_{n=0}^{\infty} x_n \mid (\forall n \ge 0) \quad x_n \in \otimes^n\mathcal{V} := \underbrace{\mathcal{V}\otimes\cdots\otimes\mathcal{V}}_{n \text{ times}}; \\ & \qquad (\exists N \ge 1)(\forall n \ge N)\ x_n = 0\Big\}\end{aligned}$$

is a unital associative algebra over $\mathbb{K}$ called the *free associative algebra* over $\mathbb{K}$ generated by $\mathcal{V}$, or the *tensor algebra* of $\mathcal{V}$. The multiplication in this algebra is given by $\otimes$. However, for the sake of simplicity, we denote $x \otimes y =: xy$.

We will also consider the larger algebra of formal series

$$\widehat{\mathcal{T}}(\mathcal{V}) := \Big\{\sum_{n=0}^{\infty} x_n \mid (\forall n \ge 0) \quad x_n \in \otimes^n\mathcal{V}\Big\}.$$

This is a unital associative $\mathbb{K}$-algebra with the multiplication defined by

$$\Big(\sum_{n=0}^{\infty} x_n\Big)\cdot\Big(\sum_{n=0}^{\infty} y_n\Big) = \sum_{n=0}^{\infty}\Big(\sum_{k=0}^{n} x_k y_{n-k}\Big).$$

Note that for every $n \ge 0$ we have $\sum\limits_{k=0}^{n} x_k y_{n-k} \in \otimes^n\mathcal{V}$.

We also note that

$$\widehat{\mathcal{T}}_0(\mathcal{V}) := \Big\{\sum_{n=0}^{\infty} x_n \in \mathcal{T}(\mathcal{V}) \mid x_0 = 0\Big\}$$

is a two-sided ideal of $\widehat{\mathcal{T}}(\mathcal{V})$.

REMARK 1.7 In the setting of Definition 1.6 it is not hard to see that $\mathbf{1} + \widehat{\mathcal{T}}_0(\mathcal{V})$ is a multiplicative group.

REMARK 1.8 In the setting of Definition 1.6, the free associative algebra generated by $\mathcal{V}$ has the following universality property: For every $\mathbb{K}$-linear mapping $\theta\colon \mathcal{V} \to \mathcal{A}$ of $\mathcal{V}$ into a unital associative algebra $\mathcal{A}$ over $\mathbb{K}$ there exists a unique homomorphism $\bar{\theta}\colon \mathcal{T}(\mathcal{V}) \to \mathcal{A}$ of unital $\mathbb{K}$-algebras such that $\bar{\theta}|_{\mathcal{V}} = \theta$.

Now we are ready to introduce the fundamental notion of universal enveloping algebra of a Lie algebra. The corresponding universality property will be noted in Remark 1.10.

DEFINITION 1.9 Let $\mathfrak{g}$ be a Lie algebra over $\mathbb{K} \in \{\mathbb{R}, \mathbb{C}\}$. Denote by $\mathcal{I}(\mathfrak{g})$ the two-sided ideal generated by the subset

$$\{xy - yx - [x, y] \mid x, y \in \mathfrak{g}\}$$

in the tensor algebra $\mathcal{T}(\mathfrak{g})$. Then the quotient

$$U(\mathfrak{g}) := \mathcal{T}(\mathfrak{g})/\mathcal{I}(\mathfrak{g})$$

has a natural structure of unital associative $\mathbb{K}$-algebra, and it is called the *universal enveloping algebra* of $\mathfrak{g}$. The mapping

$$\iota\colon \mathfrak{g} \to U(\mathfrak{g}), \quad \iota(x) = x + \mathcal{I}(\mathfrak{g}),$$

is called the *natural embedding* of $\mathfrak{g}$ into $U(\mathfrak{g})$. It easily follows by the definition of the ideal $\mathcal{I}(\mathfrak{g})$ that

$$\iota([x, y]) = \iota(x)\iota(y) - \iota(y)\iota(x)$$

for all $x, y \in \mathfrak{g}$.

REMARK 1.10 In the setting of Definition 1.9, the pair $\big(U(\mathfrak{g}), \iota\big)$ has the following universality property: If $\mathcal{A}$ is a unital associative $\mathbb{K}$-algebra and $\theta\colon \mathfrak{g} \to \mathcal{A}$ is a $\mathbb{K}$-linear mapping such that

$$(\forall x, y \in \mathfrak{g}) \quad \theta([x, y]) = \theta(x)\theta(y) - \theta(y)\theta(x),$$

then there exists a unique homomorphism of unital $\mathbb{K}$-algebras $\bar{\theta}\colon U(\mathfrak{g}) \to \mathcal{A}$ such that $\bar{\theta} \circ \iota = \theta$.

We now state one of the most basic property of universal enveloping algebras (the Poincaré-Birkhoff-Witt theorem).

THEOREM 1.11 *Let $\mathfrak{g}$ be a Lie algebra over $\mathbb{K} \in \{\mathbb{R}, \mathbb{C}\}$ and choose a basis $\{a_i\}_{i \in I}$ in $\mathfrak{g}$. Assume that the index set I is totally ordered, and denote*

$$S := \{\emptyset\} \cup \{(i_1, \ldots, i_n) \in I^n \mid n \geq 1, i_1 \leq \cdots \leq i_n\},$$

and then define

$$\nu\colon S \to U(\mathfrak{g})$$

such that $\nu(\emptyset) = \mathbf{1}$ and $\nu(i_1, \ldots, i_n) = \iota(a_{i_1}) \cdots \iota(a_{i_n})$ whenever $(i_1, \ldots, i_n) \in S$, where $\iota\colon \mathfrak{g} \to U(\mathfrak{g})$ is the natural embedding. Then ν is a bijection of S onto a basis of $U(\mathfrak{g})$.

PROOF See e.g., Theorem 1.1 in Chapter X in [Ho65]. $\square$

REMARK 1.12 Theorem 1.11 says in particular that $\{\iota(a_i) \mid i \in I\}$ is a linearly independent subset of $U(\mathfrak{g})$, hence the linear mapping $\iota\colon \mathfrak{g} \to U(\mathfrak{g})$ is always injective. Since $\iota([x, y]) = \iota(x)\iota(y) - \iota(y)\iota(x)$ for all $x, y \in \mathfrak{g}$, it thus follows that ι is an isomorphism of Lie algebras from $\mathfrak{g}$ onto the Lie subalgebra $\iota(\mathfrak{g})$ of $U(\mathfrak{g})$. That is why we shall perform the identification $\mathfrak{g} \simeq \iota(\mathfrak{g}) \subseteq U(\mathfrak{g})$, thus thinking of $\mathfrak{g}$ as a Lie subalgebra of $U(\mathfrak{g})$.

To conclude this section, we mention another connection between Lie algebras and associative algebras. It is stronger than Theorem 1.11 in the sense that it provides embeddings into *finite-dimensional* unital associative algebras. However, this theorem applies only to finite-dimensional Lie algebras. (It is called the Ado theorem).

THEOREM 1.13 *Let $\mathfrak{g}$ be a finite-dimensional Lie algebra over $\mathbb{K} \in \{\mathbb{R}, \mathbb{C}\}$. Then there exist a finite-dimensional vector space $\mathfrak{X}$ over $\mathbb{K}$ and an injective linear map*

$$\rho\colon \mathfrak{g} \to \mathrm{End}\,(\mathfrak{X})$$

such that

$$\rho([a, b]) = [\rho(a), \rho(b)] = \rho(a)\rho(b) - \rho(b)\rho(a)$$

for all $a, b \in \mathfrak{g}$.

PROOF See e.g., Theorem 2 in §7 in [Bo71b]. $\square$

1.3 The Baker-Campbell-Hausdorff series

In this section we obtain one of the most important results of the present chapter (see Corollary 1.25), concerning the way the formal expression

$$\log(\exp x \cdot \exp y)$$

can be written as a formal Baker-Campbell-Hausdorff series of iterated commutators of the variables x and y. To this end we shall need formal series belonging to certain completions of universal enveloping algebras of the free Lie algebras. As we shall see a little bit later (in Proposition 1.17 below), the tensor algebra of any vector space is the universal enveloping algebra of a free Lie algebra, which we introduce in the following definition.

DEFINITION 1.14 Let $\mathcal{V}$ be a vector space over $\mathbb{K} \in \{\mathbb{R}, \mathbb{C}\}$ with the tensor algebra $\mathcal{T}(\mathcal{V})$. We denote by $\mathcal{L}(\mathcal{V})$ the Lie subalgebra of $\mathcal{T}(\mathcal{V})$ generated by $\mathcal{V}$, and call it the *free Lie algebra* generated by $\mathcal{V}$.

We shall consider also the following subset of $\widehat{\mathcal{T}}(\mathcal{V})$:

$$\widehat{\mathcal{L}}(\mathcal{V}) := \Big\{\sum_{n=0}^{\infty} x_n \in \widehat{\mathcal{T}}(\mathcal{V}) \mid (\forall n \geq 0) \quad x_n \in \mathcal{L}(\mathcal{V}) \cap \otimes^n \mathcal{V}\Big\}.$$

It is clear that $\mathcal{L}(\mathcal{V}) \subseteq \widehat{\mathcal{T}}_0(\mathcal{V})$, and this implies that $\widehat{\mathcal{L}}(\mathcal{V}) \subseteq \widehat{\mathcal{T}}_0(\mathcal{V})$ as well. Note that for $x = \sum_{n=0}^{\infty} x_n, y = \sum_{n=0}^{\infty} y_n \in \widehat{\mathcal{L}}(\mathcal{V})$ we have $xy - yx = \sum_{n=0}^{\infty} z_n$, where $z_n := \sum_{k=0}^{n} (x_k y_{n-k} - y_{n-k} x_k) \in \mathcal{L}(\mathcal{V}) \cap \otimes^n \mathcal{V}$ for all $n \geq 0$, hence $xy - yx \in \widehat{\mathcal{L}}(\mathcal{V})$. Thus $\widehat{\mathcal{L}}(\mathcal{V})$ is a Lie subalgebra of $\widehat{\mathcal{T}}(\mathcal{V})$.

PROPOSITION 1.15 *Let $\mathcal{V}$ be a vector space over $\mathbb{K} \in \{\mathbb{R}, \mathbb{C}\}$ and $x = \sum_{n=0}^{\infty} x_n \in \mathcal{T}(\mathcal{V})$. Then $x \in \mathcal{L}(\mathcal{V})$ if and only if $x_n \in \mathcal{L}(\mathcal{V}) \cap \otimes^n \mathcal{V}$ for $n = 0, 1, 2, \ldots$ If this is the case and $\{v_i\}_{i \in I}$ is a basis of $\mathcal{V}$, then each x_n belongs to the linear subspace of $\otimes^n \mathcal{V}$ generated by $\{[v_{i_1}, \ldots, [v_{i_{n-1}}, v_{i_n}] \ldots] \mid i_1, \ldots, i_n \in I\}$.*

PROOF Denote by $\widetilde{\mathcal{L}}$ the set of all linear combinations of elements of the form

$$[v_1 \ldots v_m] := [v_1, \ldots, [v_{m-1}, v_m] \ldots] \in \mathcal{L}(\mathcal{V}) \cap \otimes^m \mathcal{V}$$

with $v_1, \ldots, v_m \in \mathcal{V}$ and $m \geq 1$ arbitrary. In particular we have $\mathcal{V} \subseteq \widetilde{\mathcal{L}} \subseteq \mathcal{L}(\mathcal{V})$.

On the other hand, for $v := [v_1 \dots v_m]$, $v' := [v_2 \dots v_m]$ and $w := [w_1 \dots w_l]$ we have $[v, w] = [[v_1, v'], w] = -[[v', w], v_1] - [[w, v_1], v']$, whence

$$[v, w] = [v_1, [v', w]] - [v', [v_1 w_1 \dots w_l]],$$

and this formula allows us to prove by induction on m that $[v, w]$ is a linear combination of elements of the form $[u_1 \dots u_k]$ with $u_1, \dots, u_k \in \mathcal{V}$. In other words, $\widetilde{\mathcal{L}}$ is a Lie subalgebra of $\mathcal{T}(\mathcal{V})$. Since $\mathcal{V} \subseteq \widetilde{\mathcal{L}} \subseteq \mathcal{L}(\mathcal{V})$ and $\mathcal{L}(\mathcal{V})$ is the Lie subalgebra of $\mathcal{T}(\mathcal{V})$ generated by $\mathcal{V}$, it then follows that $\widetilde{\mathcal{L}} = \mathcal{L}(\mathcal{V})$.

In view of the definition of $\widetilde{\mathcal{L}}$, we then deduce that for every $x \in \mathcal{L}(\mathcal{V})$ there exists $x_n \in \mathcal{L}(\mathcal{V}) \cap \otimes^n \mathcal{V}$ for $n = 0, 1, 2, \dots$ such that $x = \sum\limits_{n=0}^{\infty} x_n$ and $x_n = 0$ for n large enough. Since each $x \in \mathcal{T}(\mathcal{V})$ can be uniquely written as $x = \sum\limits_{n=0}^{\infty} x_n$ with $x_n \in \otimes^n \mathcal{V}$ for all $n \geq 0$, the desired conclusion follows. ∎

It will be convenient to have at hand the mapping introduced in the following definition.

DEFINITION 1.16 Let $\mathcal{V}$ be a vector space over $\mathbb{K} \in \{\mathbb{R}, \mathbb{C}\}$ and $\{v_i\}_{i \in I}$ a basis of $\mathcal{V}$. Then $\{\mathbf{1}\} \cup \{v_{i_1} \cdots v_{i_m} \mid i_1, \dots, i_m \in I; m \geq 1\}$ is a basis of $\mathcal{T}(\mathcal{V})$, so that there exists a unique $\mathbb{K}$-linear mapping

$$\pi_{\mathcal{V}} \colon \mathcal{T}(\mathcal{V}) \to \mathcal{T}(\mathcal{V})$$

with $\pi_{\mathcal{V}}(\mathbf{1}) = 0$, $\pi_{\mathcal{V}}|_{\mathcal{V}} = \mathrm{id}_{\mathcal{V}}$, and $\pi_{\mathcal{V}}(v_{i_1} \cdots v_{i_m}) = [v_{i_1}, \dots, [v_{i_{m-1}}, v_{i_m}] \dots]$ whenever $i_1, \dots, i_m \in I$ and $m \geq 2$. Then it is clear that for all $m \geq 2$ and $u_1, \dots, u_m \in \mathcal{V}$ we have

$$\pi_{\mathcal{V}}(u_1 \cdots u_m) = [u_1, \dots, [u_{m-1}, u_m] \dots],$$

and this remark shows that $\pi_{\mathcal{V}}$ does not depend on the choice of the basis $\{v_i\}_{i \in I}$ of $\mathcal{V}$. Moreover, we have

$$\mathrm{Ran}\, \pi_{\mathcal{V}} = \mathcal{L}(\mathcal{V})$$

according to Proposition 1.15.

Now we are able to describe the universal enveloping algebra of a free Lie algebra.

PROPOSITION 1.17 *Let $\mathcal{V}$ be a vector space over $\mathbb{K} \in \{\mathbb{R}, \mathbb{C}\}$.*

(i) *For every Lie algebra $\mathfrak{h}$ over $\mathbb{K}$ and for each $\mathbb{K}$-linear map $\theta \colon \mathcal{V} \to \mathfrak{h}$ there exists a unique Lie algebra homomorphism $\bar{\theta} \colon \mathcal{L}(\mathcal{V}) \to \mathfrak{h}$ such that $\bar{\theta}|_{\mathcal{V}} = \theta$.*

(ii) *We have an isomorphism of unital $\mathbb{K}$-algebras $\mathcal{T}(\mathcal{V}) \simeq U(\mathcal{L}(\mathcal{V}))$.*

PROOF (i) We have $\mathfrak{h} \hookrightarrow U(\mathfrak{H})$ (see Remark 1.12), hence we can write $\theta\colon \mathcal{V} \to U(\mathfrak{h})$. Thus, according to Remark 1.8, there exists a unique homomorphism of unital $\mathbb{K}$-algebras $\widetilde{\theta}\colon \mathcal{T}(\mathcal{V}) \to U(\mathfrak{h})$ such that $\widetilde{\theta}|_{\mathcal{V}} = \theta$. Then the map

$$\bar{\theta} := \widetilde{\theta}|_{\mathcal{L}(\mathcal{V})}\colon \mathcal{L}(\mathcal{V}) \to U(\mathfrak{h})$$

is a Lie algebra homomorphism. Moreover, since $\bar{\theta}(\mathcal{V}) = \theta(\mathcal{V}) \subseteq \mathfrak{h}$, $\mathfrak{h}$ is a Lie algebra and $\mathcal{L}(\mathcal{V})$ is the Lie subalgebra of $\mathcal{T}(\mathcal{V})$ generated by $\mathcal{V}$, it follows that actually $\operatorname{Ran}\bar{\theta} \subseteq \mathfrak{h}$.

Also note that $\bar{\theta}$ is uniquely determined since $\mathcal{V}$ generates $\mathcal{L}(\mathcal{V})$.

(ii) Let A be a unital associative algebra over $\mathbb{K}$ and $\theta\colon \mathcal{L}(\mathcal{V}) \to A$ a homomorphism of Lie algebras. It follows by Remark 1.8 that there exists a unique homomorphism of unital algebras $\widetilde{\theta}\colon \mathcal{T}(\mathcal{V}) \to A$ such that $\widetilde{\theta}|_{\mathcal{V}} = \theta|_{\mathcal{V}}$. Then both mappings $\theta, \widetilde{\theta}|_{\mathcal{L}(\mathcal{V})}\colon \mathcal{L}(\mathcal{V}) \to A$ are Lie algebra homomorphisms that coincide on $\mathcal{V}$. Since $\mathcal{V}$ generates $\mathcal{L}(\mathcal{V})$, it follows that $\theta = \widetilde{\theta}|_{\mathcal{V}}$.

Now use this fact for $A = U(\mathcal{L}(\mathcal{V}))$ and $\theta\colon \mathcal{L}(\mathcal{V}) \hookrightarrow U(\mathcal{L}(\mathcal{V}))$ (see Remark 1.12) to get a (uniquely determined) homomorphism of unital algebras $\widetilde{\theta}\colon \mathcal{T}(\mathcal{V}) \to U(\mathcal{L}(\mathcal{V}))$ such that $\widetilde{\theta}|_{\mathcal{L}(\mathcal{V})} = \mathrm{id}_{\mathcal{L}(\mathcal{V})}$. Theorem 1.11 implies that $\widetilde{\theta}$ maps some basis of $\mathcal{T}(\mathcal{V})$ onto a basis of $U(\mathcal{L}(\mathcal{V}))$. Hence $\widetilde{\theta}$ is bijective, and thus we get an isomorphism of unital algebras $\mathcal{T}(\mathcal{V}) \simeq U(\mathcal{L}(\mathcal{V}))$. ∎

Our next aim is to prove the Specht-Wever theorem (Theorem 1.19 below) which gives a useful characterization of elements of free Lie algebras within the corresponding universal enveloping algebras. To this end we need the following auxiliary fact.

LEMMA 1.18 *If $\mathcal{V}$ is a vector space over $\mathbb{K} \in \{\mathbb{R}, \mathbb{C}\}$, then the linear mapping*

$$\pi_{\mathcal{V}}|_{\mathcal{L}(\mathcal{V})}\colon \mathcal{L}(\mathcal{V}) \to \mathcal{L}(\mathcal{V})$$

is a derivation of the Lie algebra $\mathcal{L}(\mathcal{V})$.

PROOF Denote $\mathfrak{g} = \mathcal{L}(\mathcal{V})$ and then consider the adjoint representation $\mathrm{ad}_{\mathfrak{g}}\colon \mathfrak{g} \to \mathrm{End}\,(\mathfrak{g})$, which is a Lie algebra homomorphism (see Definition 1.1 and Example 1.3). We have $U(\mathfrak{g}) = \mathcal{T}(\mathcal{V})$ by Proposition 1.17, hence Remark 1.10 shows that there exists a unique homomorphism of unital algebras $\theta\colon \mathcal{T}(\mathcal{V}) \to \mathrm{End}\,(\mathfrak{g})$ such that $\theta|_{\mathfrak{g}} = \mathrm{ad}_{\mathfrak{g}}$.

Now let $x_1, \dots, x_m, y_1, \dots, y_s \in \mathcal{V}$ arbitrary and $u := x_1 \cdots x_m \in \otimes^m \mathcal{V}$, $v := y_1 \cdots y_s \in \otimes^s \mathcal{V}$. Then we have

$$\begin{aligned}\pi_{\mathcal{V}}(uv) &= [x_1, [x_2, \dots, [x_m, [y_1, \dots, [y_{s-1}, y_s] \dots] \\ &= (\mathrm{ad}_{\mathfrak{g}} x_1) \cdots (\mathrm{ad}_{\mathfrak{g}} x_m) \pi_{\mathcal{V}}(v) = \theta(u) \pi_{\mathcal{V}}(v).\end{aligned}$$

It then follows that for all $u, v \in \mathfrak{g} = \mathcal{L}(\mathcal{V})$ we have

$$\pi_{\mathcal{V}}([u,v]) = \pi_{\mathcal{V}}(uv) - \pi_{\mathcal{V}}(vu) = \theta(u)\pi_{\mathcal{V}}(v) - \theta(v)\pi_{\mathcal{V}}(u) = [u, \pi_{\mathcal{V}}(v)] + [\pi_{\mathcal{V}}(u), v].$$

(To prove the latter equality, note that $\theta(u) = \mathrm{ad}_{\mathfrak{g}} u$ and $\theta(v) = \mathrm{ad}_{\mathfrak{g}} v$ since $u, v \in \mathfrak{g}$.) ∎

Now we are ready to prove the Specht-Wever theorem.

THEOREM 1.19 *Let $\mathcal{V}$ be a vector space over $\mathbb{K} \in \{\mathbb{R}, \mathbb{C}\}$, $n \geq 1$ and $x \in \otimes^n \mathcal{V} \subseteq \mathcal{T}(\mathcal{V})$. Then $x \in \mathcal{L}(\mathcal{V})$ if and only if $\pi_{\mathcal{V}}(x) = nx$.*

PROOF We have already noted in Definition 1.16 that $\mathrm{Ran}\, \pi_{\mathcal{V}} = \mathcal{L}(\mathcal{V})$ hence, if $\pi_{\mathcal{V}}(x) = nx$, then $x = \frac{1}{n}\pi_{\mathcal{V}}(x) \in \mathcal{L}(\mathcal{V})$.

Conversely, let $x \in \mathcal{L}(\mathcal{V}) \cap \otimes^n \mathcal{V}$. It follows by Proposition 1.15 that x is a linear combination of elements of the form $[v_1, \dots, [v_{n-1}, v_n] \dots]$ with $v_1, \dots, v_n \in \mathcal{V}$, hence it suffices to prove that $\pi_{\mathcal{V}}(x) = nx$ provided $x = [v_1, \dots, [v_{n-1}, v_n] \dots]$ with $v_1, \dots, v_n \in \mathcal{V}$. We prove this fact by induction on n. For $n = 1$ it holds since $\pi_{\mathcal{V}}|_{\mathcal{V}} = \mathrm{id}|_{\mathcal{V}}$. In the general case, denoting $x' = [v_2, \dots, [v_{n-1}, v_n] \dots]$, we have by Lemma 1.18

$$\pi_{\mathcal{V}}(x) = \pi_{\mathcal{V}}([v_1, x']) = [\pi_{\mathcal{V}}(v_1), x'] + [v_1, \pi_{\mathcal{V}}(x')] = [v_1, x'] + (n-1)[v_1, x'] = nx,$$

and the proof is finished. ∎

In the next statement we build a sort of completion of a tensor algebra.

PROPOSITION 1.20 *Let $\mathcal{V}$ be a vector space over $\mathbb{K} \in \{\mathbb{R}, \mathbb{C}\}$, $\mathfrak{h}$ a Banach-Lie algebra over $\mathbb{K}$ and $\theta \colon \mathcal{V} \to \mathfrak{h}$ a linear map. Consider the corresponding homomorphism of Lie algebras $\bar{\theta} \colon \mathcal{L}(\mathcal{V}) \to \mathfrak{h}$ and denote*

$$\overline{\mathcal{L}}^{\theta}(\mathcal{V}) = \Big\{ x = \sum_{n=0}^{\infty} x_n \in \widehat{\mathcal{L}}(\mathcal{V}) \mid \sum_{n=0}^{\infty} \|\bar{\theta}(x_n)\| < \infty \Big\}.$$

Then $\overline{\mathcal{L}}^{\theta}(\mathcal{V})$ is a subalgebra of the Lie algebra $\widehat{\mathcal{L}}(\mathcal{V})$, we have $\mathcal{L}(\mathcal{V}) \subseteq \overline{\mathcal{L}}^{\theta}(\mathcal{V})$, and the map

$$\widetilde{\theta} \colon \overline{\mathcal{L}}^{\theta}(\mathcal{V}) \to \mathfrak{h}, \qquad \widetilde{\theta}\Big(\sum_{n=0}^{\infty} x_n\Big) = \sum_{n=0}^{\infty} \bar{\theta}(x_n),$$

is a homomorphism of Lie algebras such that $\widetilde{\theta}|_{\mathcal{L}(\mathcal{V})} = \bar{\theta}$.

PROOF Let $M > 0$ such that $\|[a,b]\| \leq M \|a\| \cdot \|b\|$ for all $a, b \in \mathfrak{h}$. Consider $x = \sum\limits_{n=0}^{\infty} x_n, y = \sum\limits_{n=0}^{\infty} y_n \in \overline{\mathcal{L}}^{\theta}(\mathcal{V})$ and denote $z := [x, y] = xy - yx = \sum\limits_{n=0}^{\infty} z_n \in$

$\widehat{\mathcal{L}}(\mathcal{V})$. Then for all $n \geq 0$ we have $z_n = \sum\limits_{k=0}^{n} (x_k y_{n-k} - y_{n-k} x_k) = \sum\limits_{k=0}^{n} [x_k, y_{n-k}]$, hence

$$\begin{aligned}
\sum_{n=0}^{\infty} \|\bar{\theta}(z_n)\| &= \sum_{n=0}^{\infty} \| \sum_{k=0}^{n} [\bar{\theta}(x_k), \bar{\theta}(y_{n-k})] \| \\
&\leq M \sum_{n=0}^{\infty} \sum_{k=0}^{n} \|\bar{\theta}(x_k)\| \cdot \|\bar{\theta}(y_{n-k})\| \\
&= M \Big(\sum_{k=0}^{\infty} \|\bar{\theta}(x_k)\| \Big) \cdot \Big(\sum_{j=0}^{\infty} \|\bar{\theta}(y_j)\| \Big) < \infty.
\end{aligned}$$

Consequently $z = [x, y] \in \overline{\mathcal{L}}^{\theta}(\mathcal{V})$. Moreover,

$$\begin{aligned}
\widetilde{\theta}(z) &= \sum_{n=0}^{\infty} \bar{\theta}(z_n) = \sum_{n=0}^{\infty} \sum_{k=0}^{n} [\bar{\theta}(x_k), \bar{\theta}(y_{n-k})] = [\sum_{k=0}^{\infty} \bar{\theta}(x_k), \sum_{j=0}^{\infty} \bar{\theta}(y_j)] \\
&= [\widetilde{\theta}(x), \widetilde{\theta}(y)],
\end{aligned}$$

where the terms can be permuted since all of these series are absolutely convergent.

It is clear that $\overline{\mathcal{L}}^{\theta}(\mathcal{V})$ is a $\mathbb{K}$-linear subspace of $\widehat{\mathcal{L}}(\mathcal{V})$ with $\mathcal{L}(\mathcal{V}) \subseteq \overline{\mathcal{L}}^{\theta}(\mathcal{V})$ and $\widetilde{\theta}|_{\mathcal{L}(\mathcal{V})} = \bar{\theta}$, hence the proof is complete. ∎

The following definition describes the setting we need in order to deal with exponentials and logarithms on a purely algebraic level.

DEFINITION 1.21 Let $\mathcal{V}$ be a vector space over $\mathbb{K} \in \{\mathbb{R}, \mathbb{C}\}$. For every $x \in \widehat{\mathcal{T}}_0(\mathcal{V})$ we define

$$\exp x = \sum_{n=0}^{\infty} \frac{1}{n!} x^n \in \mathbf{1} + \widehat{\mathcal{T}}_0(\mathcal{V}) \subseteq \mathcal{T}(\mathcal{V}),$$

while for $y \in \mathbf{1} + \widehat{\mathcal{T}}_0(\mathcal{V})$ we denote

$$\log y = \sum_{n=1}^{\infty} \frac{(-1)^{n-1}}{n} (y - 1)^n \in \widehat{\mathcal{T}}_0(\mathcal{V}).$$

Moreover, we define

$$a \star b = \log(\exp a \cdot \exp b) \in \widehat{\mathcal{T}}_0(\mathcal{V})$$

whenever $a, b \in \widehat{\mathcal{T}}_0(\mathcal{V})$.

REMARK 1.22 For every vector space $\mathcal{V}$ over $\mathbb{K} \in \{\mathbb{R}, \mathbb{C}\}$, the mappings

$$\exp\colon \widehat{\mathcal{T}}_0(\mathcal{V}) \to \mathbf{1} + \widehat{\mathcal{T}}_0(\mathcal{V}) \text{ and } \log\colon \mathbf{1} + \widehat{\mathcal{T}}_0(\mathcal{V}) \to \widehat{\mathcal{T}}_0(\mathcal{V})$$

are bijections inverse to one another.

PROPOSITION 1.23 *Let $\mathcal{V}$ be a vector space over $\mathbb{K} \in \{\mathbb{R}, \mathbb{C}\}$. Then $\widehat{\mathcal{T}}_0(\mathcal{V})$ be a group with respect to the composition law $\star$. The identity element of this group is 0, and the inverse of any $x \in \widehat{\mathcal{T}}_0(\mathcal{V})$ is $-x$.*

PROOF Since $\mathbf{1} + \widehat{\mathcal{T}}_0(\mathcal{V})$ is a multiplicative group and the bijection

$$\exp\colon \widehat{\mathcal{T}}_0(\mathcal{V}) \to \mathbf{1} + \widehat{\mathcal{T}}_0(\mathcal{V})$$

has the property

$$(\forall a, b \in \widehat{\mathcal{T}}_0(\mathcal{V})) \quad \exp(a \star b) = \exp a \cdot \exp b,$$

it clearly follows that $(\widehat{\mathcal{T}}_0(\mathcal{V}), \star)$ is a group, isomorphic to $(\mathbf{1} + \widehat{\mathcal{T}}_0(\mathcal{V}), \cdot)$ by means of the bijection exp. The other assertions then follow since we have $\exp 0 = \mathbf{1}$ and $\big(\exp x\big) \cdot \big(\exp(-x)\big) = \mathbf{1}$ for all $x \in \widehat{\mathcal{T}}_0(\mathcal{V})$. □

Now we come to a very deep property of the Lie algebra $\widehat{\mathcal{L}}(\mathcal{V})$ introduced in Definition 1.14. This property is very close to the central principle of Lie theory, which predicts a close relationship between Lie algebras and groups (see Chapters 2 and 3).

THEOREM 1.24 *Let $\mathcal{V}$ be a vector space over $\mathbb{K} \in \{\mathbb{R}, \mathbb{C}\}$. Then $\widehat{\mathcal{L}}(\mathcal{V})$ is a subgroup of the group $(\widehat{\mathcal{T}}_0(\mathcal{V}), \star)$.* ■

PROOF The proof has several stages.

1° Denote $\mathcal{R} := \widehat{\mathcal{T}}(\mathcal{V})$,

$$A := \mathcal{R}[[t]] := \Big\{\sum_{n=0}^{\infty} r_n t^n \mid r_0, r_1, \ldots \in \mathcal{R}\Big\}$$

the $\mathbb{K}$-algebra of formal power series with coefficients in $\mathcal{R}$, and

$$A_+ := \Big\{\sum_{n=1}^{\infty} r_n t^n \in A \mid r_0 = 0\Big\}.$$

Since $\mathbb{K} = \mathbb{K} \cdot 1 \subseteq \mathcal{R}$, we have $\mathbb{K}[[t]] \subseteq \mathcal{R}[[t]] = A$. Moreover, it is easy to check that every element of $\mathbb{K}[[t]]$ commutes with every element of A.

2° For every $a = \sum\limits_{n=0}^{\infty} r_n t^n \in A$, denote

$$\omega(a) = \min\{n \geq 0 \mid a_n \neq 0\}$$

if $a \neq 0$, and $\omega(a) = \infty$. Then, since $\mathcal{R}$ has no zero-divisors, we have

$$(\forall a, b \in A) \quad \omega(ab) = \omega(a) + \omega(b) \text{ and } \omega(a+b) \geq \min\{\omega(a), \omega(b)\}.$$

Moreover, if $\left\{a_k = \sum\limits_{n=0}^{\infty} r_n^{(k)} t^n\right\}_{k \geq 0}$ is a sequence in A such that $\lim\limits_{k \to \infty} \omega(a_k) = \infty$, then for each $n \geq 0$ we have $r_n^{(k)} = 0$ for all but finitely many indices k, so that we may define $r_n := \sum\limits_{k=0}^{\infty} r_n^{(k)} \in \mathcal{R}$ and then

$$\sum_{k=0}^{\infty} a_k := \sum_{n=0}^{\infty} r_n t^n \in A.$$

For example, if $a \in A_+$, then $\omega(a) < \omega(a^2) < \cdots$, so that we may consider "power series" $\sum\limits_{n=0}^{\infty} \alpha_n a^n (\in A)$ for arbitrary $\alpha_0, \alpha_1, \ldots \in \mathbb{K}$, and it is clear that the mapping

$$\mathbb{K}[[s]] \to \mathcal{R}[[t]], \quad \sum_{n=0}^{\infty} \alpha_n s^n \mapsto \sum_{n=0}^{\infty} \alpha_n a^n,$$

is a homomorphism of unital $\mathbb{K}$-algebras.

In particular, we can define the maps

$$\exp\colon A_+ \to 1 + A_+, \quad \exp a := \mathrm{e}^a := \sum_{n=0}^{\infty} \frac{1}{n!} a^n$$

and

$$\log\colon 1 + A_+ \to A_+, \quad \log(1+a) = \sum_{n=1}^{\infty} \frac{(-1)^{n-1}}{n} a^n.$$

3° We denote by $\mathrm{Lin}_{\mathbb{K}}(A)$ the unital associative $\mathbb{K}$-algebra of all $\mathbb{K}$-linear mappings $A \to A$, and consider $D, L_u, R_u, \mathrm{ad}\, u \in \mathrm{Lin}_{\mathbb{K}}(A)$ (where $u \in A$ is arbitrary) defined by

$$\begin{aligned} D\Big(\sum_{n=0}^{\infty} r_n t^n\Big) &= \sum_{n=0}^{\infty} (n+1) r_{n+1} t^n, \\ L_u(a) &= ua, \\ R_u(a) &= au, \\ (\mathrm{ad}\, u)(a) &= L_u(a) - R_u(a) = ua - au, \end{aligned}$$

for all $a = \sum\limits_{n=0}^{\infty} r_n t^n \in A$.

Then it is easy to check that for all $a, b \in A$ we have

$$D(ab) = (Da) \cdot b + a \cdot (Db) \text{ and } D(\mathrm{e}^{ta}) = a \cdot \mathrm{e}^{ta} = \mathrm{e}^{ta} \cdot a. \tag{1.1}$$

Moreover, for all $u \in A$ we have $L_u R_u = R_u L_u$, so $(\operatorname{ad} u)^n = (L_u - R_u)^n = \sum_{k=0}^{n} (-1)^k \binom{n}{k} (L_u)^{n-k} (R_u)^k$, whence

$$(\forall u, a \in A) \quad (\operatorname{ad} u)^n a = \sum_{k=0}^{n} (-1)^k \binom{n}{k} u^{n-k} a u^k$$

for every positive integer n.

4° We show at this stage that for every $c \in A_+$ there exists a unique homomorphism of unital $\mathbb{K}$-algebras

$$\Phi\colon \mathbb{K}[[s]] \to \operatorname{Lin}_{\mathbb{K}}(A)$$

such that $\Phi(s) = \operatorname{ad} c$. In fact, for $\lambda = \sum_{n=0}^{\infty} \lambda_n s^n \in \mathbb{K}[[s]]$ and $a \in A$, we define

$$\Phi(\lambda)a := \sum_{n=0}^{\infty} \lambda_n (\operatorname{ad} c)^n a.$$

Note that

$$\begin{aligned} \omega\big(\lambda_n (\operatorname{ad} c)^n a\big) &\ge \omega\big((\operatorname{ad} c)^n a\big) \\ &\ge \min_{0 \le k \le n} \omega(c^{n-k} a c^k) && \text{(see stages 2° and 3°)} \\ &\ge \min_{0 \le k \le n} \big((n-k)\omega(c) + \omega(a) + k\omega(c)\big) && \text{(see stage 2°)} \\ &\ge n\omega(c) + \omega(a), \end{aligned}$$

hence $\lim_{n\to\infty} \omega\big(\lambda_n (\operatorname{ad} c)^n a\big) = \infty$, and thus the definition of $\Phi(\lambda)a$ makes sense (see stage 2°). The properties required for Φ are clear.

It will be convenient to denote

$$\lambda(\operatorname{ad} c) := \Phi(\lambda)$$

whenever $\lambda \in \mathbb{K}[[s]]$.

For instance, we now show that, with this notation, for all $c \in A_+$ and $a \in A$ we have $(\mathrm{e}^{\operatorname{ad} c})a = \mathrm{e}^c a \mathrm{e}^{-c}$. In fact,

$$\begin{aligned} (\mathrm{e}^{\operatorname{ad} c})a &= \sum_{n=0}^{\infty} \frac{1}{n!} (\operatorname{ad} c)^n a = \sum_{n=0}^{\infty} \sum_{k=0}^{n} \frac{(-1)^k}{n!} \binom{n}{k} c^{n-k} a c^k \\ &= \sum_{j,k=0}^{\infty} \frac{1}{j!} c^j \cdot a \cdot \frac{1}{k!} (-c)^k = \mathrm{e}^c a \mathrm{e}^{-c}. \end{aligned}$$

(The second of the above equalities follows by the formula established at the end of stage 3°.)

5° We now come back to the conclusion of Theorem 1.24. We have to prove that, if $x, y \in \widehat{\mathcal{L}}(\mathcal{V})$, then $x \star y \in \widehat{\mathcal{L}}(\mathcal{V})$. Denote $x = \sum_{n=0}^{\infty} x_n$, $y = \sum_{n=0}^{\infty} y_n$, and $x \star y = \sum_{n=0}^{\infty} z_n$. With log and $\exp \cdot = \mathrm{e}^{\cdot}$ as at stage 2°, it is clear that

$$h := \log(\mathrm{e}^{tx}\mathrm{e}^{ty}) = \sum_{n=0}^{\infty} t^n z_n.$$

We note that according to (1.1), we have

$$\big(D(\mathrm{e}^h)\big)\cdot \mathrm{e}^{-h} = \big(D(\mathrm{e}^{tx}\mathrm{e}^{ty})\big)\cdot \mathrm{e}^{-h} = (x\mathrm{e}^{tx}\mathrm{e}^{ty} + \mathrm{e}^{tx}\mathrm{e}^{ty}y)\cdot \mathrm{e}^{-h} = (x\mathrm{e}^h + \mathrm{e}^h y)\cdot \mathrm{e}^{-h},$$

whence

$$\big(D(\mathrm{e}^h)\big)\cdot \mathrm{e}^{-h} = x + \mathrm{e}^h y \mathrm{e}^{-h} = x + (\mathrm{e}^{\mathrm{ad}\, h})y, \tag{1.2}$$

where the latter equality follows by the formula established in the last part of stage 4°.

6° Now denote $f := \sum_{n=0}^{\infty} \frac{1}{(n+1)!} s^n \in \mathbb{K}[[s]]$. Since the free term of f is different from 0, it follows that there exists $g \in \mathbb{K}[[s]]$ with $f \cdot g = g \cdot f = 1$. We prove at this stage that

$$D(h) = g(\mathrm{ad}\, h)(x + \mathrm{e}^{\mathrm{ad}\, h} y). \tag{1.3}$$

(In the above equality we have $g(\mathrm{ad}\, h) \in \mathrm{Lin}_{\mathbb{K}}(A)$ defined as in stage 4°, and $x + \mathrm{e}^{\mathrm{ad}\, h} y \in A$.) To prove formula (1.3) we compute

$$\begin{aligned} D(\mathrm{e}^h) \cdot \mathrm{e}^{-h} &= \Big(\sum_{j=0}^{\infty} \frac{1}{j!} D(h^j)\Big)\Big(\sum_{k=0}^{\infty} \frac{(-1)^k}{k!} h^k\Big) \\ &= \sum_{j\geq 1,\ k\geq 0} \frac{(-1)^k}{j!k!} D(h^j)\cdot h^k \\ &= \sum_{n=1}^{\infty} \frac{1}{n!} \sum_{j=1}^{n} (-1)^{n-j} \binom{n}{j} D(h^j)\cdot h^{n-j}. \end{aligned}$$

By using repeatedly the first of the equalities in (1.1), we get further

$$\begin{aligned} D(\mathrm{e}^h) \cdot \mathrm{e}^{-h} &= \sum_{n=1}^{\infty} \frac{1}{n!} \sum_{j=1}^{n} \sum_{i=0}^{j-1} (-1)^{n-j} \binom{n}{j} h^i \cdot D(h) \cdot h^{j-i-1} \cdot h^{n-j} \\ &= \sum_{n=1}^{\infty} \frac{1}{n!} \sum_{i=0}^{n-1} \Big(\sum_{j=i+1}^{n} (-1)^{n-j} \binom{n}{j} \Big) h^i \cdot D(h) \cdot h^{n-i-1}. \end{aligned}$$

We now note that $\sum_{j=i+1}^{n} (-1)^{n-j}\binom{n}{j} = \sum_{l=0}^{n-i-1} (-1)^l \binom{n}{l} = (-1)^{n-i-1}\binom{n-1}{i}$, where the latter equality follows easily by induction on n. Then the above equalities lead us to

$$\begin{aligned} D(\mathrm{e}^h)\cdot \mathrm{e}^{-h} &= \sum_{n=1}^{\infty}\frac{1}{n!}\sum_{i=0}^{n-1}(-1)^{n-i-1}\binom{n-1}{i}h^i\cdot D(h)\cdot h^{n-i-1} \\ &= \sum_{n=1}^{\infty}\frac{1}{n!}(\mathrm{ad}\,h)^{n-1}(Dh) \\ &= f(\mathrm{ad}\,h)(Dh), \end{aligned}$$

where the second equality follows by the formula established at the end of stage 3°. Now formula (1.2) in stage 5° implies that

$$x + \mathrm{e}^{\mathrm{ad}\,h}y = f(\mathrm{ad}\,h)(Dh).$$

Since $g\cdot f = 1$ in $\mathbb{K}[[s]]$, it follows by the homomorphism $\mathbb{K}[[s]] \to \mathrm{Lin}_{\mathbb{K}}(A)$ constructed at stage 4° (for $c = h$) that $g(\mathrm{ad}\,h)\cdot f(\mathrm{ad}\,h) = \mathrm{id}_A$, hence the above equality implies equality (1.3).

7° We now prove by induction on n that $z_n \in \mathcal{L}(\mathcal{V})$ for all $n \geq 0$ (see the notation from the beginning of stage 5°), whence it will follow that $x \star y \in \widehat{\mathcal{L}}(\mathcal{V})$, as desired.

We have $z_0 = 0 \in \mathcal{L}(\mathcal{V})$. Now assume that $z_0, \ldots, z_n \in \mathcal{L}(\mathcal{V})$. Consider the coefficients of t^n in both sides of equation (1.3) obtained in stage 6°. In the left-hand side, the coefficient is $(n+1)z_{n+1}$, while the coefficient in the right-hand side is a linear combination of expressions of the form

$$(\mathrm{ad}\,z_{n_1})\cdots(\mathrm{ad}\,z_{n_k})(u_{m_k}) = [z_{n_1},\ldots,[z_{n_k},u_{m_k}]\ldots],$$

where $u_{m_k} \in \{x_{m_k}, y_{m_k}\}$ and $n_1 + \cdots + n_k + m_k = n+1$. Since $m_k \geq 1$, the induction hypothesis shows that the coefficient of t^n in the right-hand side of (1.3) belongs to $\mathcal{L}(\mathcal{V})$. Then the coefficient of t^n in the left-hand side of (1.3), that is $(n+1)z_{n+1}$, in turn belongs to $\mathcal{L}(\mathcal{V})$. Thus $z_{n+1} \in \mathcal{L}(\mathcal{V})$, and the induction is complete. ∎

The series that shows up in the following statement will be called the *Baker-Campbell-Hausdorff series.*

COROLLARY 1.25 *If $\mathcal{V}$ is a vector space over $\mathbb{K} \in \{\mathbb{R}, \mathbb{C}\}$, then for all $x, y \in \mathcal{V}$ we have*

$$x\star y = \sum_{k=1}^{\infty}\frac{(-1)^{k-1}}{k}\sum_{(p_1+q_1)\cdots(p_k+q_k)>0}\frac{\pi_{\mathcal{V}}(x^{p_1}y^{q_1}\cdots x^{p_k}y^{q_k})}{p_1!q_1!\cdots p_k!q_k!(p_1+q_1+\cdots+p_k+q_k)}$$

PROOF We have

$$\begin{aligned}
x \star y &= \log(\exp x \cdot \exp y) \\
&= \sum_{k=1}^{\infty} \frac{(-1)^{k-1}}{k} (\exp x \cdot \exp y - 1)^k \\
&= \sum_{k=1}^{\infty} \frac{(-1)^{k-1}}{k} \Big(\sum_{p+q>0} \frac{1}{p!q!} x^p y^q \Big)^k \\
&= \sum_{k=1}^{\infty} \frac{(-1)^{k-1}}{k} \sum_{(p_1+q_1)\cdots(p_k+q_k)>0} \frac{1}{p_1!q_1!\cdots p_k!q_k!} x^{p_1} y^{q_1} \cdots x^{p_k} y^{q_k} \\
&= \sum_{n=1}^{\infty} z_n,
\end{aligned}$$

where, for each $n \geq 1$, we denoted

$$z_n := \sum_{k=1}^{\infty} \frac{(-1)^{k-1}}{k} \sum_{\substack{p_1+q_1+\cdots+p_k+q_k=n \\ (p_1+q_1)\cdots(p_k+q_k)>0}} \frac{1}{p_1!q_1!\cdots p_k!q_k!} x^{p_1} y^{q_1} \cdots x^{p_k} y^{q_k} \in \otimes^n \mathcal{V}.$$

Since $x, y \in \mathcal{V} \subseteq \widehat{\mathcal{L}}(\mathcal{V})$, it follows by Theorem 1.24 that $x \star y \in \widehat{\mathcal{L}}(\mathcal{V})$, hence Proposition 1.15 implies that $z_n \in \mathcal{L}(\mathcal{V})$ for all $n \geq 1$. Then Theorem 1.19 shows that $\pi_{\mathcal{V}}(z_n) = n z_n$, whence

$$\begin{aligned}
z_n &= \frac{1}{n} \pi_{\mathcal{V}}(z_n) \\
&= \sum_{k=1}^{\infty} \frac{(-1)^{k-1}}{k} \sum_{\substack{p_1+q_1+\cdots+p_k+q_k=n \\ (p_1+q_1)\cdots(p_k+q_k)>0}} \frac{1}{p_1!q_1!\cdots p_k!q_k!n} \pi_{\mathcal{V}}(x^{p_1} y^{q_1} \cdots x^{p_k} y^{q_k}),
\end{aligned}$$

and the desired formula follows since $x \star y = \sum_{n=1}^{\infty} z_n$. ∎

1.4 Convergence of the Baker-Campbell-Hausdorff series

This section includes some estimates of the terms of the Baker-Campbell-Hausdorff series that shows up in Corollary 1.25 in the case when x and y belong to a Banach-Lie algebra. Eventually we shall be able to prove that the Baker-Campbell-Hausdorff series is absolutely convergent provided $\|x\| + \|y\| < \log 2$ (see Proposition 1.33).

In order to find the aforementioned estimates it will be convenient to make the following definition.

DEFINITION 1.26 A *contractive Banach-Lie algebra* is a (real or complex) Banach-Lie algebra $\mathfrak{g}$ with a distinguished norm that defines the topology of $\mathfrak{g}$ and satisfies $\|[x,y]\| \le \|x\| \cdot \|y\|$ for all $x, y \in \mathfrak{g}$.

REMARK 1.27 Every real Banach-Lie algebra $\mathfrak{g}$ can be made into a contractive Banach-Lie algebra in the following way: Let $\|\cdot\|_1$ be an arbitrary norm defining the topology of $\mathfrak{g}$. Since the bracket $[\cdot,\cdot]\colon \mathfrak{g}\times\mathfrak{g} \to \mathfrak{g}$ is a bounded bilinear map, it follows that there exists $M > 0$ such that

$$(\forall x, y \in \mathfrak{g}) \qquad \|[x,y]\|_1 \le M \cdot \|x\|_1 \cdot \|y\|_1.$$

Defining $\|\cdot\| := (1/M)\|\cdot\|_1$, we get a norm that turns $\mathfrak{g}$ into a contractive Banach-Lie algebra.

DEFINITION 1.28 Let $\mathfrak{g}$ be a real Banach-Lie algebra and consider the real Lie algebra $\overline{\mathcal{L}}^{\mathrm{id}_\mathfrak{g}}(\mathfrak{g})$ and the homomorphism of Lie algebras $\widetilde{\mathrm{id}}_\mathfrak{g}\colon \overline{\mathcal{L}}^{\mathrm{id}_\mathfrak{g}}(\mathfrak{g}) \to \mathfrak{g}$ associated to $\theta := \mathrm{id}_\mathfrak{g}\colon \mathfrak{g} \to \mathfrak{g}$ as in Proposition 1.20. We will denote

$$\widetilde{\mathfrak{g}} := \overline{\mathcal{L}}^{\mathrm{id}_\mathfrak{g}}(\mathfrak{g}),$$

so that we have a homomorphism of Lie algebras $\widetilde{\mathrm{id}}_\mathfrak{g}\colon \widetilde{\mathfrak{g}} \to \mathfrak{g}$.

PROPOSITION 1.29 *If $\mathfrak{g}$ is a contractive Banach-Lie algebra, then we have*

$$x \star y \in \widetilde{\mathfrak{g}} \text{ and } \|\widetilde{\mathrm{id}}_\mathfrak{g}(x \star y)\| < -\log\bigl(2 - \mathrm{e}^{\|x\|+\|y\|}\bigr)$$

whenever $x, y \in \mathfrak{g}$ and $\|x\| + \|y\| < \log 2$.

PROOF Let $x, y \in \mathfrak{g}$ with $\|x\| + \|y\| < \log 2$. If $k \ge 1$ and $p_1, \dots, p_k \ge 0$, $q_1, \dots, q_k \ge 0$, with $(p_1+q_1)\cdots(p_k+q_k) > 0$ and $p_1+q_1+\cdots+p_k+q_k = n$, then we clearly have

$$\pi_\mathfrak{g}(x^{p_1}y^{q_1}\cdots x^{p_k}y^{q_k}) \in \otimes^n \mathfrak{g},$$

where $\pi_\mathfrak{g}\colon \mathcal{T}(\mathfrak{g}) \to \mathcal{T}(\mathfrak{g})$ is constructed in Definition 1.16. It then follows by the definition of $\widetilde{\mathfrak{g}}$ (see Definition 1.28) along with Corollary 1.25 that, in order to prove that $x \star y \in \widetilde{\mathfrak{g}}$, it suffices to show that the series

$$\sum_{k=1}^{\infty} \frac{1}{k} \sum_{(p_1+q_1)\cdots(p_k+q_k)>0} \frac{\|\overline{\mathrm{id}}_\mathfrak{g}\bigl(\pi_\mathcal{V}(x^{p_1}y^{q_1}\cdots x^{p_k}y^{q_k})\bigr)\|}{p_1!q_1!\cdots p_k!q_k!(p_1+q_1+\cdots+p_k+q_k)} \tag{1.4}$$

is convergent, where $\overline{\mathrm{id}}_{\mathfrak{g}}\colon \mathcal{L}(\mathfrak{g}) \to \mathfrak{g}$ is given by Proposition 1.17(i) for $\mathcal{V} = \mathfrak{h} = \mathfrak{g}$ and $\theta = \mathrm{id}_{\mathfrak{g}}$. Since $\overline{\mathrm{id}}_{\mathfrak{g}}$ is a homomorphism of Lie algebras, it follows by Definition 1.16 that

$$\begin{aligned}\overline{\mathrm{id}}_{\mathfrak{g}}\big(\pi_{\mathfrak{g}}(v_1\cdots v_m)\big) &= \overline{\mathrm{id}}_{\mathfrak{g}}\big([v_1,\dots,[v_{m-1},v_m]\cdots]\big)\\ &= [\overline{\mathrm{id}}_{\mathfrak{g}}(v_1),\dots,[\overline{\mathrm{id}}_{\mathfrak{g}}(v_{m-1}),\overline{\mathrm{id}}_{\mathfrak{g}}(v_m)]\dots]\\ &= [v_1,\dots,[v_{m_1},v_m]\dots]\end{aligned}$$

for all $v_1,\dots,v_m \in \mathfrak{g}$ and $m \ge 1$. Since $\mathfrak{g}$ is a contractive Banach-Lie algebra (see Definition 1.26), we get

$$(\forall m \ge 1)\,(\forall v_1,\dots,v_m \in \mathfrak{g}) \qquad \|\overline{\mathrm{id}}_{\mathfrak{g}}\big(\pi_{\mathfrak{g}}(v_1\cdots v_m)\big)\| \le \|v_1\|\cdots\|v_m\|.$$

Consequently, series (1.4) is dominated by

$$\begin{aligned}&\sum_{k=1}^{\infty}\frac{1}{k}\sum_{(p_1+q_1)\cdots(p_k+q_k)>0}\frac{\|x\|^{p_1}\|y\|^{q_1}\cdots\|x\|^{p_k}\|y\|^{q_k}}{p_1!q_1!\cdots p_k!q_k!(p_1+q_1+\cdots+p_k+q_k)}\\ &\le \sum_{k=1}^{\infty}\frac{1}{k}\sum_{(p_1+q_1)\cdots(p_k+q_k)>0}\frac{\|x\|^{p_1}}{p_1!}\frac{\|y\|^{q_1}}{q_1!}\cdots\frac{\|x\|^{p_k}}{p_k!}\frac{\|y\|^{q_k}}{q_k!}\\ &= \sum_{k=1}^{\infty}\frac{1}{k}\Big(\sum_{p+q>0}\frac{\|x\|^{p}}{p!}\frac{\|y\|^{q}}{q!}\Big)^k = \sum_{k=1}^{\infty}\frac{1}{k}\big(\mathrm{e}^{\|x\|}\cdot\mathrm{e}^{\|y\|}-1\big)^k\\ &= -\sum_{k=1}^{\infty}\frac{(-1)^{k-1}}{k}\big(1-\mathrm{e}^{\|x\|}\cdot\mathrm{e}^{\|y\|}\big)^k = -\log\big(2-\mathrm{e}^{\|x\|+\|y\|}\big),\end{aligned}$$

where the latter follows since $0 < \mathrm{e}^{\|x\|+\|y\|} - 1 < 1$ by the hypothesis. $\square$

DEFINITION 1.30 Let $\mathfrak{g}$ be a Lie algebra over $\mathbb{K} \in \{\mathbb{R},\mathbb{C}\}$ and

$$\pi_{\mathfrak{g}}\colon \mathcal{T}(\mathfrak{g}) \to \mathcal{L}(\mathfrak{g}), \quad v_1\cdots v_m \mapsto [v_1,\dots,[v_{m-1},v_m]\dots],$$

as in Definition 1.16. Moreover consider the homomorphism of Lie algebras

$$\overline{\mathrm{id}}_{\mathfrak{g}}\colon \mathcal{L}(\mathfrak{g}) \to \mathfrak{g}$$

given by Proposition 1.17(i) for $\mathcal{V} = \mathfrak{h} = \mathfrak{g}$ and $\theta = \mathrm{id}_{\mathfrak{g}}$. Then for every integer $n \ge 1$ we define $\beta_n\colon \mathfrak{g}\times\mathfrak{g} \to \mathfrak{g}$ by

$$\begin{aligned}\beta_n(x,y) &= \sum_{k=1}^{\infty}\sum_{\substack{p_1+q_1+\cdots+p_k+q_k=n\\(p_1+q_1)\cdots(p_k+q_k)>0}}\frac{1}{p_1!q_1!\cdots p_k!q_k!n}\overline{\mathrm{id}}_{\mathfrak{g}}\big(\pi_{\mathfrak{g}}(x^{p_1}y^{q_1}\cdots x^{p_k}y^{q_k})\big)\\ &= \sum_{k=1}^{n}\sum_{\substack{p_1+q_1+\cdots+p_k+q_k=n\\(p_1+q_1)\cdots(p_k+q_k)>0}}\frac{1}{p_1!q_1!\cdots p_k!q_k!n}\overline{\mathrm{id}}_{\mathfrak{g}}\big(\pi_{\mathfrak{g}}(x^{p_1}y^{q_1}\cdots x^{p_k}y^{q_k})\big).\end{aligned}$$

In the case when $\mathfrak{g}$ is a Banach-Lie algebra, we also define

$$\mathcal{H}(x,y) := \widetilde{\mathrm{id}}_{\mathfrak{g}}(x \star y) = \sum_{n=1}^{\infty} \beta_n(x,y) \in \mathfrak{g}$$

whenever $x \star y \in \widetilde{\mathfrak{g}}$. (Compare Corollary 1.25.)

REMARK 1.31

(a) If $\mathfrak{g}$ is a Lie algebra over $\mathbb{K} \in \{\mathbb{R}, \mathbb{C}\}$, then it turns out that

$$\beta_1(x,y) = x+y,\ \beta_2(x,y) = \frac{1}{2}[x,y],\ \beta_3(x,y) = \frac{1}{12}[x,[x,y]]+\frac{1}{12}[y,[y,x]],$$

for all $x, y \in \mathfrak{g}$.

(b) If $\mathfrak{h}$ is another Lie algebra over $\mathbb{K}$ and $\varphi\colon \mathfrak{g} \to \mathfrak{h}$ is a Lie algebra homomorphism, then $\varphi(\beta_n(x,y)) = \beta_n(\varphi(x), \varphi(y))$ whenever $x, y \in \mathfrak{g}$ and $n \geq 1$.

REMARK 1.32 Let $\mathfrak{g}$ be a Banach-Lie algebra over $\mathbb{K} \in \{\mathbb{R}, \mathbb{C}\}$, and $n \geq 1$. Denote $\mathfrak{h} := \mathfrak{g} \times \mathfrak{g}$. Then there exists a unique continuous symmetric n-linear mapping

$$\widetilde{\beta}_n\colon \underbrace{\mathfrak{h} \times \cdots \times \mathfrak{h}}_{n \text{ times}} \to \mathfrak{g}$$

such that $\widetilde{\beta}_n(h, \dots, h) = \beta_n(h)$ for all $h \in \mathfrak{g} \times \mathfrak{g}$.

PROPOSITION 1.33 *Let $\mathfrak{g}$ be a (real or complex) contractive Banach-Lie algebra and denote*

$$D := \{(x,y) \in \mathfrak{g} \times \mathfrak{g} \mid \|x\| + \|y\| < \log 2\}.$$

Then $\mathcal{H}\colon D \to \mathfrak{g}$ is a (real or complex) analytic mapping and

$$\mathcal{H}(x, \mathcal{H}(y,z)) = \mathcal{H}(\mathcal{H}(x,y), z)$$

whenever $\|x\| + \|y\| + \|z\| < \log(4/3)$.

PROOF For all $(x,y) \in D$ we have

$$\mathcal{H}(x,y) = \widetilde{\mathrm{id}}_{\mathfrak{g}}(x \star y) = \sum_{n=1}^{\infty} \beta_n(x,y), \tag{1.5}$$

and

$$\|H(x,y)\| \leq \sum_{n=1}^{\infty} \|\beta_n(x,y)\| \leq -\log\big(2 - \mathrm{e}^{\|x\|+\|y\|}\big),$$

by Proposition 1.29 and its proof. Now, by Remark 1.32 and Theorem A.34 we easily deduce that $H\colon D \to \mathfrak{g}$ is smooth and $\mathcal{H}^{(n)}_{(0,0)} = \widetilde{\beta}_n$, whence

$$\mathcal{H}^{(n)}_{(0,0)}(h,\dots,h) = \beta_n(h) \tag{1.6}$$

for all $n \geq 1$ and $h \in \mathfrak{g} \times \mathfrak{g}$. Now it easily follows by (1.5) and Theorem A.37 that $\mathcal{H}$ is (real or complex) analytic.

Next note that for $x, y, z \in \mathfrak{g}$ with $\|x\| + \|y\| + \|z\| < \log(4/3)$ we have

$$\begin{aligned}\|x\| + \|\mathcal{H}(y,z)\| &\leq \|x\| - \log\bigl(2 - \mathrm{e}^{\|y\|+\|z\|}\bigr) \\ &= \log \frac{\mathrm{e}^{\|x\|}}{2 - \mathrm{e}^{-\|x\|}\cdot \mathrm{e}^{\|x\|+\|y\|+\|z\|}} \\ &< \log \frac{\mathrm{e}^{\|x\|}}{2 - (4/3)\mathrm{e}^{-\|x\|}} < \log \frac{4/3}{2 - (4/3)\mathrm{e}^{-\|x\|}} < \log 2,\end{aligned}$$

and similarly $\|\mathcal{H}(x,y)\| + \|z\| < \log 2$. Hence $\bigl(x, \mathcal{H}(y,z)\bigr), \bigl(\mathcal{H}(x,y), z\bigr) \in D$. Thus we can define

$$g_1, g_2\colon \Omega \to \mathfrak{g}, \quad g_1(x,y,z) = \mathcal{H}(x, \mathcal{H}(y,z)),\ g_2(x,y,z) = \mathcal{H}(\mathcal{H}(x,y),z),$$

where $\Omega = \{(x,y,z) \in \mathfrak{g}\times\mathfrak{g}\times\mathfrak{g} \mid \|x\| + \|y\| + \|z\| < \log(4/3)\}$. Since we have seen that $\mathcal{H}$ is analytic, it follows by Proposition A.38 that both g_1 and g_2 are analytic mappings on the connected subset Ω of $\mathfrak{g}\times\mathfrak{g}\times\mathfrak{g}$.

On the other hand, we have

$$(\forall x, y, z \in \mathfrak{g}) \qquad x \star (y \star z) = (x \star y) \star z \in \widehat{\mathcal{L}}(\mathfrak{g})$$

(see Proposition 1.23 and Theorem 1.24). It follows by this equality along with (1.5) and (1.6) that

$$(\forall n \geq 0) \qquad (g_1)^{(n)}_{(0,0,0)} = (g_2)^{(n)}_{(0,0,0)}.$$

Since both g_1 and g_2 are analytic, it then follows by Theorem A.37 that there exists an open neighborhood V of $(0,0,0) \in \mathfrak{g}\times\mathfrak{g}\times\mathfrak{g}$ such that $V \subseteq \Omega$ and $g_1|_V = g_2|_V$. Since both g_1 and g_2 are analytic and Ω is connected, it then follows that $g_1 = g_2$ throughout on Ω. In fact, denote

$$\Omega_0 = \{\omega \in \Omega \mid (\forall n \geq 0) \quad (g_1)^{(n)}_\omega = (g_2)^{(n)}_\omega\}.$$

Then Ω_0 is clearly closed, and $\emptyset \neq V \subseteq \Omega_0$. Moreover, Ω_0 is open as a consequence of Theorem A.37. Since Ω is connected, we get $\Omega_0 = \Omega$. In particular, $g_1 = g_2$ on Ω, as claimed. ∎

REMARK 1.34 Let $\mathfrak{g}$ and $\mathfrak{h}$ be (real or complex) contractive Banach-Lie algebras and $\varphi\colon \mathfrak{g} \to \mathfrak{h}$ a homomorphism of topological Lie algebras with $\|\varphi\| \leq 1$. Then it is not hard to see that $\mathcal{H}(\varphi(x), \varphi(y)) = \varphi(\mathcal{H}(x,y))$ whenever $x, y \in \mathfrak{g}$ and $\|x\| + \|y\| < \log 2$.

Notes

There exist a lot of books devoted to the theory of Lie algebras. Two classical references for finite-dimensional Lie algebras are [Ja62] and [Bo71b]. See the books [AS74], [Kac90], and [MP95] for various classes of infinite-dimensional Lie algebras. The relationship between the theory of Lie algebras and operator theory is explored in [BS01]. See also the paper [Be02a] for a bird's-eye view on this subject.

We refer to [Bo71b] for a nice introduction to universal enveloping algebras. One of the classical treatises in this area is [Di74].

Some classical references for the Baker-Campbell-Hausdorff series are Chapter II in [Bo72] and Chapter X in [Ho65]. See also [Ja62]. A modern treatment of free Lie algebras and Baker-Campbell-Hausdorff series can be found in [Rt93].

Our reference for the proof of the Specht-Wever Theorem (Theorem 1.19) is [Ja62]. The method of proof of Theorem 1.24 is taken from [Dj75]. See also [Ho65]. The convergence of the Baker-Campbell-Hausdorff series is treated e.g., in [Bo72]. See also the papers [Go56], [Th82], [NT82], and [Th89].

Chapter 2

Lie Groups and Their Lie Algebras

Abstract. In this chapter we introduce the notion of Lie group and construct the Lie algebra of a Lie group. We are mainly interested in the study of Lie groups, and the Lie algebras prove to be a powerful tool in this connection. A Lie group is a group possessing a manifold structure that makes the group operations into smooth maps. The corresponding Lie algebra is just the tangent space at the unit element, with a bracket reflecting the group structure of the Lie group under consideration. The second part of the chapter concerns the notions of logarithmic derivatives and exponential map. We then discuss the smoothness of the exponential map in the case of general Lie groups, and present an example of a Lie group whose exponential map is not defined throughout on the corresponding Lie algebra. In the last section we consider the important special case of Banach-Lie groups. In this case, the exponential map is smooth and throughout defined, and defines a local chart around $\mathbf{1}$. As a consequence, the group operations are real analytic. We conclude by formulas for exponentials of sums and commutators.

2.1 Definition of Lie groups

One of the thoughts underlying the present book is that the structure of "nonlinear sets" can be understood by investigating their symmetry groups. And maybe the best description of the idea of symmetry group is provided by the notion of Lie group, which we are going to define now. Afterwards we shall be discussing some of the very basic properties of Lie groups.

DEFINITION 2.1 A *Lie group* is a group G equipped with a structure of smooth manifold such that both the multiplication map

$$m \colon G \times G \to G, \qquad (a, b) \mapsto ab,$$

and the inversion map

$$\eta \colon G \to G, \qquad a \mapsto a^{-1},$$

are smooth.

If H is another Lie group, then a mapping $\psi \colon H \to G$ is called a *homomorphism* (*of Lie groups*) if it is both a group homomorphism and a smooth mapping. We say that ψ is an *isomorphism* (*of Lie groups*) if ψ is a group isomorphism and a diffeomorphism.

A Lie group modeled on a Fréchet, Banach, or Hilbert space will be called, respectively, a *Fréchet-*, *Banach-*, or *Hilbert-Lie group.*

REMARK 2.2 In the category **Man** of smooth manifolds, the Lie groups can be characterized as the manifolds G equipped with a fixed element $\mathbf{1} \in G$ and some smooth maps $m \colon G \times G \to G$ and $\eta \colon G \to G$ such that the following group properties hold.

(i) Associativity: The diagram

$$\begin{array}{ccc} G \times G \times G & \xrightarrow{m \times \mathrm{id}_G} & G \times G \\ {\scriptstyle \mathrm{id}_G \times m}\downarrow & & \downarrow{\scriptstyle m} \\ G \times G & \xrightarrow{m} & G \end{array}$$

is commutative.

(ii) Unit element: The diagram

$$\begin{array}{ccc} G & \xrightarrow{\mathrm{id}_G \times \{\mathbf{1}\}} & G \times G \\ {\scriptstyle \{\mathbf{1}\} \times \mathrm{id}_G}\downarrow & \searrow^{\mathrm{id}_G} & \downarrow{\scriptstyle m} \\ G \times G & \xrightarrow{m} & G \end{array}$$

is commutative.

(iii) Inverse elements: The diagram

$$\begin{array}{ccc} G & \xrightarrow{\mathrm{id}_G \times \eta} & G \times G \\ {\scriptstyle \eta \times \mathrm{id}_G}\downarrow & \searrow^{\mathrm{id}_G} & \downarrow{\scriptstyle m} \\ G \times G & \xrightarrow{m} & G \end{array}$$

is commutative.

REMARK 2.3 Let $F \colon \mathbf{Man} \to \mathbf{Man}$ be a correspondence with the following properties:

(i) F associates to every manifold M another manifold denoted FM.

(ii) F associates to every smooth mapping $\psi\colon M \to N$ another smooth mapping $F\psi\colon FM \to FN$.

(iii) If $M \xrightarrow{\psi} N \xrightarrow{\varphi} P$ are smooth maps between smooth manifolds, then we have $F(\varphi\circ\psi) = (F\varphi)\circ(F\psi)$. In other words, F maps every commutative diagram

$$\begin{array}{ccc} N & \xrightarrow{\varphi} & P \\ \uparrow \psi & \nearrow_{\varphi\circ\psi} & \\ M & & \end{array}$$

into another commutative diagram

$$\begin{array}{ccc} FN & \xrightarrow{F\varphi} & FP \\ \uparrow F\psi & \nearrow_{F(\varphi\circ\psi)} & \\ FM & & \end{array}$$

(iv) For every smooth manifold M we have $F(\mathrm{id}_M) = \mathrm{id}_{FM}$.

(v) If M and N are smooth manifolds, then $F(M \times N) = (FM) \times (FN)$.

(vi) If $\psi\colon M \to N$ is a constant map, then $F\psi\colon FM \to FN$ is in turn a constant map.

In these conditions, it follows by Remark 2.2 that if G is a Lie group with the multiplication $m\colon G \times G \to G$, the inversion $\eta\colon G \to G$, and the unit element $\mathbf{1}_G \in G$, then the manifold FG is in turn a Lie group, with the multiplication

$$Fm\colon FG \times FG = F(G \times G) \to FG$$

(see condition (v)), the inversion

$$F\eta\colon FG \to FG$$

and the unit $\mathbf{1}_{FG}$, defined such that $\{\mathbf{1}_{FG}\}$ is the image of the constant map

$$F(\{\mathbf{1}_G\} \hookrightarrow G)\colon F(\{\mathbf{1}_G\}) \to FG$$

(see condition (vi)).

PROPOSITION 2.4 *If G is a Lie group with the multiplication $m\colon G \times G \to G$ and the inversion $\eta\colon G \to G$, then the tangent manifold TG is in turn a Lie group, with the multiplication $Tm\colon TG \times TG \to TG$, the inversion $T\eta\colon TG \to TG$, and the unit element $0 \in T_1 \subseteq TG$.*

PROOF Use Remark 2.3 for the correspondence $F = T$ defined by constructing tangent manifolds and tangent maps. (See Remark A.62.) ∎

In the following remark we describe in detail the multiplication in the tangent group that was pointed out in Proposition 2.4.

REMARK 2.5 Let G be a Lie group, $g, h \in G$, and the tangent vectors $v_g \in T_g G$ and $w_h \in T_h G$. In particular we have $v_g, w_h \in TG$, and we want to show how the product $v_g \cdot w_h$ in the group TG can be computed. According to Proposition 2.4, the multiplication in TG is just $Tm\colon TG \times TG \to TG$. Thus,

$$v_g \cdot w_h = Tm(v_g, w_h).$$

To compute the right-hand side, let $\alpha\colon \mathbb{R} \to G$ smooth with $\alpha(0) = g$ and $\dot{\alpha}(0) = v_g$, and similarly $\beta\colon \mathbb{R} \to G$ smooth with $\beta(0) = h$ and $\dot{\beta}(0) = w_h$. Then define

$$(\forall t \in \mathbb{R}) \qquad \gamma(t) := \alpha(t) \cdot \beta(t)$$

(where the product $\alpha(t) \cdot \beta(t)$ is computed in G).

Thus we get a map $\gamma\colon \mathbb{R} \to G$ which is smooth because of the commutative diagram

$$\begin{array}{ccc} \mathbb{R} \times \mathbb{R} & \xrightarrow{\alpha \times \beta} & G \times G \\ {\scriptstyle\Delta}\uparrow & & \downarrow{\scriptstyle m} \\ \mathbb{R} & \xrightarrow{\gamma} & G \end{array}$$

where m, $\alpha \times \beta$, and Δ are smooth, with $\Delta(t) = (t, t)$ for all $t \in \mathbb{R}$. Moreover $\gamma(0) = \alpha(0)\beta(0) = gh$, hence $\dot{\gamma}(0) \in T_{gh}G$.

To compute $\dot{\gamma}(0)$, we also use the above commutative diagram. We have

$$\begin{aligned} T\gamma &= T\big(m \circ (\alpha \times \beta) \circ \Delta\big) = (Tm) \circ \big(T(\alpha \times \beta)\big) \circ (T\Delta) \\ &= (Tm) \circ \big((T\alpha) \times (T\beta)\big) \circ (T\Delta). \end{aligned}$$

Since the mapping

$$T\Delta\colon T\mathbb{R} = \mathbb{R} \times \mathbb{R} \to T(\mathbb{R} \times \mathbb{R}) = T\mathbb{R} \times T\mathbb{R} = (\mathbb{R} \times \mathbb{R}) \times (\mathbb{R} \times \mathbb{R})$$

is given by $(t, \xi) \mapsto \big((t, \xi), (t, \xi)\big)$, while $\dot{\gamma}(0) = (T_0\gamma)(1) = (T\gamma)(0, 1)$, we get

$$\dot{\gamma}(0) = (T\gamma)(0,1) = (Tm)\big((T\alpha)(0,1), (T\beta)(0,1)\big) = (Tm)\big(\dot{\alpha}(0), \dot{\beta}(0)\big).$$

Thus $\dot{\gamma}(0) = (Tm)(v_g, w_h)$ is just the desired product $v_g \cdot w_h$ in the group TG.

Conclusion: The way to compute the product (in TG) of two tangent vectors $v_g \in T_g G$ and $w_h \in T_h G$ is the following: pick two smooth paths $\alpha, \beta\colon \mathbb{R} \to G$ representing the vectors v_g and w_h, respectively. That is, $\alpha(0) = g$, $\dot{\alpha}(0) = v_g$, $\beta(0) = h$, $\dot{\beta}(0) = w_h$. Then multiply them pointwise and thus get a smooth path $\gamma(\cdot) = \alpha(\cdot)\beta(\cdot)\colon \mathbb{R} \to G$ which represents the desired product $v_g \cdot w_h (\in T_{gh}G \subseteq TG)$. That is, we have $\gamma(0) = gh$ and $\dot{\gamma} = v_g \cdot w_h$.

REMARK 2.6

(a) If G is a Lie group, then the additive group $(T_1G, +)$ is a subgroup of G.

(b) If $f : G \to H$ is a homomorphism of Lie groups, then $Tf : TG \to TH$ is in turn a homomorphism of Lie groups.

REMARK 2.7 Let G be a Lie group, $\pi : TG \to G$ the natural projection mapping, and $z : G \to TG$ the mapping which associates to each $g \in G$ the zero vector in the vector space T_gG. Then

(a) both π and z are homomorphisms of Lie groups,

(b) $\operatorname{Ker} \pi = T_1G$, and

(c) $\pi \circ z = \mathrm{id}_G$.

Thus we have a short exact sequence

$$0 \to T_1G \hookrightarrow TG \xrightarrow{\pi} G \to \mathbf{1}$$

in the category of Lie groups, and this sequence splits. Hence the group TG can be expressed as the semi-direct product

$$TG = G \ltimes T_1G,$$

where the additive group T_1G is a normal subgroup of TG.

DEFINITION 2.8 If G is a Lie group with the tangent space $TG = T_1G \rtimes G$, then the *adjoint action of* G is the smooth action

$$G \times T_1G \to T_1G, \qquad (g, v) \mapsto \mathrm{Ad}(g)v := g \cdot v \cdot g^{-1},$$

which makes sense since T_1G is a normal subgroup of TG by Remark 2.7.

When we have to deal with adjoint actions of several groups, we denote the adjoint action of a group G by $\mathrm{Ad}_G(g)v$ for $g \in G$ and $v \in T_1G$.

REMARK 2.9 In Definition 2.8, the mapping $\mathrm{Ad}(g) : T_1G \to T_1G$ equals the derivative of the smooth map

$$G \to G, \qquad h \mapsto g \cdot h \cdot g^{-1},$$

at $h = 1$, for each $g \in G$. To see this, use Remark 2.5, which shows how the products in the expression $g \cdot v \cdot g^{-1}$ (where $g \in G$ and $v \in T_1 G$) can be computed.

DEFINITION 2.10 Let G be a Lie group. Consider the smooth mapping

$$G \times T_1 G \to T_1 G, \qquad (g, w) \mapsto \mathrm{Ad}(g) w,$$

and denote its first-order partial derivative with respect to g at $g = 1$ by

$$T_1 G \times T_1 G \to T_1 G, \qquad (v, w) \mapsto \mathrm{ad}(v) w,$$

which is a bilinear map.

We call this bilinear map the *adjoint action of* $T_1 G$. It is indeed an *action* in an appropriate sense, as a consequence of Theorem 2.12. For the moment, it is clear that the adjoint action of $T_1 G$ is continuous and also that it is linear with respect to the first variable.

REMARK 2.11 It is clear that if the Lie group G is commutative, then

$$(\forall g \in G) \qquad \mathrm{Ad}(g) = \mathrm{id}_{T_1 G}$$

whence

$$(\forall v \in T_1 G) \qquad \mathrm{ad}(v) = 0.$$

In this sense, we can say that in the case of general Lie groups G, the adjoint actions measure the non-commutativity of G.

2.2 The Lie algebra of a Lie group

The theorem we prove in this section (see Theorem 2.12 below) points out the fundamental phenomenon that each Lie group gives rise to a Lie algebra in a canonical way. It is the leading principle of Lie theory that the Lie algebra of a Lie group plays a critical role for that group.

THEOREM 2.12 *If G is a Lie group, then $T_1 G$ equipped with the bracket*

$$T_1 G \times T_1 G \to T_1 G, \quad (v, w) \mapsto [v, w] := \mathrm{ad}(v) w,$$

is a locally convex Lie algebra.

DEFINITION 2.13 If G is a Lie group, then the locally convex Lie algebra $(T_1 G, [\cdot, \cdot])$ constructed in Theorem 2.12 is denoted by $\mathbf{L}(G)$ and is called the *Lie algebra of G.*

We now turn to the proof of Theorem 2.12, which will be achieved by means of several lemmas.

LEMMA 2.14 *If G is a Lie group, then the mapping*

$$[\cdot,\cdot]\colon T_{\mathbf{1}}G \times T_{\mathbf{1}}G \to T_{\mathbf{1}}G$$

defined in Theorem 2.12 *is continuous, bilinear, and skew-symmetric.*

PROOF Recall that the model space of G is $X = T_{\mathbf{1}}G$ and let X_0 be an open neighborhood of $0 \in X$ such that there exists a diffeomorphism $\varphi\colon X_0 \to G_0$ onto some open neighborhood G_0 of $\mathbf{1} \in G$ with $\varphi(0) = \mathbf{1}$. By replacing G_0 by $G_0 \cap G_0^{-1}$ and X_0 by $X_0 \cap \varphi^{-1}(G_0^{-1})$, we may assume that $G_0 = G_0^{-1}$. Moreover, since φ is a diffeomorphism, $\varphi(0) = \mathbf{1}$ and $\mathbf{1} \cdot \mathbf{1} = \mathbf{1}$ in G, it follows that there exists a convex open neighborhood X_1 of $0 \in X$ such that $X_1 \subseteq X_0$ and $\varphi(X_1) \cdot \varphi(X_1) \subseteq G_0$. Then define

$$\eta\colon X_1 \to X, \quad \eta(x) := \varphi^{-1}\big(\varphi(x)^{-1}\big),$$

and

$$\mu\colon X_1 \times X_1 \to X, \quad \mu(x,y) := \varphi^{-1}\big(\varphi(x) \cdot \varphi(y)\big).$$

Next pick an open neighborhood X_2 of $0 \in X$ satisfying the conditions $\eta(X_2) \subseteq X_1$ and $\varphi(X_2) \cdot \varphi(X_2) \subseteq \varphi(X_1)$. Then note that Remark 2.9 along with Definition 2.10 imply that the mapping

$$\psi\colon X_2 \times X_2 \to X, \quad \psi(x,y) = \mu(\mu(x,y),\eta(x)) = \varphi^{-1}(\varphi(x)\varphi(y)\varphi(x)^{-1}),$$

has the property that

$$(\forall x,y \in X_2) \quad [x,y] = \mathrm{ad}(x)y = \partial_1\partial_2\psi(0,0)(x,y).$$

Then Proposition B.14 shows that

$$(\forall x,y \in X_2) \quad [x,y] = \partial_1\partial_2\mu(0,0)(y,x) - \partial_1\partial_2\mu(0,0)(x,y).$$

Since X_2 is an open neighborhood of $0 \in X = T_{\mathbf{1}}G$ and we have already noted in Definition 2.10 that $[\cdot,\cdot]$ is bilinear on $T_{\mathbf{1}}G \times T_{\mathbf{1}}G$, the above formula clearly shows that $[\cdot,\cdot]$ is moreover continuous and skew-symmetric. □

DEFINITION 2.15 If G is a Lie group, then we say that a vector field $v \in \mathcal{V}(G)$ is *left-invariant* whenever for all $g \in G$ the diagram

$$\begin{array}{ccc} TG & \xrightarrow{T(l_g)} & TG \\ {\scriptstyle v}\uparrow & & \uparrow{\scriptstyle v} \\ G & \xrightarrow{l_g} & G \end{array}$$

is commutative, where $l_g(h) = gh$ for all $g, h \in G$. Similarly, we say that $w \in \mathcal{V}(G)$ is *right-invariant* if for all $g \in G$ we have a commutative diagram

$$\begin{array}{ccc} TG & \xrightarrow{T(r_g)} & TG \\ {\scriptstyle w}\uparrow & & \uparrow{\scriptstyle w} \\ G & \xrightarrow{r_g} & G \end{array}$$

where $r_g(h) = hg$ for all $g, h \in G$.

We denote by $\mathcal{V}^l(G)$ the set of all left-invariant vector fields on G, and by $\mathcal{V}^r(G)$ the set of the right-invariant ones.

REMARK 2.16 In the setting of Definition 2.15, it easily follows by Proposition A.68 that both $\mathcal{V}^l(G)$ and $\mathcal{V}^r(G)$ are subalgebras of the Lie algebra $\mathcal{V}(G)$. (See also Theorem A.67.)

LEMMA 2.17 *Let G be a Lie group and define*

$$\iota\colon T_{\mathbf{1}}G \to \mathcal{V}(G)$$

by

$$\iota(v)(h) := T_{\mathbf{1}}(l_h)(v) \in T_hG,$$

whenever $v \in T_{\mathbf{1}}G$ and $h \in G$. Then ι is a linear isomorphism of $T_{\mathbf{1}}G$ onto $\mathcal{V}^l(G)$, having the inverse defined by

$$\iota^{-1}\colon \mathcal{V}^l(G) \to T_{\mathbf{1}}G, \quad x \mapsto x(\mathbf{1}).$$

PROOF The proof has several stages.

1° At this stage we check that for each $v \in T_{\mathbf{1}}G$ we have $\iota(v) \in \mathcal{V}^l(G)$. To this end, let $g, h \in G$ arbitrary. Then

$$\begin{aligned} \big(T(l_g) \circ \iota(v)\big)(h) &= T(l_g)\big(\iota(v)(h)\big) = T(l_g)\big(T_{\mathbf{1}}(l_h)(v)\big) \\ &= \big(T_h(l_g) \circ T_{\mathbf{1}}(l_h)\big)(v) = \big(T_{\mathbf{1}}(l_g \circ l_h)\big)(v) \\ &= \big(T_{\mathbf{1}}(l_{gh})\big)(v) = \iota(v)(gh) \\ &= \big(\iota(v) \circ l_g\big)(h), \end{aligned}$$

whence $T(l_g) \circ \iota(v) = \iota(v) \circ l_g$, as desired in order for $\iota(v)$ to be left-invariant.

2° The map ι is clearly linear, and the fact that $\operatorname{Ker} \iota = \{0\}$ follows since we have $\iota(v)(\mathbf{1}) = v$ for all $v \in T_{\mathbf{1}}G$. The latter equation also implies that $x(\mathbf{1}) = \iota^{-1}(x)$ whenever $x \in \operatorname{Ran} \iota$.

3° To prove that $\operatorname{Ran}\iota \supseteq \mathcal{V}^l(G)$, let $x \in \mathcal{V}^l(G)$ arbitrary and denote $v := x(\mathbf{1}) \in T_{\mathbf{1}}G$. Then $x = \iota(v)$. In fact, for each $h \in G$ we have

$$\begin{aligned}\iota(v)(h) &= T_{\mathbf{1}}(l_h)(v) = T_{\mathbf{1}}(l_h)(x(\mathbf{1}))\\ &= T(l_h)(x(\mathbf{1})) = (T(l_h)\circ x)(\mathbf{1})\\ &= (x\circ l_h)(\mathbf{1}) && \big(\text{since } x \in \mathcal{V}^l(G)\big)\\ &= x(h),\end{aligned}$$

and the proof ends. ∎

REMARK 2.18 Similarly to Lemma 2.17, one can prove that, if G is a Lie group and

$$\tilde{\iota}\colon T_{\mathbf{1}}G \to \mathcal{V}^r(G)$$

is defined by

$$\tilde{\iota}(v)(h) = T_{\mathbf{1}}(r_h)(v) \text{ whenever } v \in T_{\mathbf{1}}G,\ h \in G,$$

then $\tilde{\iota}$ is a linear isomorphism of $T_{\mathbf{1}}G$ onto $\mathcal{V}^r(G)$.

PROOF (of Theorem 2.12) In view of Lemmas 2.14 and 2.17, it suffices to check that, with ι as in Lemma 2.17, we have

$$\iota([x,y]) = [\iota(x),\iota(y)], \tag{2.1}$$

where $[\cdot,\cdot]$ in the left-hand side is the bracket defined in the statement of Theorem 2.12, while $[\cdot,\cdot]$ in the right-hand side is the bracket in the Lie algebra $\mathcal{V}(G)$ (see Theorem A.67). After we check formula (2.1), the fact that the Jacobi identity holds in $T_{\mathbf{1}}G$ will follow from the similar property in the Lie algebra $\mathcal{V}(G)$ along with the fact that $\iota\colon T_{\mathbf{1}}G \to \mathcal{V}(G)$ is linear and injective (see Lemma 2.17).

Using the notation introduced in the proof of Lemma 2.14, we have by Proposition B.14 that

$$(\forall x, y \in X_2)\quad [x,y] = (\tilde{x})'_0 y - (\tilde{y})'_0 x,$$

where for $x \in X_2$ we define

$$\tilde{x}\colon X_1 \to X,\quad \tilde{x}(\cdot) = \partial_2\mu(\cdot,0)x.$$

It is clear by the definition of ι in Lemma 2.17 that for each $x \in X_2$ we have a commutative diagram

$$\begin{array}{ccc} T(X_2) & \xrightarrow{T\varphi} & T(G_2)\\ \uparrow{\scriptstyle \underline{x}} & & \uparrow{\scriptstyle \iota(x)|_{G_2}}\\ X_2 & \xrightarrow{\varphi} & G_2\end{array}$$

where $G_2 = \varphi(X_2)$ and $\underline{x}\colon X_2 \to T(X_2) = X_2 \times X$ is given by $\underline{x}(\cdot) = \big(\cdot, \tilde{x}(\cdot)\big)$. Then the above formula for $[x, y]$ shows that

$$(\forall x, y \in X_2) \quad [x, y] = [\iota(x), \iota(y)](\mathbf{1})$$

(see formula (A.2) in the proof of Lemma A.66). Since Remark 2.16 shows that for $x, y \in X_2$ we have $[\iota(x), \iota(y)] \in \mathcal{V}^l(G)$, it follows by the above equality along with Lemma 2.17 that $[x, y] = \iota^{-1}([\iota(x), \iota(y)])$. Hence formula (2.1) holds whenever $x, y \in X_2$. Since both sides of that formula are bilinear as functions of the pair (x, y), and X_2 is a neighborhood of $0 \in X$, it follows that formula (2.1) actually holds for all $x, y \in X = T_{\mathbf{1}}G$. ∎

PROPOSITION 2.19 *If $f\colon G \to H$ is a homomorphism of Lie groups, then*

$$T_{\mathbf{1}}f\colon \mathbf{L}(G) \to \mathbf{L}(H)$$

is a homomorphism of real topological Lie algebras.

PROOF By Remark 2.6(b) we have a homomorphism of Lie groups

$$Tf\colon TG \to TH.$$

We also recall that $TG = G \ltimes \mathbf{L}(G)$, $TH = H \ltimes \mathbf{L}(H)$ (see Remark 2.7), and then $Tf|_G = f\colon G \to H$ and $Tf|_{\mathbf{L}(G)} = T_{\mathbf{1}}f\colon T_{\mathbf{1}}G \to T_{\mathbf{1}}H$. Since Tf is a group homomorphism, it then follows that

$$(\forall g \in G)\,(\forall v \in \mathbf{L}(G)) \quad T_{\mathbf{1}}f(g \cdot v \cdot g^{-1}) = f(g) \cdot (T_{\mathbf{1}}f)(v) \cdot f(g)^{-1}. \tag{2.2}$$

Now, for $w \in \mathbf{L}(G)$ arbitrary and $p\colon I \to G$ a smooth path with $p(0) = \mathbf{1}$ and $\dot{p}(0) = w$, applying (2.2) for $g = p(t)$ and then differentiating at $t = 0$ we get $(T_{\mathbf{1}}f)(\mathrm{ad}(w)v) = \big(\mathrm{ad}((T_{\mathbf{1}}f)w)\big)(T_{\mathbf{1}}f)(v)$. (See Definition 2.10.) In other words, according to Theorem 2.12 and Definition 2.13 we have

$$(T_{\mathbf{1}}f)[w, v] = [(T_{\mathbf{1}}f)w, (T_{\mathbf{1}}f)v]$$

for all $v, w \in \mathbf{L}(G)$, and thus $T_{\mathbf{1}}f\colon \mathbf{L}(G) \to \mathbf{L}(H)$ is a homomorphism of Lie algebras, as desired. ∎

DEFINITION 2.20 For every homomorphism of Lie groups $f\colon G \to H$, we denote by

$$\mathbf{L}(f) := T_{\mathbf{1}}f\colon \mathbf{L}(G) \to \mathbf{L}(H)$$

the corresponding homomorphism of topological Lie algebras given by Proposition 2.19.

If $\mathbf{Lie_{gr}}$ is the category of Lie groups (with the homomorphisms of Lie groups as morphisms) and $\mathbf{Lie_{alg}}$ is the category of locally convex Lie algebras (with the homomorphisms of topological Lie algebras as morphisms), then the correspondence

$$\mathbf{L}\colon \mathbf{Lie_{gr}} \to \mathbf{Lie_{alg}}$$

is a functor called the *Lie functor.*

Examples of Lie groups out of topological algebras

EXAMPLE 2.21 Let A be a real continuous inverse algebra (see Definition A.17). Moreover we assume that the topological vector space underlying A is locally convex. We are going to show that the multiplicative group

$$A^\times = \{a \in A \mid a \text{ is invertible in } A\}$$

has a natural structure of Lie group with

$$\mathbf{L}(A^\times) = A,$$

the bracket being defined by

$$(\forall x, y \in A) \quad [x, y] = xy - yx.$$

Since A is a continuous inverse algebra, it follows that $A^\times$ is an open subset of A, hence $A^\times$ is naturally a smooth manifold modeled on A.

Furthermore, since the multiplication mapping $A \times A \to A$, $(x, y) \mapsto xy$, is smooth (being actually bilinear), it follows that its restriction $m\colon A^\times \times A^\times \to A^\times$ is also smooth. On the other hand, we know from Lemma A.44 that the inversion mapping

$$\eta\colon A^\times \to A^\times, \quad a \mapsto a^{-1},$$

is also smooth, hence $A^\times$ is indeed a Lie group. Since it is modeled on A, it follows that moreover $\mathbf{L}(A^\times) = A$.

It only remains to compute the bracket of $\mathbf{L}(A^\times)$. To this end, first note that, for every $a \in A^\times$, if we define

$$I_a\colon A^\times \to A^\times, \quad I_a(b) = aba^{-1},$$

then we clearly have

$$T_{\mathbf{1}}(I_a)\colon T_{\mathbf{1}}(A^\times) = A \to T_{\mathbf{1}}(A^\times) = A, \quad x \mapsto axa^{-1}.$$

Next fix $x \in A$ and define

$$\theta\colon A^\times \to A, \quad \theta(a) = I_a(x) = axa^{-1}.$$

Then we have by Definition 2.10 and Theorem 2.12 that

$$[y, x] = \mathrm{ad}(y)x = \theta'_{\mathbf{1}}y \text{ whenever } y \in A = T_{\mathbf{1}}(A^\times).$$

To compute $\theta'_{\mathbf{1}}y$, let us fix $y \in A$. Since $\mathbf{1} \in A^\times$ and $A^\times$ is open in A, it follows that there exists $\varepsilon > 0$ such that $\mathbf{1} + ty \in A^\times$ whenever $|t| < \varepsilon$. Then we have a smooth path

$$p\colon (-\varepsilon, \varepsilon) \to A^\times, \quad p(t) = \mathbf{1} + ty$$

with the properties $p(0) = \mathbf{1}$ and $\dot{p}(0) = y \in A = T_{\mathbf{1}}(A^\times)$. Consequently, if we consider the smooth path $\gamma = \theta \circ p\colon (-\varepsilon, \varepsilon) \to A$, then we get

$$\dot{\gamma}(0) = \theta'_{\mathbf{1}}y.$$

On the other hand, we have

$$\gamma(t) = \theta(p(t)) = (\mathbf{1} + ty)x(\mathbf{1} + ty)^{-1} = (\mathbf{1} + ty) \cdot x \cdot \eta(\mathbf{1} + ty),$$

hence

$$\begin{aligned}\dot{\gamma}(t) &= y \cdot x \cdot \eta(1 + ty) + (1 + ty) \cdot x \cdot \eta'_{1+ty}y \\ &= y \cdot x \cdot \eta(1 + ty) - (1 + ty) \cdot x \cdot (1 + ty)^{-1}y(1 + ty)^{-1},\end{aligned}$$

where the latter equality follows by Lemma A.44. In particular $\dot{\gamma}(0) = yx - xy$, and according to the previous remarks we get

$$yx - xy = \dot{\gamma}(0) = \theta'_{\mathbf{1}}y = [y, x]$$

whenever $x, y \in A$, as claimed.

EXAMPLE 2.22 Let $\mathcal{X}$ be a real Banach space and $A = \mathcal{B}(\mathcal{X})$ the unital associative real Banach algebra of all bounded linear operators on $\mathcal{X}$. Then A is in particular a continuous inverse algebra (see Remark A.20). Hence Example 2.21 shows that the group

$$\mathrm{GL}(\mathcal{X}) = \{T \in \mathcal{B}(\mathcal{X}) \mid T \text{ is invertible}\}$$

is a (Banach-)Lie group whose Lie algebra is $\mathcal{B}(\mathcal{X})$ with the bracket defined by $[T, S] = TS - ST$ whenever $T, S \in \mathcal{B}(\mathcal{X})$.

In the special case when $X = \mathbb{K}^n$, where $\mathbb{K} \in \{\mathbb{R}, \mathbb{C}\}$ and n is a positive integer, we denote

$$\mathrm{GL}(n, \mathbb{K}) := \mathrm{GL}(\mathbb{K}^n)$$

and call this Lie group the *general linear group* corresponding to n and $\mathbb{K}$. It is just the group of the invertible matrices of size $n \times n$ with entries from $\mathbb{K}$. One usually denotes

$$\mathfrak{gl}(n, \mathbb{K}) := \mathbf{L}(\mathrm{GL}(n, \mathbb{K})).$$

Note that this is just the vector space of all matrices of size $n \times n$ with entries from $\mathbb{K}$, equipped with the bracket defined as above.

2.3 Logarithmic derivatives

Our aim in this section is to discuss the notion of logarithmic derivatives on a Lie group. The main consequence of that discussion concerns uniqueness for group homomorphism to which a Lie algebra homomorphism integrates (see Proposition 2.26 below).

DEFINITION 2.23 Let G be a Lie group with the tangent group $TG = G \ltimes T_1G = G \ltimes \mathbf{L}(G)$ (see Remark 2.7), I an open interval in $\mathbb{R}$ and $p\colon I \to G$ a smooth path. For every $t \in I$ we have

$$\dot{p}(t) \in T_{p(t)}G \simeq \{p(t)\} \times \mathbf{L}(G) \subseteq G \ltimes \mathbf{L}(G) = TG,$$

and, computing in the group TG, we define

$$\delta^l p(t) := p(t)^{-1} \cdot \dot{p}(t) \in T_1 G \simeq \{\mathbf{1}\} \times \mathbf{L}(G) \subseteq G \ltimes \mathbf{L}(G) = TG.$$

The smooth path

$$\delta^l p\colon I \to \mathbf{L}(G)$$

defined in this way is called the *left logarithmic derivative* of the path p. The *right logarithmic derivative* of p is the smooth path

$$\delta^r : I \to \mathbf{L}(G), \quad \delta^r(t) = \mathrm{Ad}(p(t))v(t) = \dot{p}(t) \cdot p(t)^{-1}.$$

Moreover, we say that $p(\cdot)$ is a *left* (respectively *right*) *indefinite product integral* of the path $\delta^l p(\cdot)$ (respectively of the path $\delta^r p(\cdot)$).

REMARK 2.24

(a) In the setting of Definition 2.23 we have

$$\dot{p}(t) = p(t) \cdot \delta^l p(t) = \delta^r p(t) \cdot p(t)$$

whenever $t \in I$.

(b) The left logarithmic integral allows us to compute the velocity vector of the path $\gamma(\cdot) = p(\cdot)^{-1}$ in the following way:

$$(\forall t \in I) \quad \dot{\gamma}(t) = -\delta^l p(t) \cdot p(t)^{-1}.$$

To see this, differentiate the equation $\gamma(t) \cdot p(t) = \mathbf{1}$ to get (by means of Remark 2.5) $\dot{\gamma}(t) \cdot p(t) + \gamma(t) \cdot \dot{p}(t) = 0$, where the computations in the left-hand side are performed in the group TG. Thus

$$\dot{\gamma}(t) = -\gamma(t) \cdot \dot{p}(t) \cdot p(t)^{-1} = -p(t)^{-1} \cdot \dot{p}(t) \cdot p(t)^{-1} = -\delta^l p(t) \cdot p(t)^{-1},$$

as claimed.

LEMMA 2.25 *Let G be a Lie group, I an open interval in $\mathbb{R}$, and $p, q\colon I \to G$ two smooth paths such that $\delta^l p = \delta^l q$. Then there exists $g_0 \in G$ such that $q(t) = g_0 \cdot p(t)$ whenever $t \in I$.*

PROOF Consider the smooth path

$$\alpha\colon I \to G, \quad \alpha(t) = q(t) \cdot p(t)^{-1}.$$

Denoting $\gamma(\cdot) = p(\cdot)^{-1}$ and computing in the group TG, we then have

$$\dot{\alpha}(t) = \dot{q}(t) \cdot \gamma(t) + q(t) \cdot \dot{\gamma}(t) = q(t) \cdot \delta^l q(t) \cdot p(t)^{-1} - q(t) \cdot \delta^l p(t) \cdot p(t)^{-1} = 0,$$

where the first equality follows by means of Remark 2.5, while the second one follows by Remark 2.24(b). Since $\dot{\alpha}(t) = 0$ for all $t \in I$, it follows that α is a constant. (See e.g., Proposition A.30.) Thus there exists $g_0 \in G$ such that for all $t \in I$ we have $\alpha(t) = g_0$, that is, $q(t) = g_0 \cdot p(t)$. ∎

PROPOSITION 2.26 *Let G and H be two Lie groups and $\varphi\colon \mathbf{L}(G) \to \mathbf{L}(H)$ a homomorphism of topological Lie algebras. If G is connected, then there exists at most one homomorphism of Lie groups $f\colon G \to H$ with $\mathbf{L}(f) = \varphi$.*

PROOF Let $f_1, f_2\colon G \to H$ be homomorphisms of Lie groups with $\mathbf{L}(f_1) = \mathbf{L}(f_2) = \varphi$. We have to prove that, for $g \in G$ arbitrary, we have $f_1(g) = f_2(g)$.

Since G is connected, there exist an open interval $I \subseteq \mathbb{R}$ with $0, 1 \in I$ and a smooth path $p\colon I \to G$ with $p(0) = \mathbf{1}$ and $p(1) = g$. For $i = 1, 2$ we can compute the left logarithmic derivative of the path $\gamma_i := f_i \circ p\colon I \to H$ in the following way:

$$\begin{aligned}
\delta^l(\gamma_i)(t) &= (f_i \circ p)(t)^{-1} \cdot \dot{\gamma}_i(t) = f_i(p(t)^{-1}) \cdot (Tf_i)(\dot{p}(t)) \\
&= Tf_i\big(p(t)^{-1} \cdot \dot{p}(t)\big) = Tf_i(\delta^l p(t)) \\
&= \mathbf{L}(f_i)(\delta^l p(t)) = \varphi(\delta^l p(t)),
\end{aligned}$$

where the third equality follows by the fact that $Tf_i\colon TH \to TH$ is a group homomorphism (see Remark 2.6(b)), while $TG = G \ltimes \mathbf{L}(G)$ and $Tf_i|_G = f_i$.

Consequently, both paths $f_1 \circ p, f_2 \circ p\colon I \to G$ have the same left logarithmic derivative (namely $\varphi \circ \delta^l p\colon I \to \mathbf{L}(H)$). Then Lemma 2.25 shows that there exists $h_0 \in H$ such that

$$(\forall t \in I) \quad (f_2 \circ p)(t) = h_0 \cdot (f_1 \circ p)(t).$$

Since $(f_2 \circ p)(0) = (f_1 \circ p)(0) = \mathbf{1}$, we must have $h_0 = \mathbf{1}$, hence $(f_2 \circ p)(t) = (f_1 \circ p)(t)$ whenever $t \in I$. In particular, for $t = 1$ we get $f_2(g) = f_1(g)$, as desired. ∎

DEFINITION 2.27 Let G be a Lie group and $v\colon \mathbb{R} \to \mathbf{L}(G)$ a smooth path. If there exists a left indefinite product integral $p\colon \mathbb{R} \to G$ of $v(\cdot)$ (that is, $\delta^l p = v$), then the element $\Pi(v) := p(0)^{-1}p(1) \in G$ is called the *left definite product integral* of $v(\cdot)$. Note that, according to Lemma 2.25, the element $\Pi(v) \in G$ does not depend on the choice of the left indefinite product integral $p(\cdot)$ of $v(\cdot)$.

DEFINITION 2.28 A Lie group G is said to be *regular* if it satisfies the following two conditions:

(i) Every smooth path $v\colon \mathbb{R} \to \mathbf{L}(G)$ admits left indefinite product integrals.

(ii) The left definite product integral

$$\Pi\colon \mathcal{C}^\infty(\mathbb{R}, \mathbf{L}(G)) \to G, \quad v \mapsto \Pi(v),$$

is a smooth mapping, provided we think of $\mathcal{C}^\infty(\mathbb{R}, \mathbf{L}(G))$ as a locally convex space in the usual way.

REMARK 2.29 Every Banach-Lie group is regular. (See e.g., [Mi84] or the discussion at the end of section 38.4 in [KM97].)

It is an open question whether or not there exist non-regular Lie groups modeled on complete locally convex spaces. In the case of non-complete locally convex spaces, an example of a non-regular Lie group will be described below in Example 2.38.

2.4 The exponential map

The exponential map is the main tool that allows us to use Lie algebras in the study of Lie groups. In this section we establish the basic properties of exponential maps for general Lie groups (that is, Lie groups modeled on locally convex spaces). To this end we use the properties of the logarithmic derivatives that were introduced in Definition 2.23, along with the notion of one-parameter subgroup, which we define here.

DEFINITION 2.30 If G is a Lie group, then a *one-parameter subgroup* of G is a homomorphism of Lie groups $f\colon \mathbb{R} \to G$, where we think of $\mathbb{R}$ as the additive group $(\mathbb{R}, +)$.

REMARK 2.31 According to Proposition 2.26, a one-parameter subgroup $f\colon \mathbb{R} \to G$ of the Lie group G is uniquely determined by the linear map

$$\mathbf{L}(f)\colon \mathbf{L}(\mathbb{R}) = \mathbb{R} \to \mathbf{L}(G),$$

or, equivalently, by the vector $\dot{f}(0) \in \mathbf{L}(G)$.

REMARK 2.32 Let G be a Lie group, $\delta > 0$ and $f_1\colon (-\delta, \delta) \to G$ a smooth path such that $f_1(t+s) = f_1(t)f_1(s)$ whenever $\max\{|t|, |s|\} < \delta/2$. Then it is not hard to see that there exists a unique one-parameter subgroup of G whose restriction to $(-\delta, \delta)$ coincides with f_1.

DEFINITION 2.33 Let G be a Lie group and denote by $\mathcal{D}(\exp_G)$ the set of all $v \in \mathbf{L}(G)$ such that there exists $f_v\colon \mathbb{R} \to G$, a one-parameter subgroup of G, with $\dot{f}_v(0) = v$. Then the *exponential map* of G is defined by

$$\exp_G\colon \mathcal{D}(\exp_G) \to G, \quad v \mapsto f_v(1).$$

Note that the exponential map is correctly defined according to Remark 2.31.

REMARK 2.34 It is easy to see that if G and H are Lie groups and $f\colon G \to H$ is a Lie group homomorphism, then $\mathbf{L}(f)(\mathcal{D}(\exp_G)) \subseteq \mathcal{D}(\exp_H)$ and the diagram

$$\begin{array}{ccc} \mathcal{D}(\exp_G) & \xrightarrow{\mathbf{L}(f)} & \mathcal{D}(\exp_H) \\ {\scriptstyle \exp_G}\downarrow & & \downarrow{\scriptstyle \exp_H} \\ G & \xrightarrow{f} & H \end{array}$$

is commutative.

PROPOSITION 2.35 *If G is a regular Lie group, then $\mathcal{D}(\exp_G) = \mathbf{L}(G)$ and the exponential map $\exp_G\colon \mathbf{L}(G) \to G$ is smooth.*

PROOF To prove that $\mathcal{D}(\exp_G) = \mathbf{L}(G)$, let $v \in \mathbf{L}(G)$ arbitrary and define the constant path

$$\tilde{v}\colon \mathbb{R} \to \mathbf{L}(G), \quad \tilde{v}(t) = v.$$

It then easily follows by condition (i) in Definition 2.28 (along with Proposition 2.26) that there exists a unique smooth path $f_v\colon \mathbb{R} \to G$ such that $f_v(0) = \mathbf{1}$ and $\dot{f}_v(t) = f_v(t) \cdot \tilde{v}(t)$ for all $t \in \mathbb{R}$.

Now fix $f_0 \in \mathbb{R}$ and consider the smooth paths

$$p_1, p_2\colon \mathbb{R} \to G, \quad p_1(t) = f_v(t + t_0), \quad p_2(t) = f_v(t_0)f_v(t).$$

Then $\dot{p}_1(t) = f_v(t+t_0)\cdot\tilde{v}(t+t_0) = f_v(t+t_0)\cdot v = p_1(t)\cdot v$. On the other hand, by Remark 2.5, we get $\dot{p}_2(t) = f_v(t_0)\cdot\dot{f}_v(t) = f_v(t_0)f_v(t)\cdot\tilde{v}(t) = p_2(t)\cdot v$. Thus p_1 and p_2 have the same left logarithmic derivative (namely the constant path $\tilde{v}$). Since $p_1(0) = p_2(0)$, it follows by Lemma 2.25 that $p_1(t) = p_2(t)$ for all $t \in \mathbb{R}$. Thus $f_v(t+t_0) = f_v(t_0)f_v(t)$ for all $t, t_0 \in \mathbb{R}$, that is, f is a one-parameter subgroup of G.

Now, since the mapping

$$\theta\colon \mathbf{L}(G) \to \mathcal{C}^\infty(\mathbb{R}, \mathbf{L}(G)), \quad v \mapsto \tilde{v},$$

is clearly smooth (being linear and continuous, and $\Pi(\tilde{v}) = f_v(1)$ for all $v \in \mathbf{L}(G)$, it follows that $\exp_G = \Pi\circ\theta\colon \mathbf{L}(G) \to G$ is smooth. ∎

PROPOSITION 2.36 *If G is a Lie group such that $\mathcal{D}(\exp_G) = \mathbf{L}(G)$ and the exponential map $\exp_G\colon \mathbf{L}(G) \to G$ is smooth, then*

$$T_0(\exp_G) = \mathrm{id}_{\mathbf{L}(G)}.$$

PROOF For $v = 0 \in T_{\mathbf{1}}G$, the corresponding one-parameter subgroup of G is just the constant one, $f_0\colon \mathbb{R} \to G$, $f_0(\cdot) = \mathbf{1}$, so that $\exp_G 0 = f_0(1) = \mathbf{1}$. Since the tangent space of $\mathbf{L}(G)$ at $0 \in \mathbf{L}(G)$ is just $\mathbf{L}(G)$, while $T_{\mathbf{1}}G = \mathbf{L}(G)$, we have

$$T_0(\exp_G)\colon \mathbf{L}(G) \to \mathbf{L}(G).$$

Next let $v \in \mathbf{L}(G)$ arbitrary. We will show that

$$(\forall t \in \mathbb{R}) \quad \exp_G(tv) = f_v(t), \tag{2.3}$$

which will imply that $T_0(\exp_G)v = \dot{f}_v(0) = v$, hence $T_0(\exp_G) = \mathrm{id}_{\mathbf{L}(G)}$, as desired.

To check (2.3), fix $t_0 \in \mathbb{R}$ and define $l_{t_0}\colon \mathbb{R} \to \mathbb{R}$, $s \mapsto t_0 s$. Then $t_0 v = t_0\dot{f}_v(0) = (f_v\circ l_{t_0})'(0)$, and $f_v\circ l_{t_0}\colon \mathbb{R} \to G$ is clearly a one-parameter subgroup of G, hence $\exp_G(t_0 v) = (f_v\circ l_{t_0})(1) = f_v(t_0)$, which is just (2.3). ∎

Example of non-regular Lie group

We introduce some notation to be used in Proposition 2.37 and Example 2.38 below. Let $\mathbb{K} \in \{\mathbb{R}, \mathbb{C}\}$ and consider the unital commutative $\mathbb{K}$-algebra of $\mathbb{K}$-valued polynomial functions on $[0,1]$:

$$\mathcal{P} := \{a\colon [0,1] \to \mathbb{K} \mid a \in \mathbb{K}[X]\}.$$

Also consider the $\mathbb{K}$-algebra of rational functions with poles off $[0,1]$:

$$A := \Big\{\frac{f}{g} \mid f, g \in \mathcal{P} \text{ and } g(s) \neq 0 \text{ whenever } s \in [0,1]\Big\}.$$

We think of A equipped with the topology it inherits as a unital subalgebra of the unital Banach algebra of the continuous $\mathbb{K}$-valued functions on $[0,1]$. Then it is easy to deduce by Remarks A.20 and A.21 that A is a continuous inverse algebra. Thus its group of invertible elements $A^\times$ is a Lie group with $\mathbf{L}(A^\times) = A$, according to Example 2.21.

PROPOSITION 2.37 *If $p\colon (\mathbb{R},+) \to (A^\times,\cdot)$ is a group homomorphism, then $p(\mathbb{R}) \subseteq \mathbb{K}^* \cdot \mathbf{1}$.*

PROOF We may suppose that $\mathbb{K} = \mathbb{C}$, since the other case clearly follows by this one. Next note that for every $a \in A \setminus \{0\}$ there exist uniquely determined polynomials $f, g \in \mathbb{C}[X]$ such that f and g have no common zero, $g(0) = 1$, $g([0,1]) \subseteq \mathbb{C}^*$ and $a(s) = f(s)/g(s)$ for all $s \in [0,1]$.

In particular, for all $t \in \mathbb{R}$, there exist uniquely determined $f_t, g_t \in \mathbb{C}[X]$ having no common zero, with $g_t(0) = 1$, $g_t([0,1]) \subseteq \mathbb{C}^*$ and $p(t) = f_t/g_t$ on $[0,1]$. Moreover, for all $t \in \mathbb{R}$ denote

$$U_t := g_t^{-1}(\mathbb{C}^*) \supseteq [0,1]$$

and $h_t := f_t/g_t\colon U_t \to \mathbb{C}$, so that $h_t|_{[0,1]} = p(t)$. (Note that since g_t is a polynomial, U_t is just $\mathbb{C}$ with finitely many points—the roots of g_t—removed.)

We now fix $t_0 \in \mathbb{R}$ and show that both polynomials f_{t_0} and g_{t_0} are constant, whence $p(t_0) \in \mathbb{C}^* \cdot \mathbf{1}$, as desired.

To this end, assume the contrary. Then the rational function $p(t_0) = f_{t_0}/g_{t_0}$ has either a pole or a zero at some $z_0 \in \mathbb{C}$. Then there exist $n \in \mathbb{Z} \setminus \{0\}$ and a holomorphic function $k\colon U_{t_0} \cup \{z_0\} \to \mathbb{C}$ such that $k(z_0) \neq 0$ and

$$(\forall z \in U_{t_0}) \quad h_{t_0}(z) = (z - z_0)^n k(z).$$

Consider

$$W := U_{t_0} \cap U_{t_0/(2|n|)},$$

which is an open connected subset of $\mathbb{C}$. We have $p(t_0) = p(t_0/(2|n|))^{2|n|}$, that is, $h_{t_0} = (h_{t_0/(2|n|)})^{2|n|}$ on $[0,1]$. Since both sides of the latter equality are holomorphic functions on the open connected set W, it follows that $h_{t_0} = (h_{t_0/(2|n|)})^{2|n|}$ throughout on W. Consequently, for $\epsilon = n/|n| \in \{-1,+1\}$ we have

$$\lim_{z\to z_0} \frac{|h_{t_0/(2|n|)}(z)|}{|z - z_0|^{\epsilon/2}} = \lim_{z\to z_0} \frac{|h_{t_0}(z)|^{1/(2n)}}{|z - z_0|^{\epsilon/2}} = |k(z_0)|^{1/(2|n|)} \neq 0,$$

which is impossible since $h_{t_0/(2|n|)}$ is a rational function. Consequently both polynomials f_{t_0} and g_{t_0} are constant, and the proof ends. ∎

EXAMPLE 2.38 For the Lie group $A^\times$ we have

$$\mathcal{D}(\exp_{A^\times}) = \mathbb{K}^\times \mathbf{1} \neq A = \mathbf{L}(A^\times).$$

Indeed, $v \in \mathcal{D}(\exp_{A^\times})$ if and only if there exists a one-parameter subgroup $f\colon \mathbb{R} \to A^\times$ such that $\dot{f}(0) = v$. According to Proposition 2.37, the latter condition implies that $f(\mathbb{R}) \subseteq \mathbb{K} \cdot \mathbf{1}$, whence $v = \dot{f}(0) \in \mathbb{K} \cdot \mathbf{1}$. Conversely, for $v \in \mathbb{K} \cdot \mathbf{1}$, we can define

$$f_v\colon \mathbb{R} \to \mathbb{K}^* \cdot \mathbf{1} \hookrightarrow A^\times, \quad f_v(t) = \mathrm{e}^{tv}.$$

Then clearly f_v is a one-parameter subgroup of G and $\dot{f}_v(0) = v$, hence $v \in \mathcal{D}(\exp_{A^\times})$.

We note that, since $\mathcal{D}(\exp_{A^\times}) \neq \mathbf{L}(A^\times)$, it follows by Proposition 2.35 that the Lie group $A^\times$ is non-regular (see Definition 2.28).

2.5 Special features of Banach-Lie groups

For the applications we want to make in the following chapters of this book we need a number of special properties of Banach-Lie groups, to which the present section is devoted. These properties can be summarized as follows: the exponential map of a Banach-Lie group is defined throughout the corresponding Lie algebra and defines a local chart around the unit element (Theorems 2.39 and 2.42 below). By using that local chart we can prove that every Banach-Lie group is real analytic and obtain some useful formulas for exponentials of sums and brackets (see Theorem 2.47).

Here is the main result of this section.

THEOREM 2.39 *For every Banach-Lie group G we have $\mathcal{D}(\exp_G) = \mathbf{L}(G)$ and the exponential map $\exp_G\colon \mathbf{L}(G) \to G$ is smooth.*

The proof of Theorem 2.39 needs several lemmas.

LEMMA 2.40 *If G is a Banach-Lie group with the Lie algebra $\mathfrak{g}$, then there exist $r_0 > 0$ and a smooth mapping $\theta\colon B_{\mathfrak{g}}(0, r_0) \to G$ satisfying the following conditions.*

(i) *We have $\theta(0) = \mathbf{1}$ and $T_0\theta = \mathrm{id}_{\mathfrak{g}}$.*

(ii) *If $0 \neq v \in B_{\mathfrak{g}}(0, r_0)$, then*

$$\theta((t+s)v) = \theta(tv) \cdot \theta(sv)$$

whenever $\max\{|t|, |s|\} < r_0/(2\|v\|)$.

PROOF Denote by $m\colon G \times G \to G$ the multiplication mapping and consider an open subset V of G such that $\mathbf{1} \in V$ and there exists a local chart

$\varphi\colon V \to B \subseteq \mathfrak{g}$, where B is open in $\mathfrak{g}$, $\varphi(\mathbf{1}) = 0 \in B$ and $T_{\mathbf{1}}\varphi = \mathrm{id}_{\mathfrak{g}}$. Next pick $r_1 > 0$ such that $B_1 := B_{\mathfrak{g}}(0, r_1) \subseteq B$ and $\varphi^{-1}(B_1) \cdot \varphi^{-1}(B_1) \subseteq V$, and define

$$\mu\colon B_1 \times B_1 \to \mathfrak{g}, \quad (x, y) \mapsto \varphi\big(\varphi^{-1}(x)\varphi^{-1}(y)\big) = \varphi\big(m(\varphi^{-1}(x)\varphi^{-1}(y))\big).$$

In view of the properties of the multiplication in G, it then follows that we can use Corollary B.7 to get $r_0 \in (0, r_1)$ and a smooth mapping $\chi\colon B_{\mathfrak{g}}(0, r_0) \to B_1$ such that $\chi(0) = 0$, $\chi_0' = \mathrm{id}_{\mathfrak{g}}$ and

$$\chi\big((t+s)v\big) = \mu\big(\chi(tv), \chi(sv)\big)$$

whenever $0 \neq v \in B_{\mathfrak{g}}(0, r_0)$ and $\max\{|t|, |s|\} < r_0/(2\|v\|)$.

Now it is easy to check that the mapping

$$\theta := \varphi^{-1} \circ \chi\colon B_{\mathfrak{g}}(0, r_0) \to G$$

has the desired properties. For instance,

$$T_0\theta = T_0(\varphi^{-1} \circ \chi) = T_0(\varphi^{-1}) \circ T_0\chi = (T_{\mathbf{1}}\varphi)^{-1} \circ \mathrm{id}_{\mathfrak{g}} = \mathrm{id}_{\mathfrak{g}},$$

where the latter equality follows by the fact that $T_{\mathbf{1}}\varphi = \mathrm{id}_{\mathfrak{g}}$. ∎

LEMMA 2.41 *If G is a Banach-Lie group with the Lie algebra $\mathfrak{g}$, then there exists $r_0 > 0$ such that $B_{\mathfrak{g}}(0, r_0) \subseteq \mathcal{D}(\exp_G)$ and $\exp_G|_{B_{\mathfrak{g}}(0,r_0)}\colon B_{\mathfrak{g}}(0, r_0) \to G$ is smooth.*

PROOF Let $\tilde{r}_0 > 0$ and $\theta\colon B_{\mathfrak{g}}(0, \tilde{r}_0) \to G$ be given by Lemma 2.40, and denote $r_0 = \tilde{r}_0/2$. We are going to prove that $B_{\mathfrak{g}}(0, r_0) \subseteq \mathcal{D}(\exp_G)$ and $\exp_G|_{B_{\mathfrak{g}}(0,r_0)} = \theta$, whence the desired conclusion will follow since θ is smooth.

It suffices to show that if $0 \neq v \in B_{\mathfrak{g}}(0, r_0)$, then $v \in \mathcal{D}(\exp_G)$ and $\exp_G v = \theta(v)$. By condition (ii) in Lemma 2.40 and Remark 2.32, we get a one-parameter subgroup of G, say $f\colon \mathbb{R} \to G$, such that $f(t) = \theta(tv)$ whenever $|t| < r_0/\|v\|$. Moreover,

$$\dot{f}(0) = \theta_0'(v) = v$$

since $\theta_0' = \mathrm{id}_{\mathfrak{g}}$ by property (i) in Lemma 2.40. On the other hand, since $0 < \|v\| < r_0$, we have $1 < r_0/\|v\|$, hence $f(1) = \theta(v)$, and then

$$\exp_G v = f(1) = \theta(v),$$

as desired. ∎

PROOF of Theorem 2.39 Fix r_0 as in Lemma 2.41 and $v_0 \in \mathfrak{g}$ arbitrary. Also let a positive integer n and $\varepsilon > 0$ such that $\|v_0\| + \varepsilon < nr_0$, and define

$$\omega\colon B_{\mathfrak{g}}(v_0, \varepsilon) \to G, \quad \omega(w) := \Big(\exp_G\big((1/n)w\big)\Big)^n.$$

Note that for $w \in B_{\mathfrak{g}}(v_0, \varepsilon)$ we have $\|w\| \le \|w - v_0\| + \|v_0\| < \varepsilon + \|v_0\| < nr_0$, hence $(1/n)w \in B_{\mathfrak{g}}(0, r_0) \subseteq \mathcal{D}(\exp_G)$, according to the choice of r_0. Thus ω is smooth in view of the commutative diagram

$$\begin{array}{ccc} B_{\mathfrak{g}}(0, r_0) & \xrightarrow{\exp_G|_{B_{\mathfrak{g}}(0,r_0)}} & G \\ {\scriptstyle\lambda_{1/n}}\uparrow & & \downarrow{\scriptstyle m_n} \\ B_{\mathfrak{g}}(v_0, \varepsilon) & \xrightarrow{\omega} & G \end{array}$$

where $\lambda_{1/n}(w) = (1/n)w$ for all $w \in B_{\mathfrak{g}}(v_0, \varepsilon)$ and $m_n(g) = g^n$ for all $g \in G$. (The mapping $\lambda_{1/n}$ is obviously smooth, $\exp_G|_{B_{\mathfrak{g}}(0,r_0)}$ is smooth by Lemma 2.41 and m_n is smooth since G is a Lie group.)

We are going to show that for arbitrary $w \in B_{\mathfrak{g}}(v_0, \varepsilon)$ we have $w \in \mathcal{D}(\exp_G)$ and $\exp_G w = \omega(w)$. This will imply that $\exp_G$ is defined and smooth on the neighborhood $B_{\mathfrak{g}}(v_0, \varepsilon)$ of v_0, which leads to the desired conclusion since $v_0 \in \mathfrak{g}$ is arbitrary.

We have seen above that $(1/n)w \in B_{\mathfrak{g}}(0, r_0) \subseteq \mathcal{D}(\exp_G)$, hence there exists a one-parameter subgroup of G, let's call it $f\colon \mathbb{R} \to G$, such that $\dot{f}(0) = (1/n)w$. Define

$$h\colon \mathbb{R} \to G, \quad g := f(nt).$$

It is clear that h is in turn a one-parameter subgroup of G and

$$\dot{h}(0) = n\dot{f}(0) = n \cdot (1/n)w = w,$$

hence indeed $w \in \mathcal{D}(\exp_G)$. Moreover,

$$\exp_G w = h(1) = f(n) = f(1)^n = \Big(\exp_G\big((1/n)w\big)\Big)^n = \omega(w),$$

and the proof is finished. □

THEOREM 2.42 *If G is a Banach-Lie group, then the following assertions hold.*

(i) *There exist an open neighborhood V_0 of $0 \in T_{\mathbf{1}}G$ and an open neighborhood $U_{\mathbf{1}}$ of $\mathbf{1} \in G$ such that $\exp_G(V_0) = U_{\mathbf{1}}$ and $\exp_G|_{V_0}\colon V_0 \to U_{\mathbf{1}}$ is a diffeomorphism.*

(ii) *All of the mappings*

$$\begin{aligned} \exp_G\colon \mathbf{L}(G) &\to G && \text{(the exponential mapping)} \\ m\colon G \times G &\to G && \text{(the multiplication mapping)} \\ \eta\colon G &\to G && \text{(the inversion mapping)} \end{aligned}$$

are real analytic.

PROOF (i) Since G is a Banach-Lie group, it follows that $T_{\mathbf{1}}G$ is a Banach space. Now let $\varphi\colon V \to U$ be a local coordinate chart on G, where V is an open subset of the model (Banach) space X of G, and U is an open neighborhood of $\mathbf{1} \in U$. Denote $\psi := \varphi^{-1}\colon U \to V$ and $v_0 := \psi(\mathbf{1}) \in V$.

On the other hand, the exponential map $\exp_G\colon T_{\mathbf{1}}G \to G$ is smooth by Theorem 2.39 and $\exp_G 0 = \mathbf{1}$. Then there exists an open neighborhood V' of $0 \in T_{\mathbf{1}}G$ such that $\exp_G(V') \subseteq U$, and the mapping

$$\chi = \psi \circ (\exp_G|_{V'})\colon V' \to V$$

is smooth. Moreover, by Proposition 2.36 along with Theorem 2.39 we get

$$T_0\chi = T_{\mathbf{1}}\psi \circ T_0(\exp_G) = T_{\mathbf{1}}\psi\colon T_{\mathbf{1}}G \to X,$$

which is an isomorphism of the Banach space $T_{\mathbf{1}}G$ onto the Banach space X (see Definition A.55). Since χ is a smooth map from an open subset of $T_{\mathbf{1}}G$ into X, it then follows by the inverse function theorem that there exists an open subset V_0 of X such that $0 \in V_0$, $\chi(V_0)$ is an open subset of V and $\chi|_{V_0}\colon V_0 \to \chi(V_0)$ is a diffeomorphism. Since

$$\exp_G|_{V_0} = \varphi|_{\chi(V_0)} \circ \chi|_{V_0},$$

and φ is a local coordinate chart, it then follows that $U_1 := \exp_G(V_0)$ is an open neighborhood of $\mathbf{1} \in G$, and $\exp_G|_{V_0}\colon V_0 \to U_1$ is a diffeomorphism, as desired.

(ii) Define

$$\tilde{\varphi}\colon V_0 \to U_1, \quad \tilde{\varphi} := \exp_G|_{V_0}.$$

It follows by (i) that $\tilde{\varphi}$ is a local coordinate chart on G around $\mathbf{1} \in G$. Then denote $\mathfrak{g} = \mathbf{L}(G)$, pick $r_1 > 0$ such that $\tilde{\varphi}(B_{\mathfrak{g}}(0,r_1)) \cdot \tilde{\varphi}(B_{\mathfrak{g}}(0,r_1)) \subseteq U_1$, and define $\mu\colon B_{\mathfrak{g}}(0,r_1) \times B_{\mathfrak{g}}(0,r_1) \to \mathfrak{g}$ by

$$\mu(x,y) = \tilde{\varphi}^{-1}(\tilde{\varphi}(x), \tilde{\varphi}(y)) = \tilde{\varphi}^{-1}(m(\tilde{\varphi}(x), \tilde{\varphi}(y))).$$

For $0 \neq x \in B_{\mathfrak{g}}(0,r_1)$, $t, s \in \mathbb{R}$ and $\max\{|t|,|s|\} < \frac{r_1}{2\|x\|}$, we have $\mu(tx,sx) = \tilde{\varphi}^{-1}(\exp_G(tx) \cdot \exp_G(sx)) = \tilde{\varphi}^{-1}(\exp_G((t+s)x)) = (t+s)x$. Hence we may make use of Theorem B.12 to deduce that μ is real analytic on some neighborhood of $(0,0) \in \mathfrak{g} \times \mathfrak{g}$. That is, the multiplication mapping m is real analytic on some neighborhood of $(\mathbf{1},\mathbf{1}) \in G \times G$. Now, by means of Remark 3.4 and Remark 3.6, it is easy to see that $m\colon G \times G \to G$ is real analytic throughout the group G.

The analyticity of η and $\exp_G$ then follows at once. ∎

For the statement of the following lemma we recall that, if G is a Lie group, then we have an isomorphism of Lie algebras

$$\iota\colon \mathbf{L}(G) \to \mathcal{V}^l(G)$$

from the Lie algebra $\mathbf{L}(G)$ onto the Lie algebra $\mathcal{V}^l(G)$ of all left-invariant vector fields on G. (See Lemma 2.17.)

LEMMA 2.43 *Let G be a Banach-Lie group, U an open subset of G, Y a real Banach space, and $\alpha\colon U \to Y$ a smooth mapping. Also let $g \in G$, $r \in (0,\infty)$, and $v \in \mathbf{L}(G)$ such that $g\exp_G(tv) \in U$ whenever $t \in (-r,r)$.*

Then the function

$$(-r/2, r/2) \to Y, \quad t \mapsto \alpha\big(g\exp_G(tv)\big),$$

has the following properties:

(i) *For every positive integer n and all $t \in (-r/2, r/2)$ we have*

$$\Big((D_{\iota(v)})^n\alpha\Big)\big(g\exp_G(tv)\big) = \frac{d^n}{dt^n}\Big(\alpha\big(g\exp_G(tv)\big)\Big).$$

(ii) *If the mapping α is real analytic on some neighborhood of $g \in U$, then there exists $\varepsilon > 0$ such that*

$$\alpha\big(g\exp_G(tv)\big) = \sum_{n=0}^{\infty}\frac{t^n}{n!}\Big((D_{\iota(v)})^n\alpha\Big)(g)$$

whenever $t \in \mathbb{R}$, $v \in \mathbf{L}(G)$ and $\|tv\| < \varepsilon$.

PROOF (i) Let $\pi\colon TY = Y \times Y \to Y$, $(y_1, y_2) \mapsto y_2$, be the second projection (as in Definition A.64), and consider the smooth path

$$\gamma\colon (-r,r) \to G, \quad t \mapsto \exp_G(tv).$$

Then $\gamma(0) = \mathbf{1} \in G$ and $\dot\gamma(0) = v \in T_{\mathbf{1}}G = \mathbf{L}(G)$. For every smooth mapping $\beta\colon U \to Y$ and $h \in G$ in some sufficiently small neighborhood of g we have $h\exp_G(tv) \in U$ for all $t \in (-r/2, r/2)$, hence

$$\Big(D_{\iota(v)}\beta\Big)\big(h\exp_G(tv)\big) = \big(\pi \circ T\beta \circ \iota(v)\big)\big(h\exp_G(tv)\big).$$

Denoting as usual $l_h\colon G \to G$, $k \mapsto hk$, we get

$$\Big(D_{\iota(v)}\beta\Big)\big(h\exp_G(tv)\big) = \big(\pi \circ T\beta \circ \iota(v) \circ l_h\big)\big(\gamma(t)\big).$$

But $\iota(v) \in \mathcal{V}(G)$ is a left-invariant vector field (see Lemma 2.17), hence

$$\begin{aligned}\Big(D_{\iota(v)}\beta\Big)\big(h\exp_G(tv)\big) &= \big(\pi \circ T\beta \circ T(l_h) \circ \iota(v)\big)\big(\gamma(t)\big)\\ &= \big(\pi \circ T(\beta \circ l_h) \circ \iota(v)\big)\big(\gamma(t)\big).\end{aligned}$$

Since $\gamma(0) = \mathbf{1}$, $\dot{\gamma}(0) = v$ and $\iota(v)(\mathbf{1}) = v$, we get by Definition A.60

$$(D_{\iota(v)}\beta)(h) = \big(\pi \circ T(\beta \circ l_h)\big)(v) = \frac{d}{dt}\beta(h\gamma(t))|_{t=0}.$$

Using the above formula for $h = g\gamma(s)$ with $s \in (-r/2, r/2)$, we get

$$(D_{\iota(v)}\beta)(g\gamma(s)) = \frac{d}{dt}\beta(g\gamma(s)\gamma(t))|_{t=0} = \frac{d}{dt}\beta(g\gamma(t+s))|_{t=0} = \frac{d}{dt}\beta(g\gamma(t))|_{t=s}.$$

Now the desired formula follows by applying the latter formula for $\beta = \alpha$, $\beta = D_{\iota(v)}\alpha$, $\beta = (D_{\iota(v)})^2\alpha$, etc.

(ii) If α is real analytic on some neighborhood of $g \in U$, then the function $f\colon (-r/2, r/2) \to Y$, $f(t) = \alpha\big(g \exp_G(tv)\big)$, is real analytic on some neighborhood of 0 (see Proposition A.38 and Theorem 2.42(ii)), hence

$$f(t) = \sum_{n=0}^{\infty} \frac{t^n}{n!}\frac{d^n f}{dt^n}(0).$$

Since $f(t) = \alpha(g \exp_G(tv))$, the desired formula then follows by (i). ∎

We now use Lemma 2.43 to prove the following version of the Taylor formula on Lie groups. (Compare Theorems A.32 and A.37.)

THEOREM 2.44 *Let G be a Banach-Lie group, U an open neighborhood of $\mathbf{1} \in G$, and $v_1, \dots, v_k$ in $\mathbf{L}(G)$. Then there exists $r > 0$ such that for all $t_1, \dots, t_k \in (-r, r)$ we have $\exp_G(t_1 v_1) \cdots \exp_G(t_k v_k) \in U$ and, for every real Banach space Y and every real analytic function $\alpha\colon U \to Y$ we have*

$$\alpha\big(\exp_G(t_1v_1)\cdots\exp_G(t_kv_k)\big) = \sum_{n_1,\dots,n_k \geq 0} \frac{t_1^{n_1}\cdots t_k^{n_k}}{n_1!\cdots n_k!}\big((D_{\iota(v_1)})^{n_1}\cdots(D_{\iota(v_k)})^{n_k}\alpha\big)(\mathbf{1}).$$

PROOF It follows by making repeated use of Lemma 2.43 that for every smooth function $\alpha\colon U \to Y$ we have

$$(\forall n_1, \dots, n_k \geq 0) \quad \big((D_{\iota(v_1)})^{n_1}\cdots(D_{\iota(v_k)})^{n_k}\alpha\big)(\mathbf{1}) = \frac{\partial^{n_1+\cdots+n_k} f}{\partial t_1^{n_1}\cdots\partial t_k^{n_k}}(0),$$

where $f(t_1, \dots, t_k) = \alpha\big(\exp_G(t_1v_1)\cdots\exp_G(t_kv_k)\big)$. Now, if $\alpha\colon U \to Y$ is moreover real analytic, then $f\colon (-r, r)^k \to Y$ is real analytic, hence

$$f(t_1, \dots, t_k) = \sum_{n_1,\dots,n_k \geq 0} \frac{t_1^{n_1}\cdots t_k^{n_k}}{n_1!\cdots n_k!}\frac{\partial^{n_1+\cdots+n_k} f}{\partial t_1^{n_1}\cdots\partial t_k^{n_k}}(0),$$

and thus the desired formula follows. ∎

REMARK 2.45 Let G be a Banach-Lie group with the Lie algebra $\mathfrak{g}$, V_0 an open neighborhood of $0 \in \mathfrak{g}$, and U_1 an open neighborhood of $\mathbf{1} \in G$ such that $\exp_G(V_0) = U_1$ and $\exp_G|_{V_0}: V_0 \to U_1$ is a diffeomorphism. Denote $f := (\exp_G|_{V_0})^{-1}: U_1 \to V_0$. Then for every $v \in \mathfrak{g}$ the operator $D_{\iota(v)}: \mathcal{C}^\infty(U_1, \mathfrak{g}) \to \mathcal{C}^\infty(U_1, \mathfrak{g})$ has the property $(D_{\iota(v)}f)(\mathbf{1}) = v$.

LEMMA 2.46 *Let G be a Banach-Lie group with the Lie algebra $\mathfrak{g}$ and $v, w \in \mathfrak{g}$. Then there exist $r > 0$ and a bounded function $\theta_{v,w}: (-r, r) \to \mathfrak{g}$ such that for all $t \in (-r, r)$ we have*

$$\exp_G(tv)\exp_G(tw) = \exp_G\Big(t(v+w) + \frac{t^2}{2}[v,w] + t^3\theta_{v,w}(t)\Big).$$

There exist $\delta > 0$ and $M > 0$ such that r can be chosen independently of v and w for $\max\{\|v\|, \|w\|\} < \delta$ and

$$\sup\{\|\theta_{v,w}(t)\| \mid t \in (-r,r),\ v,w \in B_{\mathfrak{g}}(0,\delta)\} \le M.$$

PROOF According to Theorem 2.42(i), there exist an open neighborhood V_0 of $0 \in \mathfrak{g}$ and an open neighborhood U_1 of $\mathbf{1} \in G$ such that $\exp_G(V_0) = U_1$ and the mapping $\exp_G|_{V_0}: V_0 \to U_1$ is a diffeomorphism. Denote

$$f := (\exp_G|_{V_0})^{-1}: U_1 \to V_0.$$

We pick $r_0 > 0$ such that $\exp_G(tv)\exp_G(tw) \in U_1$ whenever $t \in (-r_0, r_0)$, and define

$$u: (-r_0, r_0) \to \mathfrak{g}, \quad u(t) = f\big(\exp_G(tv)\exp_G(tw)\big).$$

Then u is a smooth function and $u(0) = 0$, hence there exist $r_1 \in (0, r_0)$, $u_1, u_2 \in \mathfrak{g}$ and a bounded function $\theta_1: (-r_1, r_1) \to \mathfrak{g}$ such that

$$u(t) = tu_1 + t^2u_2 + t^3\theta_1(t) \text{ whenever } t \in (-r_1, r_1)$$

(see Corollary A.33). What we have to prove is that

$$u_1 = v + w \text{ and } u_2 = \frac{1}{2}[v,w].$$

To this end, note that since $f = (\exp_G|_{V_0})^{-1}$, we have

$$u(t) = f\big(\exp_G(tu_1 + t^2u_2)\big) + t^3\theta_1(t) = \sum_{n=0}^{\infty} \frac{1}{n!}\Big(\big(D_{\iota(tu_1+t^2u_2)}\big)^n f\Big)(\mathbf{1}) + t^3\theta_1(t).$$

The latter equality follows by Lemma 2.43(ii), since f is real analytic as an easy consequence of the fact that $\exp_G|_{V_0}$ is a local chart of G around $\mathbf{1} \in G$

(see Theorem 2.42(i)). By Remark A.65 and the fact that $\iota\colon T_{\mathbf{1}}G \to \mathcal{V}(G)$ is linear (see Lemma 2.17), we further deduce that

$$u(t) = \sum_{n=0}^{\infty} \frac{1}{n!}\Big(\big(tD_{\iota(u_1)} + t^2 D_{\iota(u_2)}\big)^n f\Big)(\mathbf{1}) + t^3\theta_1(t)$$
$$= \Big(\big(tD_{\iota(u_1)} + t^2 D_{\iota(u_2)}\big) f\Big)(\mathbf{1}) + \frac{t^2}{2}\Big(\big(D_{\iota(u_1)}\big)^2 f\Big)(\mathbf{1}) + t^3\theta_2(t),$$

for some bounded function $\theta_2\colon (-r_0, r_0) \to \mathfrak{g}$, Consequently,

$$u(t) = t\big(D_{\iota(u_1)} f\big)(\mathbf{1}) + t^2\Big(\big(D_{\iota(u_2)} + \frac{1}{2}(D_{\iota(u_1)})^2\big) f\Big)(\mathbf{1}) + t^3\theta_2(t). \tag{2.4}$$

On the other hand, we get by Corollary 2.44 that, for all $t \in (-r_1, r_1)$,

$$u(t) = f\big(\exp_G(tv)\exp_G(tw)\big) = \sum_{n,m\geq 0} \frac{t^{n+m}}{n!m!}\big((D_{\iota(v)})^n (D_{\iota(w)})^m f\big)(\mathbf{1}),$$

hence

$$\begin{aligned} u(t) =& t\big((D_{\iota(v)} + D_{\iota(w)}) f\big)(\mathbf{1}) \\ &+ \frac{t^2}{2}\big(((D_{\iota(v)})^2 + 2D_{\iota(v)}D_{\iota(w)} + (D_{\iota(w)})^2) f\big)(\mathbf{1}) + \theta_{v,w}(t), \end{aligned} \tag{2.5}$$

for some $r \in (0, r_0)$ and some bounded function $\theta_{v,w}\colon (-r, r) \to \mathfrak{g}$. Now (2.4) and (2.5) imply that $\theta_1 = \theta_{v,w}$ on $(-r, r)$, and also

$$(D_{\iota(u_1)} f)(\mathbf{1}) = (D_{\iota(v+w)} f)(\mathbf{1})$$

and

$$\Big(\big(D_{\iota(u_2)} + \frac{1}{2}(D_{\iota(u_1)})^2\big) f\Big)(\mathbf{1}) = \frac{1}{2}\Big(\big((D_{\iota(v)})^2 + 2D_{\iota(v)}D_{\iota(w)} + (D_{\iota(w)})^2\big) f\Big)(\mathbf{1}).$$

Consequently, by Remark 2.45, the first of the above equalities implies that $u_1 = v + w$. Then the second of the above equalities takes on the form

$$\begin{aligned} &\Big(\big(D_{\iota(u_2)} + \frac{1}{2}(D_{\iota(v)} + D_{\iota(w)})^2\big) f\Big)(\mathbf{1}) \\ &\qquad = \frac{1}{2}\Big(\big((D_{\iota(v)})^2 + 2D_{\iota(v)}D_{\iota(w)} + (D_{\iota(w)})^2\big) f\Big)(\mathbf{1}), \end{aligned}$$

whence

$$\begin{aligned} (D_{\iota(u_2)} f)(\mathbf{1}) &= \frac{1}{2}\big((D_{\iota(v)}D_{\iota(w)} - D_{\iota(w)}D_{\iota(v)}) f\big)(\mathbf{1}) = \frac{1}{2}(D_{[\iota(v),\iota(w)]} f)(\mathbf{1}) \\ &= \frac{1}{2}(D_{\iota([v,w])} f)(\mathbf{1}) = (D_{\iota(\frac{1}{2}[v,w])} f)(\mathbf{1}) \end{aligned}$$

(where the second equality follows by Lemma A.66, and the third equality follows by formula (2.1) in the proof of Theorem 2.12). Now $u_2 = \frac{1}{2}[v, w]$ by Remark 2.45 again.

The last assertion of the lemma clearly follows by the way $\theta_{v,w}$ is constructed. In fact, $\theta_{v,w}$ only depends on the second derivative of the mapping

$$(v, w) \mapsto f(\exp_G v \cdot \exp_G w)$$

on some neighborhood of $(0,0) \in \mathfrak{g} \times \mathfrak{g}$. ∎

Now we are able to prove a couple of useful formulas for exponentials of sums and brackets.

THEOREM 2.47 *If G is a Banach-Lie group, then*

$$\exp_G(v+w) = \lim_{n\to\infty} \Big(\exp_G(v/n) \cdot \exp_G(w/n)\Big)^n$$

and

$$\exp_G[v,w] = \lim_{n\to\infty} \Big(\exp_G(-(v/n))\cdot\exp_G(-(w/n))\cdot\exp_G(v/n)\cdot\exp_G(w/n)\Big)^{n^2}$$

for all $v, w \in \mathbf{L}(G)$.

PROOF To prove the first formula, denote $u(t) = f\big(\exp_G(tv)\exp_G(tw)\big)$ as in the proof of Lemma 2.46, where $f := (\exp_G|_{V_0})^{-1}\colon U_1 \to V_0$. Then for n large enough we have by Lemma 2.46

$$f\big(\exp_G((1/n)v)\cdot\exp_G((1/n)w)\big) = u(1/n) = \frac{1}{n}(v+w) + \frac{1}{2n^2}[v,w] + \frac{1}{n^3}\theta(1/n),$$

so that

$$\lim_{n\to\infty} nf\big(\exp_G((1/n)v)\cdot\exp_G((1/n)w)\big) = v+w.$$

Since $\exp_G\colon \mathbf{L}(G) \to G$ is continuous (see Theorem 2.39), we get

$$\begin{aligned}\exp_G(v+w) &= \lim_{n\to\infty} \exp_G\Big(nf\big(\exp_G((1/n)v)\cdot\exp_G((1/n)w)\big)\Big)\\ &= \lim_{n\to\infty}\Big(\exp_G f\big(\exp_G((1/n)v)\cdot\exp_G((1/n)w)\big)\Big)^n\\ &= \Big(\exp_G((1/n)v)\cdot\exp_G((1/n)w)\Big)^n,\end{aligned}$$

as desired.

For the second formula, again use Lemma 2.46 to deduce that

$$\exp_G(tv)\exp_G(tw) = \exp_G\Big(t(v+w) + \frac{t^2}{2}[v,w] + t^3\theta_{v,w}(t)\Big)$$

and also that

$$\exp_G(-tv)\exp_G(-tw) = \exp_G\Big(-t(v+w) + \frac{t^2}{2}[v,w] - t^3\theta_{v,w}(-t)\Big).$$

Hence we deduce by Lemma 2.46 again, there exist $\tilde{r} > 0$ and a bounded function $\tilde{\theta}\colon (-\tilde{r},\tilde{r}) \to \mathbf{L}(G)$ such that

$$\exp_G(-tv)\exp_G(-tw)\exp_G(tv)\exp_G(tw) = \exp_G\big(t^2[v,w] + t^3\tilde{\theta}(t)\big).$$

Now reason as above: denote

$$\tilde{u}(t) := f\big(\exp_G(-tv)\exp_G(-tw)\exp_G(tv)\exp_G(tw)\big),$$

and then the above equality implies that $\lim_{n\to\infty} n^2\tilde{u}(1/n) = [v,w]$, whence

$$\exp_G[v,w] = \lim_{n\to\infty} n^2\big(\tilde{u}(1/n)\big) = \lim_{n\to\infty}\big(\exp_G \tilde{u}(1/n)\big)^{n^2},$$

which is just the desired formula, in view of the definition of $\tilde{u}$. ∎

EXAMPLE 2.48 Let A be a real associative unital Banach algebra. It follows by Example 2.21 (along with Remark A.20) that $A^\times$ is a (Banach-)Lie group whose Lie algebra is just A with the bracket

$$(\forall a,b \in A) \quad [a,b] := ab - ba.$$

For arbitrary $a \in A$, we define

$$f_a\colon \mathbb{R} \to A^\times, \quad f_a(t) := \sum_{n=0}^{\infty} \frac{t^n}{n!}a^n.$$

(Note that the series is absolutely convergent in A.) Then f_a is a one-parameter subgroup of $A^\times$ and $\dot{f}_a(0) = a$, hence $a \in \mathcal{D}(\exp_{A^\times})$ and

$$\exp_{A^\times} a = \mathrm{e}^a := \sum_{n=0}^{\infty} \frac{a^n}{n!}.$$

Thus $\mathcal{D}(\exp_{A^\times}) = A = \mathbf{L}(A^\times)$ (which agrees with Theorem 2.39) and the exponential map $\exp_{A^\times}\colon A \to A^\times$ is defined by the usual power series.

In particular, it follows by Theorem 2.47 that for all $a,b \in A$ we have

$$\mathrm{e}^{a+b} = \lim_{n\to\infty}\big(\mathrm{e}^{a/n}\mathrm{e}^{b/n}\big)^n \quad \text{and} \quad \mathrm{e}^{[a,b]} = \lim_{n\to\infty}\big(\mathrm{e}^{-a/n}\mathrm{e}^{-b/n}\mathrm{e}^{a/n}\mathrm{e}^{b/n}\big)^{n^2}.$$

The first of these formulas is sometimes called the *Trotter formula.*

For the following statement we recall the notation $\mathcal{H}(\cdot,\cdot)$ from Definition 1.30.

PROPOSITION 2.49 *Let G be a Banach-Lie group and assume that $\mathfrak{g} := \mathbf{L}(G)$ is equipped with a norm making it into a contractive Banach-Lie algebra. Pick an open neighborhood V of $0 \in \mathfrak{g}$ such that $U := \exp_G(V)$ is an open subset of G and $\exp_G|_V : V \to U$ is a diffeomorphism. Also, let $R \in (0, (1/2)\log 2)$ such that $\exp_G(B_{\mathfrak{g}}(0,R)) \cdot \exp_G(B_{\mathfrak{g}}(0,R)) \subseteq U$. Then there exists $r_0 \in (0,R)$ such that*

$$(\exp_G|_V)^{-1}(\exp_G x \cdot \exp_G y) = \mathcal{H}(x,y)$$

whenever $x, y \in B_{\mathfrak{g}}(0, r_0)$.

PROOF We first recall that $f := (\exp_G|_V)^{-1} : U \to \mathfrak{g}$ is real analytic. Then Theorem 2.44 shows that there exists $r \in (0,R)$ such that

$$f(\exp_G x \cdot \exp_G y) = \sum_{n,m\geq 0} \frac{1}{n!m!}\big((D_{\iota(x)})^n (D_{\iota(y)})^m f\big)(\mathbf{1}) \tag{2.6}$$

whenever $x, y \in B_{\mathfrak{g}}(0,r)$. Moreover, it follows by Proposition 1.29 and Definition 1.30 that there exists $r_0 \in (0,r)$ such that $\mathcal{H}\big(B_{\mathfrak{g}}(0,r_0) \times B_{\mathfrak{g}}(0,r_0)\big) \subseteq B_{\mathfrak{g}}(0,r)$. Since $f \circ \exp_G|_V = \mathrm{id}_V$, it then follows by Theorem 2.44 again that

$$\mathcal{H}(x,y) = f\big(\exp_G(\mathcal{H}(x,y))\big) = \sum_{k=1}^{\infty} \frac{1}{k!}\big((D_{\iota(\mathcal{H}(x,y))})^k f\big)(\mathbf{1}). \tag{2.7}$$

On the other hand, define

$$\varphi\colon \mathfrak{g} \to \mathrm{End}\,\big(\mathcal{C}^\infty(U,\mathfrak{g})\big), \qquad \varphi(v) := D_{\iota(v)}.$$

It then follows by Theorem A.67 along with formula (2.1) in the proof of Theorem 2.12 that φ is a Lie algebra homomorphism. Moreover, by the definition of ι (see Lemma 2.17) along with formula (A.3) in the proof of Lemma A.66, it follows that, if $h \in \mathcal{C}^\infty(U,\mathfrak{g})$, $x \in U$ and $\lim\limits_{n\to\infty} v_n = v$ in $\mathfrak{g}$, then

$$\lim_{n\to\infty} \big(\varphi(v_n)h\big)(x) = \big(\varphi(v)h\big)(x)$$

in $\mathfrak{g}$. Since $\mathcal{H}(x,y) = \sum\limits_{l=1}^{\infty} \beta_l(x,y)$ (see Definition 1.30), the above property of φ implies by (2.7) that

$$\begin{aligned}\mathcal{H}(x,y) &= \sum_{k=0}^{\infty} \frac{1}{k!}\big((\varphi(\mathcal{H}(x,y)))^k f\big)(\mathbf{1}) = (\mathrm{e}^{\varphi(\mathcal{H}(x,y))} f)(\mathbf{1}) = (\mathrm{e}^{\varphi(x)\star\varphi(y)} f)(\mathbf{1}) \\ &= (\mathrm{e}^{\varphi(x)}\mathrm{e}^{\varphi(y)} f)(\mathbf{1}) = f(\exp_G x \cdot \exp_G y)\end{aligned}$$

where the third equality follows by Remark 1.31, the fourth equality is a consequence of Definition 1.21, and the last equality follows by (2.6). The proof is finished. ∎

Notes

In the present book we are interested mainly in infinite-dimensional Lie groups. In particular, according to Definition A.51, the smooth manifolds used in our definition of a Lie group (Definition A.51) are in general infinite dimensional.

There are several brilliant books treating the theory of finite-dimensional Lie groups. Some of them, which might be useful in order to get a better understanding of some topics that we have only sketched here, are [Hel62], [Ho65], [Wa71], and [Kn96]. Our basic references for infinite-dimensional Lie groups are [Ma62], [Bo72], [Mi84], [Up85], and [KM97]. See also [Om74] and [Om97].

The first two sections of the present chapter are mainly based on the papers [Mi84] and [Ma62]. A detailed study of Lie groups arising from topological algebras can be found in [Gl02b], which is our reference for Example 2.21. We refer to [Mi84] and [KM97] for the notions of logarithmic derivatives and regular Lie group. Example 2.38 is taken from [Gl02b]. Theorems 2.39 and 2.42 can be found e.g., in [Bi38] and [Ma62]. Our Theorem 2.44 is an infinite-dimensional version of results like Theorem 4.3 in [Ho65]. Lemma 2.46 in the present chapter is a straightforward extension of Lemma 1.8 in Chapter II of [Hel62]. Theorem 2.47 on exponentials of sums and commutators is a classical fact; compare Proposition 6.7 in [Up85]. See e.g., Theorem 3.1 in Chapter X of [Ho65] for a version of our Proposition 2.49 in the case of finite-dimensional Lie groups.

Chapter 3

Enlargibility

Abstract. One of the central results of this chapter is the fact that the structure of a local Lie group can be enlarged to a structure of a Lie group provided the local Lie group is embeddable into a group. Other main results concern integration of Lie algebra homomorphisms and Lie group structures on closed subgroups of Lie groups. In the second part of the chapter we prove the Kuiper theorem, asserting in particular that both the unitary group and the group of invertible operators on a separable infinite-dimensional Hilbert space are simply connected. Actually any continuous map from a compact space into one of these Lie groups is homotopic to a constant map. It is only towards the end of the chapter that we are ready to approach the problem of whether a given topological Lie algebra is enlargible, in the sense that it corresponds to some Lie group. Enlargibility is a hereditary property. Every finite-dimensional Lie algebra has this property, however there exist non-enlargible infinite-dimensional Lie algebras.

3.1 Integrating Lie algebra homomorphisms

The main result of this section (Theorem 3.5) concerns the way a Lie algebra homomorphism integrates to a Lie group homomorphism. This fact turns out to be closely related to the way a local Lie group structure enlarges to a (global) Lie group structure (see Theorem 3.3 below).

To begin with, let us explain what a local Lie group is.

DEFINITION 3.1 A *local Lie group* is a smooth locally convex manifold K equipped with a distinguished element $k_0 \in K$, an open neighborhood V_0 of $k_0 \in K$, a smooth mapping (the *multiplication*)

$$\mu \colon V_0 \times V_0 \to K$$

and a smooth mapping (the *inversion*)

$$\eta \colon V_0 \to V_0$$

satisfying the following conditions:

(i) For all $x \in V_0$ we have $\mu(x, k_0) = \mu(k_0, x) = x$.

(ii) There exists an open neighborhood W_0 of $k_0 \in K$ such that $\mu(W_0 \times W_0) \subseteq V_0$ (whence, by (i), $W_0 = \mu(\{k_0\} \times W_0) \subseteq V_0$) and

$$\mu(x, \mu(y, z)) = \mu(\mu(x, y), z)$$

whenever $x, y, z \in W_0$.

(iii) We have

$$\mu(x, \eta(x)) = \mu(\eta(x), x) = k_0$$

for all $x \in V_0$.

EXAMPLE 3.2 Let G be a Lie group and K an open neighborhood of $\mathbf{1} \in G$. Pick open neighborhoods V_0 and W_0 of $\mathbf{1} \in G$ such that $V_0 = V_0^{-1}$, $V_0 \cdot V_0 \subseteq K$ and $W_0 \cdot W_0 \subseteq V_0$. Then K is a local Lie group with $k_0 = \mathbf{1}$, $\mu(x, y) = xy$ and $\eta(x) = x^{-1}$ for $x, y \in V_0$.

THEOREM 3.3 *Let G be a connected topological group and assume that K is an open neighborhood of $\mathbf{1} \in G$ such that K has a structure of smooth manifold making it into a local Lie group with respect to the multiplication and inversion mapping inherited from G. Then there exists a unique structure of manifold on G making G into a Lie group such that the original manifold structure of K coincides with the one inherited by K as an open subset of the manifold G.*

PROOF The proof has several steps.

1° Since K is a local Lie group, let $\mathbf{1} \in V_0 \subseteq K$ as in Definition 3.1 such that the multiplication mapping

$$m \colon G \times G \to G$$

has a smooth restriction $m|_{V_0 \times V_0} \colon V_0 \times V_0 \to K$. After shrinking V_0 (if necessary), we may assume that the following properties hold.

(a) There exists a locally convex space E and a mapping $\varphi \colon V_0 \to E$ such that $\varphi(V_0)$ is an open subset of E and $\varphi \colon V_0 \to \varphi(V_0)$ is a diffeomorphism.

(b) We have $V_0 = V_0^{-1}$, $V_0 \cdot V_0 \cdot V_0 \cdot V_0 \subseteq K$ and the mapping

$$m|_{(V_0 \cdot V_0) \times (V_0 \cdot V_0)} \colon (V_0 \cdot V_0) \times (V_0 \cdot V_0) \to K$$

is smooth.

2° For every $a \in G$, note that aV_0 is an open neighborhood of a and define

$$\varphi_a \colon aV_0 \to E, \quad x \mapsto \varphi(a^{-1}x).$$

Then $\varphi_a(aV_0) = \varphi(V_0)$ is an open subset of E, and $\varphi_a \colon aV_0 \to \varphi_a(aV_0)$ is a homeomorphism. Our aim at the present stage of the proof is to check that the family $\{(\varphi_a)^{-1} \colon \varphi_a(aV_0) \to aV_0\}_{a \in G}$ defines a structure of smooth manifold on G modeled on E.

To this end, fix $a_1, a_2 \in G$ such that $U := a_1V_0 \cap a_2V_0 \neq \emptyset$. What we have to show is that the mapping

$$\psi := \varphi_{a_2} \circ (\varphi_{a_1})^{-1}|_{\varphi_{a_1}(U)} \colon \varphi_{a_1}(U) \to \varphi_{a_2}(U)$$

is smooth. In fact, since $a_1V_0 \cap a_2V_0 \neq \emptyset$, there exist $v_1, v_2 \in V_0$ such that $a_1v_1 = a_2v_2$. We then have

$$a_2^{-1}a_1 = v_2v_1^{-1} \in V_0 \cdot V_0^{-1} = V_0 \cdot V_0.$$

On the other hand, note that for all $y \in \varphi_{a_1}(U) = \varphi(a_1^{-1}U)$ we have

$$\begin{aligned}\psi(y) &= \varphi_{a_2}((\varphi_{a_1})^{-1}(y)) = \varphi_{a_2}(a_1\varphi^{-1}(y)) = \varphi(a_2^{-1}a_1\varphi^{-1}(y)) \\ &= (\varphi \circ m)(a_2^{-1}a_1, \varphi^{-1}(y)).\end{aligned}$$

Since $\varphi^{-1}(y) \in a_1^{-1}U \subseteq V_0$ and the mapping $m \colon (V_0 \cdot V_0) \times V_0 \to K$ is smooth (see property (b) in stage 1°), it thus follows that ψ is smooth.

3° We have seen at stage 2° that G has a structure of smooth manifold. We will show at stages 3°–5° that the group operations of G are smooth. To this end, we first prove that, for all $a \in G$, the left translation mapping

$$l_a \colon G \to G, \quad h \mapsto ah,$$

is smooth.

In order to check this fact, fix $h \in G$ and note that the mappings $\varphi_h \colon hV_0 \to E$ and $\varphi_{ah} \colon ahV_0 \to E$ are local charts on G around h and ah, respectively. We have

$$\varphi_{ah} \circ (l_a|_{aV_0}) \circ \varphi_h^{-1} \colon \varphi(V_0) \to \varphi(V_0),$$

and, for all $y \in \varphi(V_0)$,

$$\begin{aligned}\left(\varphi_{ah} \circ (l_a|_{aV_0}) \circ \varphi_h^{-1}\right)(y) &= \varphi_{ah}(l_a(h\varphi^{-1}(y))) = \varphi_{ah}(ah\varphi^{-1}(y)) \\ &= \varphi((ah)^{-1}ah\varphi^{-1}(y)) = y.\end{aligned}$$

In particular, $\varphi_{ah} \circ (l_a|_{aV_0}) \circ \varphi_h^{-1}$ is smooth, and then l_a is smooth.

4° Our aim at this stage is to show that, for all $a \in G$, the continuous mapping

$$I_a \colon G \to G, \quad h \mapsto aha^{-1},$$

is smooth. To this end, consider the set

$$M := \{a \in G \mid I_a \text{ is smooth on some neighborhood of } \mathbf{1} \in G\}$$

and note that

$$M \cdot M \subseteq M \text{ and } V_0 \subseteq M.$$

(The first of these inclusions follows since $I_a \circ I_b = I_{ab}$ for all $a, b \in G$, while the second inclusion follows by property (b) at stage 1°.) The above inclusions show that

$$\bigcup_{n=1}^{\infty} \underbrace{V_0 \cdots V_0}_{n \text{ times}} \subseteq M.$$

Since G is connected, it then follows by Remarks C.8 and C.9(a) that $G = M$.

Consequently, for each $a \in G$, the mapping $I_a \colon G \to G$ is smooth on some neighborhood U of $\mathbf{1} \in K (\subseteq G)$. Now, for every $h \in G$ and $u \in U$ we have

$$(I_a \circ l_h)(u) = ahua^{-1} = aha^{-1} \cdot aua^{-1} = (l_{aha^{-1}} \circ I_a)(u),$$

hence the diagram

$$\begin{array}{ccc} U & \xrightarrow{l_h} & hU \\ {\scriptstyle I_a|_U}\downarrow & & \downarrow{\scriptstyle I_a|_{hU}} \\ G & \xrightarrow{l_{aha^{-1}}} & G \end{array}$$

is commutative. Since both l_h and $l_{aha^{-1}}$ are diffeomorphisms (by stage 3°) and $I_a|_U \colon U \to G$ is smooth, it follows that

$$I_a|_{hU} \colon hU \to G$$

is in turn smooth. Thus for every $h \in G$ we can find the open neighborhood hU of h such that I_a is smooth on hU. That is, $I_a \colon G \to G$ is smooth.

5° Now we are ready to prove that the group operations of G are smooth. For all $a \in G$ denote

$$r_a \colon G \to G, \quad h \mapsto ha,$$

and also

$$\eta \colon G \to G, \quad h \mapsto h^{-1}.$$

Then η is smooth on the neighborhood V_0 of $\mathbf{1} \in G$. For $h \in G$ arbitrary and $v \in V_0$ we have $\eta(hv) = v^{-1}h^{-1} = l_{h^{-1}}(I_h(\eta(v)))$, hence the diagram

$$\begin{array}{ccc} V_0 & \xrightarrow{l_h} & hV_0 \\ {\scriptstyle \eta}\downarrow & & \downarrow{\scriptstyle \eta} \\ G & \xrightarrow{l_{h^{-1}} \circ I_h} & G \end{array}$$

is commutative. Since both mappings l_h and $l_{h^{-1}} \circ I_h$ are diffeomorphisms (by stages 3° and 4°) and $\eta|_{V_0}$ is smooth, it then follows that η is smooth on the neighborhood hV_0 of h. Since $h \in G$ was arbitrary, we deduce that η is smooth throughout G.

Finally, note that for all $a_1, a_2, g, h \in G$ we have

$$m(a_1 g, a_2 h) = a_1 a_2 \cdot a_2^{-1} g a_2 \cdot h = (l_{a_1 a_2} \circ m)(I_{a_2^{-1}}(g), h),$$

hence the diagram

$$\begin{array}{ccc} W_0 \times W_0 & \xrightarrow{l_{a_1} \times l_{a_2}} & a_1 W_0 \times a_2 W_0 \\ {\scriptstyle I_{a_2^{-1}} \times \mathrm{id}_G} \downarrow & & \downarrow {\scriptstyle m} \\ V_0 \times V_0 & \xrightarrow{l_{a_1 a_2} \circ m} & G \end{array}$$

is commutative, where W_0 is a neighborhood of $\mathbf{1} \in G$ such that $\mathbf{1} \in W_0 \subseteq V_0$ and $I_{a_2^{-1}}(W_0) \subseteq V_0$. Since $l_{a_1 a_2} \circ m \colon V_0 \times V_0 \to G$ is smooth, and the mappings $l_{a_1} \times l_{a_2}$ and $I_{a_2^{-1}} \times \mathrm{id}_G$ are diffeomorphisms of $G \times G$ onto itself, it follows by the above commutative diagram that the multiplication mapping m is smooth on the neighborhood $a_1 W_0 \times a_2 W_0$ of $(a_1, a_2) \in G \times G$. Since the point $(a_1, a_2) \in G \times G$ is arbitrary, we get that $m \colon G \times G \to G$ is smooth.

6° It only remained to explain the uniqueness assertion. To this end, let V_0 be as at stage 1°, and G_1 and G_2 two structures of Lie group on the topological group G such that both G_1 and G_2 induce the initial manifold structure of K. We have already seen at stage 4° that

$$G = \bigcup_{n=1}^{\infty} \underbrace{V_0 \cdots V_0}_{n \text{ times}}. \tag{3.1}$$

For every $n \geq 2$ and $v_1, \dots, v_{n-1} \in V_0$ we have the commutative diagram

$$\begin{array}{ccc} V_0 & \xrightarrow{\mathrm{id}_{V_0}} & V_0 \\ {\scriptstyle l^{G_1}_{v_1 \cdots v_{n-1}}} \downarrow & & \downarrow {\scriptstyle l^{G_2}_{v_1 \cdots v_{n-1}}} \\ v_1 \cdots v_{n-1} V_0 & \xrightarrow{\mathrm{id}_{v_1 \cdots v_{n-1} V_0}} & v_1 \cdots v_{n-1} V_0 \end{array}$$

where the columns are the left translations $g \mapsto v_1 \cdots v_{n-1} g$ in the Lie groups G_1 and G_2, respectively. Since the upper horizontal arrow is smooth by the assumption on G_1 and G_2, it follows that the lower horizontal arrow is smooth as well. Thus $\mathrm{id}_G \colon G_1 \to G_2$ is smooth on a neighborhood of $v_1 \cdots v_{n-1} \in G$. Then (3.1) implies that $\mathrm{id}_G \colon G_1 \to G_2$ is smooth. Similarly, the mapping $\mathrm{id}_G \colon G_2 \to G_1$ is smooth, hence the manifold structures of the Lie groups G_1 and G_2 coincide. ∎

REMARK 3.4 In the setting of Theorem 3.3, if the manifold underlying the local group K is actually real analytic and the multiplication and inversion are real analytic on some neighborhood of $\mathbf{1} \in K$, then the manifold underlying the Lie group G is in turn real analytic and the group operations of G are real analytic mappings.

THEOREM 3.5 *Let H be a Banach-Lie group, $\mathfrak{g}$ a real Banach-Lie algebra, and $\varphi\colon \mathfrak{g} \to \mathbf{L}(H)$ an injective homomorphism of topological Lie algebras. Then there exists a connected Banach-Lie group G and an injective homomorphism of Lie groups $f\colon G \to H$ such that $\mathfrak{g} = \mathbf{L}(G)$ and $\varphi = \mathbf{L}(f)$.*

PROOF Denote $\mathfrak{h} = \mathbf{L}(H)$. It follows by Theorem 2.42(i) that there exist an open neighborhood V of $\mathbf{1} \in H$ and a real number $R > 0$ such that the exponential map of H induces a diffeomorphism $\exp_H|_{B_{\mathfrak{h}}(0,R)}\colon B_{\mathfrak{h}}(0,R) \to V$.

On the other hand, since $\varphi\colon \mathfrak{g} \to \mathfrak{h}$ is continuous, there exists $r > 0$ such that $\varphi(B_{\mathfrak{g}}(0,r)) \subseteq B_{\mathfrak{h}}(0,R)$. Denote

$$G := \langle(\exp_H \circ\varphi)(B_{\mathfrak{g}}(0,r))\rangle \subseteq H$$

(i.e., G is the subgroup of H generated by $(\exp_H \circ\varphi)(B_{\mathfrak{g}}(0,r))$). We will equip the group G with a structure of connected Banach-Lie group such that $\mathbf{L}(G) = \mathfrak{g}$ and the inclusion mapping $f\colon G \hookrightarrow H$ satisfies $\mathbf{L}(f) = \varphi$.

To this end, denote

$$K := (\exp_H \circ\varphi)(B_{\mathfrak{g}}(0,r)).$$

There exists a unique Banach manifold structure (in particular a Hausdorff topology) on K such that the bijective map $\theta := \exp_H\colon \varphi|_{B_{\mathfrak{g}}(0,r)}\colon B_{\mathfrak{g}}(0,r) \to K$ is a diffeomorphism. Now recall from Remark 1.27 that the Banach-Lie algebra $\mathfrak{g}$ may be assumed contractive. It then follows by Proposition 1.33 that $B_{\mathfrak{g}}(0,r)$ has a structure of local Lie group with the multiplication defined by the Baker-Campbell-Hausdorff series $\mathcal{H}(\cdot,\cdot)$, and the inversion given by $x \mapsto -x$ (see also Proposition 1.23). Then the diffeomorphism $\theta\colon B_{\mathfrak{g}}(0,r) \to K$ allows us to define a structure of local Lie group on K with respect to the multiplication and the inversion inherited from G (see Proposition 2.49). Since the subgroup G of H is generated by K, it follows by Theorem C.11 that G has a unique group topology such that the inclusion map $K \hookrightarrow G$ is an embedding and K is an open neighborhood of $\mathbf{1} \in G$. Since K is connected (being homeomorphic to the ball $B_{\mathfrak{g}}(0,r)$), we deduce by Remark C.9(b) that the topological group G is connected.

We have seen above that K has a structure of a local Lie group, hence it follows by Theorem 3.3 that G possesses a structure of a Lie group modeled on the Banach space underlying $\mathfrak{g}$. (The latter fact follows since $\theta\colon B_{\mathfrak{g}}(0,r) \to K$ is a diffeomorphism of $B_{\mathfrak{g}}(0,r)$ onto the open neighborhood K of $\mathbf{1} \in G$.)

We now show that we have even $\mathbf{L}(G) = \mathfrak{g}$. To this end, first use Remark 1.27 again to see that we may assume that the Banach-Lie algebra $\mathfrak{h} = \mathbf{L}(H)$ is contractive. Then pick $r_0, R_0 \in (0, (\log 2)/2)$ such that $r_0 < r$ and $\varphi(B_{\mathfrak{g}}(0, r_0)) \subseteq B_{\mathfrak{h}}(0, R_0)$. Then for all for all $x, y \in B_{\mathfrak{g}}(0, r_0)$ we have

$$\theta(\mathcal{H}(x, y)) = \exp_H(\varphi(\mathcal{H}(x, y))) = \exp_H(\varphi(x)) \cdot \exp_H(\varphi(y)) = \theta(x) \cdot \theta(y)$$

where the second equality follows by Remark 1.34, while the third one follows by Proposition 2.49. (Actually this is the way we defined above the multiplication and the inversion mapping in the local Lie group K.) Now, since $\theta\colon B_{\mathfrak{g}}(0, r) \to K$ is a local chart around $\mathbf{1} \in G$, it easily follows by the above equalities that the bracket of $\mathfrak{g}$ agrees with the one of $\mathbf{L}(G)$, that is, $\mathbf{L}(G) = \mathfrak{g}$. It also follows that the inclusion mapping $f\colon G \hookrightarrow H$ is smooth around $\mathbf{1} \in G$. Since f is a group homomorphism, and both G and H are Lie groups, it then follows that f is smooth throughout G.

To conclude the proof, we have to show that $\mathbf{L}(f) = \varphi$. In fact, for $x \in B_{\mathfrak{g}}(0, r)$, the path

$$t \mapsto \theta(tx)$$

can be extended to a one-parameter subgroup of G (see Remark 2.32) whose image by f is $t \mapsto \exp_H(\varphi(tx))$. Thus $\mathbf{L}(f)x = \varphi(x)$ for $x \in B_{\mathfrak{g}}(0, r)$, whence $\mathbf{L}(f) = \varphi$ on $\mathfrak{g}$. ∎

REMARK 3.6 Let G be a group and G_0 a normal subgroup of G. Assume that G_0 is equipped with a structure of Lie group such that for every $g \in G$ the map

$$G_0 \to G_0, \quad h \mapsto ghg^{-1},$$

is smooth. Then, using arguments from stages 2°, 3°, and 5° in the proof of Theorem 3.3, it is not hard to prove that there exist on G a uniquely determined topology and a manifold structure making G into a Lie group such that G_0 is at the same time closed and open in G and the original manifold structure of G_0 coincides with the manifold structure inherited by G_0 as an open subset of G. Moreover, if the manifold underlying the Lie group G_0 were real analytic and the multiplication and inversion in G_0 were real analytic maps, then the similar properties hold for the Lie group G.

COROLLARY 3.7 *Assume that H is a Banach-Lie group with the Lie algebra $\mathfrak{h}$, G is a closed subgroup of H and denote*

$$\mathfrak{g} = \{x \in \mathfrak{h} \mid (\forall t \in \mathbb{R}) \quad \exp_G(tx) \in G\}.$$

Then $\mathfrak{g}$ is a closed subalgebra of $\mathfrak{h}$ and there exist on G uniquely determined topology τ and manifold structure making G into a Banach-Lie group such *that* $\mathbf{L}(G) = \mathfrak{g}$, *the inclusion map* $G \hookrightarrow H$ *is smooth and the diagram*

$$\begin{array}{ccc} \mathfrak{g} & \longrightarrow & \mathfrak{h} \\ {\scriptstyle \exp_G}\downarrow & & \downarrow{\scriptstyle \exp_H} \\ G & \longrightarrow & H \end{array}$$

is commutative, where the horizontal arrows stand for inclusion maps.

PROOF The fact that $\mathfrak{g}$ is a closed subalgebra of $\mathfrak{h}$ follows by Theorem 2.47 since G is closed. Now denote $G_0 = \langle \exp_H \mathfrak{g} \rangle$ (the subgroup of H generated by $\exp_H \mathfrak{g}$).

It follows by Theorem 3.5 (and its proof) that G_0 has a uniquely determined structure of a connected Banach-Lie group such that $\mathbf{L}(G_0) = \mathfrak{g}$, the inclusion map $f\colon G_0 \hookrightarrow H$ is smooth and $\varphi := \mathbf{L}(f)\colon \mathfrak{g} \hookrightarrow \mathfrak{h}$ is the inclusion map.

On the other hand, for all $g \in G$, $t \in \mathbb{R}$ and $x \in \mathfrak{g}$ we have

$$\exp_H(g \cdot tx \cdot g^{-1}) = g \cdot \exp_H(tx) \cdot g^{-1} \in G$$

whence $g \cdot x \cdot g^{-1} \in \mathfrak{g}$. Then the above equality shows that $g \cdot \exp_H \mathfrak{g} \cdot g^{-1} \subseteq \exp_H \mathfrak{g}$, which implies that G_0 is a normal subgroup of G. Note that the above equality along with Theorem 2.42(i) imply that for all $g \in G$ the map $G_0 \to G_0$, $h \mapsto ghg^{-1}$ is smooth on some neighborhood of $\mathbf{1} \in G_0$, hence it is smooth throughout G_0 (since that map is actually an automorphism of the group G_0). Now we can make use of Remark 3.6 to endow G with a Lie group structure such that G_0 is at the same closed and open in G. In particular G_0 is just the connected $\mathbf{1}$-component in G and $\mathbf{L}(G) = \mathbf{L}(G_0) = \mathfrak{g}$. The other assertions then follow at once. ∎

COROLLARY 3.8 *In the setting of* Corollary 3.7, *if the Lie group* H *is finite dimensional, then the topology* τ *coincides with the topology inherited by* G *from* H.

PROOF We have $\dim \mathfrak{g} \leq \dim \mathfrak{h} < \infty$, hence the Lie group G is in turn finite dimensional. In particular, the topology τ is locally compact. Now recall that continuous bijective maps from compact spaces onto Hausdorff topological spaces are always homeomorphisms. It then easily follows that the continuous map $\mathrm{id}_G\colon (G, \tau) \to G$ is actually a homeomorphism, and we are done. ∎

For the next statement we recall the notation $\mathbb{T} = \{z \in \mathbb{C} \mid |z| = 1\}$ for the unit circle in the complex plane.

COROLLARY 3.9 *Let* $\gamma\colon \mathbb{R} \to \mathbb{T}^2$ *be a one-parameter subgroup of the 2-dimensional Lie group* $\mathbb{T}^2$. *If* $\operatorname{Ker} \gamma = \{0\}$, *then* $\gamma(\mathbb{R})$ *is dense in* $\mathbb{T}^2$.

PROOF Denote $H = \mathbb{T}^2$ and let G be the closure of $\gamma(\mathbb{R})$. It then follows by Corollaries 3.7 and 3.8 that G has a structure of Lie group with respect to the topology inherited from H, and the Lie algebra of G is a subalgebra of $\mathbf{L}(H)$. In particular, $\dim G \leq 2$ and G is compact. Moreover, G is connected, as the closure of a continuous image of the connected space $\mathbb{R}$.

If $\dim G = 0$, then G reduces to a set of isolated points in H. Since G is connected, we see that $G = \{\mathbf{1}\}$. Thus we get the contradiction $\operatorname{Ker}\gamma = \mathbb{R}$.

If $\dim G = 1$, then consider the linear map $\mathbf{L}(\gamma)\colon \mathbf{L}(\mathbb{R}) = \mathbb{R} \to \mathbf{L}(G) = \mathbb{R}$. (Note that actually $\gamma\colon \mathbb{R} \to G$.) Thus either $\mathbf{L}(\gamma) = 0$ or $\mathbf{L}(\gamma)$ is an isomorphism. In the first case we have $\{\mathbf{1}\} = (\exp_G \circ \mathbf{L}(\gamma))(\mathbf{L}(\mathbb{R})) = (\gamma \circ \exp_{\mathbb{R}})(\mathbf{L}(\mathbb{R})) = \gamma(\mathbb{R})$, a contradiction. Hence $\mathbf{L}(\gamma)\colon \mathbf{L}(\mathbb{R}) \to \mathbf{L}(G)$ is an isomorphism, and then $\gamma\colon \mathbb{R} \to G$ is a local homeomorphism at $0 \in \mathbb{R}$. This implies that the subgroup $\gamma(\mathbb{R})$ of G is open. But $\gamma(\mathbb{R})$ is dense in G, hence $\gamma(\mathbb{R}) = G$, and then $\gamma\colon \mathbb{R} \to G$ is a homeomorphism. Since G is compact while $\mathbb{R}$ is not, we again get a contradiction.

Thus we must have $\dim G = 2 = \dim H$. Then G has to be an open subgroup of H. But G is also closed and H is connected, hence $G = H$, and we are done. ∎

REMARK 3.10 Let us describe a way to construct one-parameter subgroups of the torus $\mathbb{T}^2$ to which Corollary 3.9 applies. To this end, note that the torus can be expressed as a quotient group of additive groups $\mathbb{T}^2 = \mathbb{R}^2/\mathbb{Z}^2$, and denote by

$$\pi\colon \mathbb{R}^2 \to \mathbb{T}^2, \quad (x,y) \mapsto (x,y) + \mathbb{Z}^2$$

the canonical projection, which is a homomorphism of Lie groups.

Now pick $\alpha \in \mathbb{R} \setminus \mathbb{Q}$. Then

$$\gamma_\alpha\colon \mathbb{R} \to \mathbb{T}^2, \quad \gamma_\alpha(t) = \pi(t, \alpha t),$$

is a one-parameter subgroup of $\mathbb{T}^2$ with $\ker\gamma_\alpha = \{0\}$. In fact, if $t \in \operatorname{Ker}\gamma_\alpha$, then $(t, \alpha t) \in \mathbb{Z}^2$. Thus $t, \alpha t \in \mathbb{Z}$, and then $t = 0$ since otherwise we get $\alpha = (\alpha t)/t \in \mathbb{Q}$, a contradiction. Consequently $\operatorname{Ker}\gamma_\alpha = \{0\}$, and then $\gamma_\alpha(\mathbb{R})$ is dense in $\mathbb{T}^2$ by Corollary 3.9.

A one-parameter subgroup of $\mathbb{T}^2$ of the type γ_α with $\alpha \in \mathbb{R} \setminus \mathbb{Q}$ is usually called a *dense wind*.

REMARK 3.11 In the setting of Corollary 3.7, assume that there exist an open neighborhood V of $0 \in \mathfrak{h}$ and an open neighborhood U of $\mathbf{1} \in H$ such that $\exp_H$ induces a diffeomorphism of V onto U and $\exp_H(V \cap \mathfrak{g}) = U \cap H$. Then the topology τ coincides with the topology inherited by G from H.

REMARK 3.12 Let $\mathcal{H}$ be a complex Hilbert space and consider the group $\mathrm{U}(\mathcal{H})$ of all unitary operators on $\mathcal{H}$. Also denote

$$\mathfrak{u}(\mathcal{H}) = \{A \in \mathcal{B}(\mathcal{H}) \mid A^* = -A\}.$$

Then $\mathrm{U}(\mathcal{H})$ is a closed connected subgroup of the Banach-Lie group $\mathrm{GL}(\mathcal{H})$ and the topology τ referred to in Corollary 3.7 is in this case just the norm topology of $\mathrm{U}(\mathcal{H})$, that is, the topology inherited from $\mathrm{GL}(\mathcal{H})$. Moreover $\mathfrak{u}(\mathcal{H})$ equals the Lie algebra of the Banach-Lie group $\mathrm{U}(\mathcal{H})$.

REMARK 3.13 Let G and H be Banach-Lie groups such that G is simply connected. Then Theorem C.19 can be used to show that for every homomorphism of topological Lie algebras $\varphi\colon \mathbf{L}(G) \to \mathbf{L}(H)$ there exists a homomorphism of Lie groups $f\colon G \to H$ with $\mathbf{L}(f) = \varphi$.

3.2 Topological properties of certain Lie groups

Our aim in this section is to prove that any continuous mapping from a compact space into the unitary group of an infinite-dimensional separable Hilbert space is homotopic to a constant mapping (see Corollary 3.29 below). In order to obtain this result, we first prove the similar property for the group of invertible operators (Theorem 3.15).

NOTATION 3.14 Throughout this section we denote by $\mathcal{H}$ an *infinite-dimensional* separable Hilbert space over $\mathbb{K} \in \{\mathbb{R}, \mathbb{C}\}$ and consider both groups

$$G := \mathrm{GL}(\mathcal{H}) = \{T \in \mathcal{B}(\mathcal{H}) \mid T \text{ invertible}\}$$

and

$$U := \mathrm{U}(\mathcal{H}) = \{T \in \mathcal{B}(\mathcal{H}) \mid T^*T = TT^* = \mathrm{id}_{\mathcal{H}}\}$$

equipped with the operator norm topology.

Moreover, for all $T_0 \in \mathcal{B}(\mathcal{H})$ and $\varepsilon > 0$ we denote

$$B(T_0, \varepsilon) = \{T \in \mathcal{B}(\mathcal{H}) \mid \|T - T_0\| < \varepsilon\}.$$

THEOREM 3.15 *If S is a compact topological space, then for every continuous map $f\colon S \to G$ there exists a homotopy $H\colon [0,1] \times S \to G$ such that $H(0,s) = f(s)$ and $H(1,s) = \mathrm{id}_{\mathcal{H}}$ for all $s \in S$.*

The proof of this theorem will be achieved by means of several lemmas. Until after that proof we will make use of the notation from the statement of Theorem 3.15.

LEMMA 3.16 *There exists a homotopy*

$$[0,1] \times S \to G, \quad (t,s) \mapsto f_t(s),$$

such that $f_0 = f$ *and*

$$\mathcal{W} := \mathrm{sp}_{\mathbb{K}}(f_1(S))$$

is a finite-dimensional subspace of $\mathcal{B}(\mathcal{H})$.

PROOF Denote $f_0 := f$, according to the statement. The construction of the desired homotopy has three steps.

1° Since S is compact and f_0 is continuous, it follows that $f_0(S)$ is a compact subset of G. Since G is an open subset of $\mathcal{B}(\mathcal{H})$, it then easily follows that there exist $N \geq 1$, $\varepsilon_1, \ldots, \varepsilon_N > 0$ and $Z_1, \ldots, Z_N \in f_0(S)$ such that $\varepsilon_1 \leq \cdots \leq \varepsilon_N$, and moreover

$$f_0(S) \subseteq \mathcal{D} := \bigcup_{j=1}^{N} B(Z_j, \varepsilon_j)$$

and $B(Z_j, 3\varepsilon_j) \subseteq G$ for $j = 1, \ldots, N$.

Now for $j = 1, \ldots, N$ define

$$\psi_j \colon \mathcal{D} \to [0, \infty), \quad \psi_j(Z) := \max(\varepsilon_j - \|Z - Z_j\|, 0).$$

If $Z \in B(Z_j, \varepsilon_j)$, then $\psi_j(Z) > 0$, hence $\sum\limits_{j=1}^{N} \psi_j > 0$ on $\mathcal{D}$. Then we can define the continuous functions

$$\varphi_1, \ldots, \varphi_N \colon \mathcal{D} \to [0, 1], \quad \varphi_j = \frac{\psi_j}{\psi_1 + \cdots + \psi_N} \text{ for } j = 1, \ldots, N.$$

We have $\varphi_1 + \cdots + \varphi_N = 1$ and $\operatorname{supp} \varphi_j = \operatorname{supp} \psi_j = \overline{B(Z_j, \varepsilon_j)}$ for $j = 1, \ldots, N$.

2° Now for all $t \in [0, 1]$ we define

$$g_t \colon \mathcal{D} \to \mathcal{B}(\mathcal{H}), \quad g_t(Z) = (1 - t)Z + t \sum_{j=1}^{N} \varphi_j(Z) Z_j.$$

It is clear that $g_0 = \mathrm{id}_{\mathcal{D}}$ and $g_1(Z) \subseteq \mathrm{co}\{Z_1, \ldots, Z_N\}$. We now claim that for each $t \in [0, 1]$ we have $g_t(\mathcal{D}) \subseteq G$.

In fact, for $Z \in \mathcal{D}$ arbitrary, denote

$$\{i \mid Z \in B(Z_i, \varepsilon_i)\} =: \{i_1, \ldots, i_l\}$$

with $i_1 < \cdots < i_l$. Then $B(Z_{i_1}, \varepsilon_{i_1}) \cup \cdots \cup B(Z_{i_l}, \varepsilon_{i_l}) \subseteq B(Z_{i_l}, 3\varepsilon_{i_l})$ since $\varepsilon_{i_1} \leq \cdots \leq \varepsilon_{i_l}$ and $\bigcap\limits_{k=1}^{l} B(Z_{i_k}, \varepsilon_{i_k}) \neq \emptyset$. Since $g_t(Z) \in \mathrm{co}\Big(\{Z\} \cup \bigcup\limits_{k=1}^{l} B(Z_{i_k}, \varepsilon_{i_k})\Big)$, we get $g_t(Z) \in B(Z_{i_l}, 3\varepsilon_{i_l}) \subseteq G$, as claimed.

3° According to the properties of g_t, we can define $f_t \colon S \to G$, $f_t := g_t \circ f$, to obtain a continuous map

$$[0, 1] \times S \to G, \quad (t, s) \mapsto f_t(s),$$

such that the subspace $\mathrm{sp}_{\mathbb{K}}(f_1(S))$ of $\mathcal{B}(\mathcal{H})$ is finite dimensional. In fact, $f_1(S) = g_1(S) \subseteq g_1(\mathcal{D}) \subseteq \mathrm{sp}_{\mathbb{K}}\{Z_1, \ldots, Z_N\}$. To conclude the proof, note that $f_0 = g_0 \circ f = \mathrm{id}_{\mathcal{D}} \circ f = f$, hence the notation for f_0 agrees with the one introduced at the very beginning of the proof. ∎

DEFINITION 3.17 In what follows, we make use of the notation $\mathcal{W} := \mathrm{sp}_{\mathbb{K}}(f_1(S))$ introduced in the statement of Lemma 3.16. Moreover, we denote

$$N := \dim_{\mathbb{K}} \mathcal{W}$$

and pick a basis $\{W_1, \ldots, W_N\} \subseteq f_1(S)$ $(\subseteq G)$ of $\mathcal{W}$.

On the other hand, we denote

$$M := \sup_{s \in S} \max\{\|f_1(s)\|, \|f_1(s)^{-1}\|\} \ (< \infty)$$

(note that $f_1(S)$ is a compact subset of G!) and

$$\mathcal{W}_M := \{W \in \mathcal{W} \cap G \mid \max\{\|W\|, \|W^{-1}\|\} \leq M\},$$

so that $f_1(S) \subseteq \mathcal{W}_M$.

REMARK 3.18 For every vector $x \in \mathcal{H}$ we have $\dim_{\mathbb{K}}\big(\mathrm{sp}_{\mathbb{K}}(f_1(S)x)\big) = \dim_{\mathbb{K}}(\mathcal{W}x) \leq N$.

DEFINITION 3.19 Now we are going to construct inductively three sequences of objects

$$\begin{aligned} &a_1, a_2, \ldots \in \mathcal{H}, \quad \|a_i\| = 1 \text{ for all } i \geq 1, \\ &a_1^0, a_2^0, \ldots \in \mathcal{H}, \quad \|a_i^0\| = 1 \text{ for all } i \geq 1, \\ &A_1, A_2, \ldots \text{ linear subspaces of } \mathcal{H}, \ \dim_{\mathbb{K}}(A_i) = N + 2 \text{ for all } i \geq 1 \end{aligned}$$

such that

- $\mathrm{sp}_{\mathbb{K}}(\{a_i\} \cup \mathcal{W}a_i) \perp (A_1 + \cdots + A_{i-1})$,
- $a_i^0 \perp \big(\mathrm{sp}_{\mathbb{K}}(\{a_i\} \cup \mathcal{W}a_i) + (A_1 + \cdots + A_{i-1})\big)$,
- $A_i \supseteq \mathrm{sp}_{\mathbb{K}}(\{a_i, a_i^0\} \cup \mathcal{W}a_i)$, and
- $A_i \perp (A_1 + \cdots + A_{i-1})$.

To this end, pick $a_1 \in \mathcal{H}$ arbitrary with $\|a_1\| = 1$. Then take $a_1^0 \in \mathcal{H}$ such that $\|a_1^0\| = 1$ and $a_1^0 \perp \mathrm{sp}_{\mathbb{K}}(\{a_1\} \cup \mathcal{W}a_1)$. It then follows by Remark 3.18 that we can find a linear subspace A_1 of $\mathcal{H}$ such that

$$A_1 \supseteq \mathrm{sp}_{\mathbb{K}}(\{a_1, a_1^0\} \cup \mathcal{W}a_1) \text{ and } \dim_{\mathbb{K}}(A_1) = N + 2.$$

Now assume that $a_1,\dots,a_{i-1}$, $a_1^0,\dots,a_{i-1}^0$, $A_1,\dots,A_{i-1}$ have been already constructed. Then each of the subspaces $A_k^\perp$ and $W_j^{-1}(A_k^\perp)$ has finite codimension (where the W_j's are the ones introduced in Definition 3.17), hence $\bigcap_{k=1}^{i-1}\Big(A_k^\perp\cap(\bigcap_{j=1}^{N} W_j^{-1}(A_k^\perp))\Big)\neq\{0\}$, since $\dim_{\mathbb{K}}\mathcal{H}=\infty$. Let us pick an arbitrary unit vector a_i in that nonzero intersection. Then we have $\mathrm{sp}_{\mathbb{K}}(\{a_i\}\cup\mathcal{W}a_i)\perp(A_1+\cdots+A_{i-1})$. Now let a_i^0 be an arbitrary unit vector with $a_i^0\perp\big(\mathrm{sp}_{\mathbb{K}}(\{a_i\}\cup\mathcal{W}a_i)+(A_1+\cdots+A_{i-1})\big)$. Then Remark 3.18 again shows that there exists a linear subspace A_i of $\mathcal{H}$ with the desired properties.

DEFINITION 3.20 For each $i\geq 1$ we define

$$\kappa_i\colon[0,1]\times\mathcal{W}_M\to U\ (\subseteq G)$$

in the following way. Let $W\in\mathcal{W}_M$ be fixed. If $t\in[0,1/2]$ then we define

$$\kappa_i(t,W)=\begin{pmatrix}\cos\pi t & -\sin\pi t & 0\\ \sin\pi t & \cos\pi t & 0\\ 0 & 0 & \mathrm{id}_{\{Wa_i,a_i^0\}^\perp}\end{pmatrix}\colon\mathcal{H}\to\mathcal{H} \tag{3.2}$$

With respect to the orthogonal decomposition $\mathcal{H}=\mathbb{K}Wa_i\oplus\mathbb{K}a_i^0\oplus\{Wa_i,a_i^0\}^\perp$. On the other hand, if $t\in[1/2,1]$, we define $\kappa_i(t,W)\colon\mathcal{H}\to\mathcal{H}$ such that

$$\kappa_i(t,W)=\begin{pmatrix}\cos\pi(t-\frac12) & -\sin\pi(t-\frac12) & 0\\ \sin\pi(t-\frac12) & \cos\pi(t-\frac12) & 0\\ 0 & 0 & \mathrm{id}_{\{a_i^0,a_i\}^\perp}\end{pmatrix}\kappa_i(\tfrac12,W)\colon\mathcal{H}\to\mathcal{H} \tag{3.3}$$

with respect to the orthogonal decomposition $\mathcal{H}=\mathbb{K}a_i^0\oplus\mathbb{K}a_i\oplus\{a_i^0,a_i\}^\perp$.

REMARK 3.21 Let us record some basic properties of the maps $\kappa_i\colon[0,1]\times\mathcal{W}_M\to U$ constructed in Definition 3.20.

(i) Both maps $\kappa_i|_{[0,1/2]\times\mathcal{W}_M}$ and $\kappa_i|_{[1/2,1]\times\mathcal{W}_M}$ are continuous and agree on the set $\{1/2\}\times\mathcal{W}_M$, hence $\kappa_i\colon[0,1]\times\mathcal{W}_M\to U$ is continuous for all $i\geq 1$.

(ii) We have

$$(\forall W\in\mathcal{W}_M)\quad \kappa_i(0,W)=\mathrm{id}_{\mathcal{H}}$$

for all $i\geq 1$.

(iii) Since $A_i\supseteq\mathrm{sp}_{\mathbb{K}}(\{a_i,a_i^0\}\cup\mathcal{W}a_i)$, we have $A_i^\perp\subseteq(\{a_i,a_i^0\}\cup\mathcal{W}a_i)^\perp$, and thus

$$(\forall t\in[0,1])(\forall W\in\mathcal{W}_M)\quad \kappa_i(t,W)|_{A_i^\perp}=\mathrm{id}_{A_i^\perp}$$

for all $i\geq 1$.

(iv) Let $W \in \mathcal{W}_M$ arbitrary. With respect to the orthogonal decomposition $\mathcal{H} = \mathbb{K}Wa_i \oplus \mathbb{K}a_i^0 \oplus \{Wa_i, a_i^0\}^\perp$ we have

$$\kappa_i(\tfrac{1}{2}, W) = \begin{pmatrix} 0 & -1 & 0 \\ 1 & 0 & 0 \\ 0 & 0 & \mathrm{id}_{\{Wa_i, a_i^0\}^\perp} \end{pmatrix}$$

hence $\big(\kappa_i(1/2, W)\big)Wa_i = \|Wa_i\|a_i^0$. On the other hand,

$$\kappa_i(1, W) = \begin{pmatrix} 0 & -1 & 0 \\ 1 & 0 & 0 \\ 0 & 0 & \mathrm{id}_{\{a_i^0, a_i\}^\perp} \end{pmatrix} \kappa_i(\tfrac{1}{2}, W),$$

where the matrix is computed according to the orthogonal decomposition $\mathcal{H} = \mathbb{K}a_i^0 \oplus \mathbb{K}a_i \oplus \{a_i^0, a_i\}^\perp$. Since $\big(\kappa_i(1/2, W)\big)Wa_i = \|Wa_i\|a_i^0$, we eventually get

$$\big(\kappa_i(1, W)\big)Wa_i = \|Wa_i\|a_i$$

for all $W \in \mathcal{W}_M$ and $i \geq 1$.

REMARK 3.22 If $x, y \in \mathcal{H}$, $\theta \in [0, \pi/2]$ and $u \in \mathbb{K}$ satisfy the conditions $(x|y) = \|x\| \cdot \|y\| \cdot u \cdot \cos\theta$ and $|u| = 1$, then a straightforward computation shows that $\|x - y\| \geq 2\sin(\theta/2)\min\{\|x\|, \|y\|\}$.

LEMMA 3.23 *The family $\{\kappa_i\}_{i\geq 1}$ is equi-continuous.*

PROOF First note that for all $i \geq 1$, $t, t' \in [0, 1]$ and $W, W' \in \mathcal{W}_M$ we have

$$\|\kappa_i(t', W') - \kappa_i(t, W)\| \leq \|\kappa_i(t', W') - \kappa_i(t, W')\| + \|\kappa_i(t, W') - \kappa_i(t, W)\|.$$

Since both $\kappa_i(t, W)$ and $\kappa_i(t, W')$ are unitary operators, we further obtain that

$$\begin{aligned} \|\kappa_i(t', W') - \kappa_i(t, W)\| \leq & \|\kappa_i(t', W')\kappa_i(t, W')^{-1} - \mathrm{id}_{\mathcal{H}}\| \\ & + \|\kappa_i(t, W')\kappa_i(t, W)^{-1} - \mathrm{id}_{\mathcal{H}}\|. \end{aligned} \tag{3.4}$$

On the other hand, note that for all $\theta \in [0, \pi]$ we have

$$\left\| I_2 - \begin{pmatrix} \cos\theta & -\sin\theta \\ \sin\theta & \cos\theta \end{pmatrix} \right\| = \left\| 2\sin\frac{\theta}{2} \begin{pmatrix} 2\sin\frac{\theta}{2} & -\cos\frac{\theta}{2} \\ \cos\frac{\theta}{2} & \sin\frac{\theta}{2} \end{pmatrix} \right\| = 2\sin\frac{\theta}{2} \leq \theta. \tag{3.5}$$

It then follows by (3.2) and (3.3) (in Definition 3.20) that

$$\|\kappa_i(t', W')\kappa_i(t, W')^{-1} - \mathrm{id}_{\mathcal{H}}\| \leq \pi|t - t'| \tag{3.6}$$

whenever $W' \in \mathcal{W}_M$ and either $t, t' \in [0, 1/2]$ or $t, t' \in [1/2, 1]$. We thus have an appropriate estimate of the first term in the right-hand side of (3.4).

The second term in the right-hand side of (3.4) is constant for $1/2 \leq t \leq 1$ (see (3.3)), hence it suffices to estimate it only for $0 \leq t \leq 1/2$. For such values of t, we are going to prove that

$$\|\kappa_i(t, W')\kappa_i(t, W)^{-1} - \mathrm{id}_{\mathcal{H}}\| \leq 2M\|W - W'\|. \tag{3.7}$$

To this end, let μ stand for the left-hand side of (3.7). If $\mathbb{K}Wa_i = \mathbb{K}W'a_i$, it follows by (3.2) that $\mu = 0$, hence (3.7) holds. Now assume that $\mathbb{K}Wa_i \neq \mathbb{K}W'a_i$ and denote

$$\mathcal{Y} := \mathrm{sp}_{\mathbb{K}}\{a_i^0, Wa_i, W'a_i\}.$$

Then $\dim_{\mathbb{K}} \mathcal{Y} = 3$ and $\kappa_i(t, W')|_{\mathcal{Y}^\perp} = \kappa_i(t, W)|_{\mathcal{Y}^\perp} = \mathrm{id}_{\mathcal{Y}^\perp}$ by (3.2).

Now let $\theta \in [0, \pi/2]$ and $u \in \mathbb{K}$, $|u| = 1$, with $(W'a_i|Wa_i) = \|Wa_i\| \cdot \|W'a_i\| \cdot u\cos\theta$ and define

$$y_1 := \frac{u^{-1}}{\|Wa_i\|}Wa_i, \quad y_2 := \frac{u^{-1}}{\|\tilde{y}_2\|}\tilde{y}_2,$$

where $\tilde{y}_2 := W'a_i - (W'a_i|y_1)y_1$. Then $\{a_i^0, y_1, y_2\}$ is an orthonormal basis in $\mathcal{Y}$. With respect to this orthonormal basis we define

$$R\colon \mathcal{Y} \to \mathcal{Y}, \quad R = \begin{pmatrix} 1 & 0 & 0 \\ 0 & \cos\theta & -\sin\theta \\ 0 & \sin\theta & \cos\theta \end{pmatrix}. \tag{3.8}$$

We have $Wa_i = \|Wa_i\|uy_1$ and $W'a_i = (Wa_i|y_1)y_1 + u\|\tilde{y}_2\|y_2$, hence

$$Wa_i = \begin{pmatrix} 0 \\ \|Wa_i\|u \\ 0 \end{pmatrix} \text{ and } W'a_i = \begin{pmatrix} 0 \\ (W'a_i|y_1) \\ u\|\tilde{y}_2\| \end{pmatrix}$$

with respect to the orthonormal basis $\{a_i^0, y_1, y_2\}$ of $\mathcal{Y}$. On the other hand, we have $(W'a_i|y_1) = (W'a_i|Wa_i)/\|Wa_i\| = \|W'a_i\|u\cos\theta$, hence

$$\|\tilde{y}_2\| = \left(\|W'a_i\|^2 - |(W'a_i|y_1)|\right)^{1/2} = \|W'a_i\|\sin\theta.$$

This implies that

$$R(Wa_i) = \|Wa_i\|u \cdot R\begin{pmatrix} 0 \\ 1 \\ 0 \end{pmatrix} = \begin{pmatrix} 0 \\ \|Wa_i\|u\cos\theta \\ \|Wa_i\|u\sin\theta \end{pmatrix} = \frac{\|Wa_i\|}{\|W'a_i\|} \cdot W'a_i.$$

Thus $R\colon \mathcal{Y} \to \mathcal{Y}$ is the rotation of $\mathcal{Y}$ around $\mathbb{K} \cdot a_i^0$ mapping $\mathbb{K} \cdot Wa_i$ onto $\mathbb{K} \cdot W'a_i$. It then follows by the definition (3.2) of $\kappa_i(t, W)$ that

$$\kappa_i(t, W')|_{\mathcal{Y}} = R \circ \kappa_i(t, W)|_{\mathcal{Y}} \circ R^{-1}.$$

Consequently, if μ stands for the left-hand side of (3.7), then we can estimate μ in the following way:

$$\begin{aligned}\mu &= \|R \circ \kappa_i(t, W)|_{\mathcal{Y}} \circ R^{-1} \circ \kappa_i(t, W)^{-1}|_{\mathcal{Y}} - \mathrm{id}_{\mathcal{H}}\| \\ &= \|\kappa_i(t, W)|_{\mathcal{Y}} \circ R^{-1} \circ \kappa_i(t, W)^{-1}|_{\mathcal{Y}} - R^{-1}\| \\ &\leq \|\kappa_i(t, W)|_{\mathcal{Y}} \circ R^{-1} \circ \kappa_i(t, W)^{-1}|_{\mathcal{Y}} - \mathrm{id}_{\mathcal{Y}}\| + \|\mathrm{id}_{\mathcal{Y}} - R^{-1}\|,\end{aligned}$$

where the second equality follows since R is unitary. Thence, using the fact that $\kappa_i(t, W)$ is unitary, we get

$$\begin{aligned}\mu &\leq 2\|\mathrm{id}_{\mathcal{Y}} - R^{-1}\| \\ &\leq 4 \sin \frac{\theta}{2} && \text{(by (3.5))} \\ &\leq \left(\min\{\|Wa_i\|, \|W'a_i\|\}\right)^{-1} \cdot \|Wa_i - W'a_i\| && \text{(by Remark 3.22)} \\ &\leq 2M \cdot \|(W - W')a_i\| && \text{(by } W, W' \in \mathcal{W}_M\text{)} \\ &\leq 2M \cdot \|W - W'\| && \text{(since } \|a_i\| = 1\text{).}\end{aligned}$$

Thus (3.7) is proved. Now the desired conclusion easily follows by inequalities (3.4), (3.6) and (3.7). ∎

DEFINITION 3.24 We define $\kappa \colon [0, 1] \times \mathcal{W}_M \to U$ by

$$\kappa(t, W)x = \begin{cases} \kappa_i(t, W)x & \text{if } x \in A_i \text{ for some } i \geq 1, \\ x & \text{if } x \perp A_i \text{ for all } i \geq 1, \end{cases}$$

whenever $t \in [0, 1]$, $W \in \mathcal{W}_M$ and $x \in \mathcal{H}$. Note that $\kappa(t, W)x$ is well defined since $A_i \perp A_j$ whenever $i \neq j$. Moreover, it follows by Lemma 3.23 that the mapping $\kappa \colon [0, 1] \times \mathcal{W}_M \to U$ is continuous.

REMARK 3.25 For later use, let us state some basic properties of the map $\kappa \colon [0, 1] \times \mathcal{W}_M \to U$ introduced in Definition 3.24.

(a) For all $W \in \mathcal{W}_M$ we have $\kappa(0, W) = \mathrm{id}_{\mathcal{H}}$ $(\in U)$.

(b) If $W \in \mathcal{W}_M$ and $i \geq 1$ then, by Definition 3.24 and Remark 3.21(iv),

$$\kappa(1, W)(Wa_i) = \|Wa_i\| a_i.$$

LEMMA 3.26 *The homotopy*

$$[1, 2] \times S \to G, \quad (t, s) \mapsto f_t(s) := \kappa(t - 1, f_1(s)) f_1(s)$$

leads to a continuous mapping $f_2 \colon S \to G$ *such that* $f_2(s)a_i = \|f_2(s)a_i\| a_i$ *for all* $s \in S$ *and* $i \geq 1$.

PROOF Note that the homotopy is continuous since both κ and f_1 are continuous. The property of f_2 follows by Remark 3.25(b) since $f_1(S) \subseteq \mathcal{W}_M$ and the values of κ are unitary operators. ∎

DEFINITION 3.27 We denote by $\widetilde{\mathcal{H}}$ the closed linear subspace of $\mathcal{H}$ spanned by the orthonormal sequence $\{a_i\}_{i\geq 1}$. Note that

$$\dim_{\mathbb{K}} \widetilde{\mathcal{H}} = \dim_{\mathbb{K}} \widetilde{\mathcal{H}}^{\perp} = \dim_{\mathbb{K}} \mathcal{H} = \aleph_0,$$

where the fact that $\dim_{\mathbb{K}} \widetilde{\mathcal{H}}^{\perp} = \aleph_0$ follows since $\widetilde{\mathcal{H}}^{\perp}$ contains the orthonormal sequence $\{a_i^0\}_{i\geq 1}$. (See Definition 3.19.)

LEMMA 3.28 *There exists a homotopy*

$$[2,3] \times S \to G, \quad (t,s) \mapsto f_t(s)$$

such that $f_3(s)|_{\widetilde{\mathcal{H}}} = \mathrm{id}_{\widetilde{\mathcal{H}}}$ *for all* $s \in S$.

PROOF For $t \in [2,3]$, $s \in S$ and $x \in \mathcal{H}$ we define

$$f_t(s) = \begin{cases} (3-t)f_2(s)x + (t-2)x & \text{if } x \in \widetilde{\mathcal{H}}, \\ f_2(s)x & \text{if } x \perp \widetilde{\mathcal{H}}. \end{cases}$$

Since $f_2(s)|_{\widetilde{\mathcal{H}}}$ is a positive invertible operator on $\widetilde{\mathcal{H}}$ (as an easy consequence of Lemma 3.26), it follows that $f_t(s)|_{\widetilde{\mathcal{H}}}$ has the same property. Now it is straightforward to check that $f_t(s) \in G$ for all $t \in [2,3]$ and $s \in S$.

Moreover, it is clear that $f_3(s)|_{\widetilde{\mathcal{H}}} = \mathrm{id}_{\widetilde{\mathcal{H}}}$. ∎

PROOF (of Theorem 3.15) It follows by Lemmas 3.16, 3.26 and 3.28 that there exists a homotopy

$$[0,3] \times S \to G, \quad (t,s) \mapsto f_t(s),$$

such that $f_0 = f\colon S \to G$ is the given continuous map, and $f_3\colon S \to G$ has the property $f_3(s)|_{\widetilde{\mathcal{H}}} = \mathrm{id}_{\widetilde{\mathcal{H}}}$ for all $s \in S$, where $\widetilde{\mathcal{H}}$ is a closed subspace of $\mathcal{H}$ such that $\dim_{\mathbb{K}} \widetilde{\mathcal{H}} = \dim_{\mathbb{K}} \widetilde{\mathcal{H}}^{\perp} = \dim_{\mathbb{K}} \mathcal{H} = \aleph_0$ (see Definition 3.27). To conclude the proof, it suffices to show that there exists a homotopy

$$[3,4] \times S \to G, \quad (t,s) \mapsto f_t(s),$$

such that $f_4(s) = \mathrm{id}_{\mathcal{H}}$ for all $s \in S$ (i.e., $f_4\colon S \to S \to G$ is a constant map).

To this end, first use the fact that both $\widetilde{\mathcal{H}}$ and $\widetilde{\mathcal{H}}^{\perp}$ have the dimension $\aleph_0$ to construct a sequence $\{\mathcal{H}_j\}_{j\geq 2}$ of closed subspaces of $\mathcal{H}$ such that $\widetilde{\mathcal{H}} = \bigoplus_{j\geq 2} \mathcal{H}_j$ and $\dim_{\mathbb{K}} \mathcal{H}_j = \aleph_0$ for all $j \geq 2$. Thus, by denoting $\mathcal{H}_1 := \widetilde{\mathcal{H}}^{\perp}$, we have

$$\mathcal{H} = \mathcal{H}_1 \oplus \mathcal{H}_2 \oplus \cdots \text{ and } \dim_{\mathbb{K}} \mathcal{H}_j = \aleph_0 \text{ for all } j \geq 1.$$

Actually, since all the Hilbert spaces $\mathcal{H}_j$ have the same dimension, we may assume for the sake of simplicity that $\mathcal{H}_1 = \mathcal{H}_2 = \cdots = \widetilde{\mathcal{H}}^\perp$.

Now fix $j \geq 1$ for the moment and consider the homotopy

$$H_j \colon [0, \tfrac{\pi}{2}] \times \big(\mathrm{GL}(\mathcal{H}_j) \times \mathrm{GL}(\mathcal{H}_{j+1})\big) \to \mathrm{GL}(\mathcal{H}_j \oplus \mathcal{H}_{j+1})$$

defined by

$$\begin{aligned} H_j\Big(t, \begin{pmatrix} A & 0 \\ 0 & B \end{pmatrix}\Big) &:= H_{j,t}\Big(\begin{pmatrix} A & 0 \\ 0 & B \end{pmatrix}\Big) \\ &:= \begin{pmatrix} A & 0 \\ 0 & 1 \end{pmatrix} \begin{pmatrix} \cos t & \sin t \\ -\sin t & \cos t \end{pmatrix} \begin{pmatrix} 1 & 0 \\ 0 & B \end{pmatrix} \begin{pmatrix} \cos t & \sin t \\ -\sin t & \cos t \end{pmatrix}^{-1}, \end{aligned}$$

where the scalars actually stand for the corresponding scalar multiples of the operator $\mathrm{id}_{\mathcal{H}_j} = \mathrm{id}_{\mathcal{H}_{j+1}}$. Then for all $A \in \mathrm{GL}(\mathcal{H}_j)$ and $B \in \mathrm{GL}(\mathcal{H}_{j+1})$ we have

$$H_j\Big(\frac{\pi}{2}, \begin{pmatrix} A & 0 \\ 0 & B \end{pmatrix}\Big) = H_{j,\pi/2}\Big(\begin{pmatrix} A & 0 \\ 0 & B \end{pmatrix}\Big) = \begin{pmatrix} AB & 0 \\ 0 & 1 \end{pmatrix}.$$

Now consider the elements of $\mathcal{B}(\mathcal{H})$ represented by infinite matrices with entries in $\mathcal{B}(\mathcal{H}_1)$, according to the orthogonal decomposition $\mathcal{H} = \mathcal{H}_1 \oplus \mathcal{H}_2 \oplus \cdots$, where $\mathcal{H}_1 = \mathcal{H}_2 = \cdots$. Using this convention, denote

$$\mathcal{V} = \left\{ \begin{pmatrix} T & & & \mathbf{0} \\ & 1 & & \\ & & 1 & \\ \mathbf{0} & & & \ddots \end{pmatrix} \in \mathrm{GL}(\mathcal{H}) \mid T \in \mathrm{GL}(\mathcal{H}_1) \right\} \subseteq G.$$

Also denote by $\tau \colon \mathcal{V} \hookrightarrow G$ the inclusion mapping, by $\tau_0 \colon \mathcal{V} \to G$ the mapping that is constant $\mathrm{id}_{\mathcal{H}} \in G$, and define

$$\eta_0 \colon \mathcal{V} \to G, \qquad \begin{pmatrix} T & & & & \mathbf{0} \\ & 1 & & & \\ & & 1 & & \\ & & & 1 & \\ \mathbf{0} & & & & \ddots \end{pmatrix} \mapsto \begin{pmatrix} T & & & & \mathbf{0} \\ & T^{-1} & & & \\ & & T & & \\ & & & T^{-1} & \\ \mathbf{0} & & & & \ddots \end{pmatrix}.$$

Then

$$\alpha \colon [0, \tfrac{\pi}{2}] \times \mathcal{V} \to G, \quad (t, V) \mapsto (H_{1,t} \times H_{3,t} \times \cdots)(\eta_0(V))$$

is a homotopy such that $\alpha(0, V) = \eta_0(V)$ and $\alpha(\pi/2, V) = \mathrm{id}_{\mathcal{H}}$ whenever $V \in \mathcal{V}$. On the other hand,

$$\beta \colon [0, \tfrac{\pi}{2}] \times \mathcal{V} \to G, \quad (t, V) \mapsto \big(\mathrm{id}_{\mathrm{GL}(\mathcal{H}_1)} \times H_{2,t} \times H_{4,t} \times \cdots\big)(\eta_0(V))$$

is a homotopy with $\beta(0, V) = \eta_0(V)$ and $\beta(\pi/2, V) = V$ for all $V \in \mathcal{V}$. Then it is easy to construct (from α and β) a homotopy

$$\gamma \colon [3, 4] \times \mathcal{V} \to G$$

such that $\gamma(3, V) = V$ and $\gamma(4, V) = \mathrm{id}_{\mathcal{H}}$ for all $V \in \mathcal{V}$.

Now, since $f_3(S) \subseteq \mathcal{V}$, it follows that

$$[3,4] \times S \to G, \quad (t,s) \mapsto \gamma(t, f_3(s)) =: f_t(s)$$

is a homotopy with $f_4(s) = \mathrm{id}_{\mathcal{H}}$ for all $s \in S$, and the proof ends. ∎

COROLLARY 3.29 *If S is a compact topological space, then for every continuous map $f\colon S \to U$ there exists a homotopy $H\colon [0,1] \times S \to U$ such that $H(0,s) = f(s)$ and $H(1,s) = \mathrm{id}_{\mathcal{H}}$ for all $s \in S$.*

PROOF The proof has two stages.

1° We show at this stage that the map

$$\tau\colon G \to U, \quad \tau(T) = T \cdot (T^*T)^{-1/2}$$

is continuous. To this end, first note that τ is well defined since for all $T \in G$ we have

$$\tau(T)^*\tau(T) = (T^*T)^{-1/2}T^*T(T^*T)^{-1/2} = \mathrm{id}_{\mathcal{H}}$$

and similarly $\tau(T)\tau(T)^* = \mathrm{id}_{\mathcal{H}}$. Now, to show that τ is continuous, it suffices to check the continuity of the map $G \to G$, $T \mapsto (T^*T)^{1/2}$.

To this end, assume that $\lim\limits_{n\to\infty} T_n = T$ in G. Since $\lim\limits_{n\to\infty} \|T_n\| = \|T\| > 0$, there exist $c, d > 0$ such that $c \cdot \mathrm{id}_{\mathcal{H}} \le T_n^*T_n \le d \cdot \mathrm{id}_{\mathcal{H}}$ for all $n \ge 1$. Then let $\{p_j\colon [c,d] \to \mathbb{R}\}_{j\ge 1}$ be a sequence of polynomial functions such that $\lim\limits_{j\to\infty} p_j = r$ uniformly on $[c,d]$, where

$$r\colon [c,d] \to \mathbb{R}, \quad r(t) = \sqrt{t}.$$

For all $j \ge 1$ and $S \in \{T\} \cup \{T_n \mid n \ge 1\}$ we have $\|p_j(S^*S) - r(S^*S)\| \le \sup\limits_{[c,d]} |p_j - r|$. Since $\|r(T^*T) - r(T_n^*T_n)\| \le \|r(T^*T) - p_j(T^*T)\| + \|p_j(T^*T) - p_j(T_n^*T_n)\| + \|p_j(T_n^*T_n) - r(T_n^*T_n)\|$, we easily get $\lim\limits_{n\to\infty} \|r(T^*T) - r(T_n^*T_n)\| = 0$, as desired.

2° Now let us use Theorem 3.15 to construct a homotopy

$$\widetilde{H}\colon [0,1] \times S \to G$$

with $\widetilde{H}(0,s) = f(s)$ and $\widetilde{H}(1,s) = \mathrm{id}_{\mathcal{H}}$ for all $s \in S$. Then, according to stage 1°, the map $H = \tau \circ \widetilde{H}\colon [0,1] \times S \to U$ is a homotopy with the desired properties, since $\tau(T) = T$ whenever $T \in U$. ∎

For the last statement of this section we recall from Notation 3.14 that we have denoted by G the group of invertible operators on a complex separable Hilbert space $\mathcal{H}$, and by U the subgroup of G consisting of the unitary operators.

COROLLARY 3.30 *Both Lie groups G and U are simply connected.*

PROOF It follows by Theorem 3.15 and Corollary 3.29 that the Lie groups G and U, respectively, are pathwise simply connected. Thus they are simply connected by Theorem C.17. □

3.3 Enlargible Lie algebras

In this section we introduce the notion of enlargible Lie algebra and discuss simple sufficient conditions for a Banach-Lie algebra to be enlargible. We then use Corollary 3.30 to construct a non-enlargible Banach-Lie algebra.

DEFINITION 3.31 A real topological Lie algebra $\mathfrak{g}$ is said to be *enlargible* if there exists a Lie group G with $\mathbf{L}(G) = \mathfrak{g}$.

The following statement shows that enlargibility is a hereditary property.

THEOREM 3.32 *Let $\mathfrak{g}$ and $\mathfrak{h}$ be two real Banach-Lie algebras. If $\mathfrak{h}$ is enlargible and there exists an injective homomorphism of topological Lie algebras $\varphi\colon \mathfrak{g} \to \mathfrak{h}$, then $\mathfrak{g}$ is in turn enlargible.*

PROOF This is an immediate consequence of Theorem 3.5. □

COROLLARY 3.33 *Let $\mathfrak{g}$ be a real Banach-Lie algebra. If there exists a real Banach space $\mathcal{X}$ and a continuous injective homomorphism of topological Lie algebras $\varphi\colon \mathfrak{g} \to \mathcal{B}(\mathcal{X})$, then $\mathfrak{g}$ is enlargible.*

PROOF We know from Example 2.22 that $\mathcal{B}(\mathcal{X})$ is the Lie algebra of the Banach-Lie group $\mathrm{GL}(\mathcal{X})$, hence $\mathcal{B}(\mathcal{X})$ is enlargible. Then the conclusion follows by Theorem 3.32. □

COROLLARY 3.34 *Every finite-dimensional real Lie algebra is enlargible.*

PROOF First recall that every finite-dimensional real vector space has a unique Banach space topology. Now let $\mathfrak{g}$ be a finite-dimensional real Lie algebra and think of it as a Banach-Lie algebra. It follows by Ado's theorem

(Theorem 1.13) that there exists a finite-dimensional real vector space $\mathcal{X}$ and an injective homomorphism of Lie algebras $\varphi\colon \mathfrak{g} \to \mathrm{End}\,(\mathcal{X})$. If we think of $\mathcal{X}$ as a Banach space, then $\mathrm{End}\,(\mathcal{X}) = \mathcal{B}(\mathcal{X})$, hence the desired conclusion follows by Corollary 3.33. ∎

Example of non-enlargible Banach-Lie algebra

In Example 3.35 below, we use the complex separable Hilbert space $\mathcal{H} = \ell^2(\mathbb{Z}_+)$, the group $U = \mathrm{U}(\mathcal{H})$ of all unitary operators on $\mathcal{H}$ and the real Banach-Lie algebra

$$\mathfrak{u} := \{A \in \mathcal{B}(\mathcal{H}) \mid A^* = -A\}.$$

We remark that the direct product $\mathfrak{g} := \mathfrak{u} \times \mathfrak{u}$ has a natural structure of real Banach-Lie algebra with componentwise defined operations.

EXAMPLE 3.35 Let $\alpha \in \mathbb{R} \setminus \mathbb{Q}$ and denote

$$\mathfrak{k} := \{(it \cdot \mathrm{id}_{\mathcal{H}}, it\alpha \cdot \mathrm{id}_{\mathcal{H}}) \mid t \in \mathbb{R}\} \subseteq \mathfrak{u} \times \mathfrak{u} =: \mathfrak{g}.$$

Then $\mathfrak{k}$ is a closed (one-dimensional) subalgebra of $\mathfrak{g}$ with $[\mathfrak{k}, \mathfrak{g}] = \{0\}$. Then it is easy to check that the quotient Banach space

$$\mathfrak{h} := \mathfrak{g}/\mathfrak{k}$$

has a well-defined structure of real Banach-Lie algebra with the bracket defined by

$$(\forall x, y \in \mathfrak{g}) \quad [x + \mathfrak{k}, y + \mathfrak{k}] := [x, y] + \mathfrak{k}.$$

We are going to show that the Banach-Lie algebra $\mathfrak{h}$ is non-enlargible.

In fact, let us assume that there exists a Banach-Lie group H with $\mathbf{L}(H) = \mathfrak{h}$. Then consider the canonical projection

$$\varphi\colon \mathfrak{g} \to \mathfrak{g}/\mathfrak{k} = \mathfrak{h},$$

and note that φ is a homomorphism of topological Lie algebras. On the other hand, recall from Remark 3.12 that U is a Banach-Lie group with $\mathbf{L}(U) = \mathfrak{u}$, hence $\mathbf{L}(U \times U) = \mathfrak{u} \times \mathfrak{u} = \mathfrak{g}$. Since U is simply connected by Corollary 3.30, it follows that $U \times U$ is in turn simply connected. Then Remark 3.13 shows that there exists a homomorphism of Lie groups

$$f\colon U \times U \to H$$

with $\mathbf{L}(f) = \varphi$. Now note that $\mathfrak{k} = \mathrm{Ker}\,\varphi$, hence $f(\exp_{U\times U} \mathfrak{k}) = \exp_H(\varphi(\mathfrak{k})) = \mathbf{1}$, and then $\langle \exp_{U\times U} \mathfrak{k}\rangle \subseteq \mathrm{Ker}\, f$, where $\langle \exp_{U\times U} \mathfrak{k}\rangle$ stands for the subgroup of $U \times U$ generated by $\exp_{U\times U} \mathfrak{k}$. Since Corollary 3.9 along with Remark 3.10 show that $\langle \exp_{U\times U} \mathfrak{k}\rangle$ is dense in $\mathbb{T} \cdot \mathrm{id}_{\mathcal{H}} \times \mathbb{T} \cdot \mathrm{id}_{\mathcal{H}}$ ($\subseteq U \times U$), it then follows that $\mathbb{T} \cdot \mathrm{id}_{\mathcal{H}} \times \mathbb{T} \cdot \mathrm{id}_{\mathcal{H}} \subseteq \mathrm{Ker}\, f$. The latter inclusion implies that $\mathrm{Ker}\,\varphi \supseteq \mathbf{L}(\mathbb{T} \cdot \mathrm{id}_{\mathcal{H}} \times \mathbb{T} \cdot \mathrm{id}_{\mathcal{H}}) = \mathrm{i}\mathbb{R} \cdot \mathrm{id}_{\mathcal{H}} \times \mathrm{i}\mathbb{R} \cdot \mathrm{id}_{\mathcal{H}}$, which contradicts the fact that $\mathrm{Ker}\,\varphi = \mathfrak{k}$ is 1-dimensional. Consequently, there cannot exist a Lie group H with $\mathbf{L}(H) = \mathfrak{h}$.

Notes

The present chapter is an introduction to some of the very basic ideas of enlargibility, in order to supply a motivation for the enlargibility criterion that will be proved in Chapter 8 (more precisely, see Corollary 8.36). We emphasize that we made no attempt to survey here the results that were obtained in the last forty years on enlargibility questions. Some recent references where such questions are addressed are [Pe88], [Pe92], [GN03], [Ne02d], and [Be04]. On the other hand, some of the first papers treating on enlargibility of local groups are [EK64] and [Sw65].

Theorem 3.3 appears as Lemma IV.1 in [Ne02d]. See also [Ne04]. Our Corollary 3.7 appears as Corollary 7.8 in [Up85]. The special case for finite-dimensional Lie groups (see Corollary 3.8) is known as "Cartan's theorem on closed subgroups." In this special case, other proofs can be found in [Ho65] and [Wa71].

Corollary 3.29 is the main result of N. Kuiper's paper [Ku65]. In the present chapter we essentially follow the method of proof of that paper. We refer to [Ne02b] for the extension of Corollary 3.29 to arbitrary Hilbert spaces. A version of Kuiper's theorem in the framework of purely infinite von Neumann algebras can be found in the paper [BW76]. Related results can be found in the papers [Pa65], [Pa66], and [AV02]. See also [Do65].

The converse to Corollary 3.33 is discussed in [ES73]. The result of Corollary 3.34 is known as "Lie's Third Theorem." Another proof of that fact can be found in [Ho65]. The first example of non-enlargible Lie algebra was described in [EK64]. Example 3.35 is taken from [DL66].

Homogeneous spaces

Chapter 4

Smooth Homogeneous Spaces

Abstract. In the first part of this chapter we explain the notion of Banach-Lie subgroup. A Banach-Lie subgroup of a Banach-Lie group is in particular a closed subgroup. Conversely, the question when a given closed subgroup is actually a Banach-Lie group in general has no simple answer in the case of infinite-dimensional groups. However, in the case of closed subgroups of invertible elements in a unital associative Banach algebra, there exists the notion of algebraic subgroup, which helps us to deal with a large number of interesting examples of Banach-Lie subgroups. In the second part of Section 4.1 we prove the basic theorem on existence of smooth structures on homogeneous spaces of Banach-Lie groups. We then introduce the symplectic manifolds of various types and describe the symplectic homogeneous spaces in Lie algebraic terms. In the last part of the chapter we present a list of smooth homogeneous spaces that show up in the theory of operator algebras: unitary orbits of self-adjoint operators, unitary orbits in preduals of W^*-algebras, and then unitary orbits of spectral measures, conditional expectations, group representations, and algebra representations, respectively.

4.1 Basic facts on smooth homogeneous spaces

In this section we prove the main theorems concerning constructions of smooth structures on homogeneous spaces of general Banach-Lie groups. The most important result is that a quotient of a Banach-Lie group by a Banach-Lie subgroup is always a smooth homogeneous space (Theorem 4.19). We also discuss the notion of algebraic subgroup, which is helpful in order to produce concrete examples of Banach-Lie subgroups.

Banach-Lie subgroups

DEFINITION 4.1 Let G be a Banach-Lie group and H a subgroup of G. We say that H is a *Banach-Lie subgroup* of G if the following conditions are satisfied.

(i) The subgroup H is endowed with a structure of Banach-Lie group whose underlying topology is the topology which H inherits from G.

(ii) The inclusion map $\iota\colon H \hookrightarrow G$ is smooth and $\mathbf{L}(\iota)\colon \mathbf{L}(H) \to \mathbf{L}(G)$ is an injective operator with closed range.

(iii) There exists a closed linear subspace $\mathfrak{M}$ of $\mathbf{L}(G)$ such that $\operatorname{Ran}\mathbf{L}(\iota) \dot{+} \mathfrak{M} = \mathbf{L}(G)$.

Because of condition (ii) we always identify $\mathbf{L}(H)$ to $\operatorname{Ran}\mathbf{L}(\iota)$, so that we think of $\mathbf{L}(H)$ as a closed subalgebra of the Banach-Lie algebra $\mathbf{L}(G)$. In this way $\mathbf{L}(\iota)$ is just the inclusion map $\mathbf{L}(H) \hookrightarrow \mathbf{L}(G)$.

EXAMPLE 4.2 If G be a finite-dimensional Lie group and H is a closed subgroup of G, then H is a Banach-Lie subgroup of G.

In fact, Corollary 3.8 shows that H has a structure of a finite-dimensional Lie group satisfying condition (i) in Definition 4.1, and the inclusion map $\iota\colon H \hookrightarrow G$ is smooth and $\mathbf{L}(\iota)\colon \mathbf{L}(H) \to \mathbf{L}(G)$ is injective. The other conditions in (ii) and (iii) in Definition 4.1 follow since $\dim \mathbf{L}(G) < \infty$.

REMARK 4.3 With the notation of Definition 4.1, there exists a basis of open neighborhoods $\mathcal{W}$ of $\mathbf{1} \in G$ such that for each $W \in \mathcal{W}$ there exist open neighborhoods $U_{\mathfrak{M},W}$ of $0 \in \mathfrak{M}$ and $U_{\mathfrak{h},W}$ of $0 \in \mathfrak{h}$ such that the mapping

$$\Phi_W\colon U_{\mathfrak{M},W} \times U_{\mathfrak{h},W} \to W, \quad (x,y) \mapsto \exp_G x \cdot \exp_G y,$$

is a real analytic diffeomorphism.

In fact, consider the mapping

$$\Phi\colon \mathfrak{M} \times \mathfrak{h} \to G, \quad (x,y) \mapsto \exp_G x \cdot \exp_G y.$$

Then Φ is a real analytic mapping by Theorem 2.42, and its differential at $(0,0) \in \mathfrak{M} \times \mathfrak{h}$ is

$$T_{(0,0)}\colon \mathfrak{M} \times \mathfrak{h} \to T_{\mathbf{1}}G = \mathfrak{g}, \quad (x,y) \mapsto x + y,$$

which is an isomorphism of Banach spaces according to the hypothesis $\mathfrak{M}\dot{+}\mathfrak{h} = \mathfrak{g}$. Now the claim follows by the local inversion theorem.

Additionally, it is easy to see that $\{U_{\mathfrak{M},W}\}_{W\in\mathcal{W}}$ and $\{U_{\mathfrak{h},W}\}_{W\in\mathcal{W}}$ are basis of neighborhoods of $0 \in \mathfrak{M}$ and $0 \in \mathfrak{h}$, respectively.

The following statement supplies a very useful characterization of Banach-Lie subgroups.

PROPOSITION 4.4 *Let G be a Banach-Lie group and $\mathfrak{h}$ a closed subalgebra of the Lie algebra $\mathfrak{g}$ of G. Assume that H is a subgroup of G that has a structure of Banach-Lie group such that the inclusion mapping $\iota\colon H \hookrightarrow G$ is smooth and $\mathbf{L}(\iota)$ equals the inclusion mapping $\mathfrak{h} \hookrightarrow \mathfrak{g}$. Then the following assertions are equivalent:*

(i) *The topology which H inherits from G equals the topology underlying the Banach-Lie group structure of H.*

(ii) *There exist an open neighborhood U of $0 \in \mathfrak{g}$ and an open neighborhood W of $\mathbf{1} \in G$ such that $\exp_G|_U\colon U \to W$ is a diffeomorphism and $\exp_G(U \cap \mathfrak{h}) = W \cap H$.*

PROOF "(i) $\Rightarrow$ (ii)" Since G is a Banach-Lie group we can use Theorem 2.42 to construct an open neighborhood U_0 of $0 \in \mathfrak{g}$ and an open neighborhood W_0 of $\mathbf{1} \in G$ such that $\exp_G|_{U_0}\colon U_0 \to W_0$ is a diffeomorphism. Similarly, for the Banach-Lie group H we can construct the open neighborhoods U_0' of $0 \in \mathfrak{h}$ and W_0' of $\mathbf{1} \in H$ such that $\exp_H|_{U_0'}\colon U_0' \to W_0'$ is a diffeomorphism. Since the topology of $\mathfrak{h}$ is the one inherited from $\mathfrak{g}$, we may additionally assume that $U_0' \subseteq \mathfrak{h} \cap U_0$.

On the other hand, we can use hypothesis (i) to find an open subset W_1 of $\mathbf{1} \in G$ such that $W_0' = W_1 \cap H$. Now pick an open neighborhood U of $0 \in \mathfrak{g}$ such that

$$U \subseteq U_0, \quad U \cap \mathfrak{h} \subseteq U_0' \quad \text{and} \quad \exp_G(U) \subseteq W_1.$$

We claim that

$$\exp_G(U) \cap H = \exp_G(U \cap \mathfrak{h}).$$

The inclusion $\supseteq$ is obvious. To prove the converse inclusion, let $x \in U$ arbitrary such that $\exp_G x \in H$. Then $\exp_G x \in H \cap W_1 = W_0' = \exp_H(U_0')$, so that there exists $x' \in U_0'$ such that $\exp_G x = \exp_H x'$. Since $x, x' \in U_0$, and $\exp_G|_H = \exp_H$ by Remark 2.34, we get $x = x' \in U_0' \subseteq \mathfrak{h} \cap U$. Consequently $\exp_G X \in \exp_G(\mathfrak{h} \cap U)$, as desired.

"(ii) $\Rightarrow$ (i)" Since the topology of $\mathfrak{h}$ is the one inherited from $\mathfrak{g}$, it follows by condition (ii) that for $\mathbf{1} \in H$ we can find a family $\mathcal{W}_1$ of subsets of H constituting a basis of neighborhoods in both the topology inherited by H from G and the Lie group topology of H. Then for each $h \in H$ the family $\mathcal{W}_h := \{hW\}_{W \in \mathcal{W}_1}$ will be a similar neighborhood basis of $h \in H$ in both aforementioned topologies. Now, with the neighborhood basis $\{\mathcal{W}_h \mid h \in H\}$ at hand, it is easy to show that (i) holds. $\square$

PROPOSITION 4.5 *Let G be a Banach-Lie group and H a Banach-Lie subgroup of G. Then H is a closed subset of G.*

PROOF It follows by Proposition 4.4 that there exists an open neighborhood $W_{\mathbf{1}}$ of $\mathbf{1} \in G$ such that $W_{\mathbf{1}} \cap H$ is closed in the relative topology of $W_{\mathbf{1}}$. Then for every $h \in H$ we can find the open neighborhood

$$W_h := hW_{\mathbf{1}}$$

of $h \in G$ such that $W_h \cap H = h(W_{\mathbf{1}} \cap H)$ is closed in W_h. In particular, denoting by $\mathrm{cl}\, H$ the closure of H in G, we have

$$(\forall h \in H) \qquad W_h \cap \mathrm{cl}\, H = W_h \cap H.$$

(In fact, the inclusion $\supseteq$ is obvious, while $W_h \cap \mathrm{cl}\, H$ is contained in the closure of $W_h \cap H$ in the relative topology of W_h, and the latter closure equals $W_h \cap H$.) Consequently, for the open subset $W := \bigcup_{h \in H} W_h$ of G we have

$$H = H \cap \bigcup_{h \in H} W_h = \bigcup_{h \in H} (H \cap W_h) = \bigcup_{h \in H} (\mathrm{cl}\, H \cap W_h) = \mathrm{cl}\, H \cap \bigcup_{h \in H} W_h,$$

that is,

$$H = \mathrm{cl}\, H \cap W.$$

With this equality at hand, we now come back to the proof of the fact that H is closed in G, that is, $H = \mathrm{cl}\, H$.

The inclusion $\subseteq$ is obvious. To prove the converse inclusion, let $g \in \mathrm{cl}\, H$ arbitrary. Denote as usually $W_{\mathbf{1}}^{-1} = \{k^{-1} \mid k \in W_{\mathbf{1}}\}$. Then $gW_{\mathbf{1}}^{-1}$ is a neighborhood of g, so that there exists $h \in H \cap gW_{\mathbf{1}}^{-1}$. Then $h^{-1}g \in W_{\mathbf{1}} \subseteq W$. On the other hand clearly $\mathrm{cl}\, H$ is a subgroup of G, and $h, g \in \mathrm{cl}\, H$, hence $h^{-1}g \in W \cap \mathrm{cl}\, H = H$. Since $h \in H$, it then follows that $g \in H$, as desired. Consequently $\mathrm{cl}\, H \subseteq H$, and we are done. ∎

REMARK 4.6 It follows by Corollary 4.5 and Example 4.2 that, for a finite-dimensional Lie group, the closed subgroups and the Banach-Lie subgroups are the same.

EXAMPLE 4.7 It follows by Remark 4.6 that the subgroup $\gamma(\mathbb{R})$ involved in Corollary 3.9 is *not* a Banach-Lie subgroup of $\mathbb{T}^2$.

The next statement is often helpful in order to prove that the pull-back of a Banach-Lie subgroup is again a Banach-Lie subgroup.

PROPOSITION 4.8 *Let $f\colon G_1 \to G_2$ be a Lie group homomorphism between the Banach-Lie groups G_1 and G_2. Assume that $\mathfrak{h}_2$ is closed subalgebra of the Lie algebra $\mathfrak{g}_2$ of G_2 and H_2 is a subgroup of G_2 that has a structure of Banach-Lie group with respect to the topology inherited from G_2 such that* *the inclusion mapping $\iota_2\colon H_2 \hookrightarrow G_2$ is smooth, $\mathbf{L}(H_2) = \mathfrak{h}_2$, and $\mathbf{L}(\iota_2)$ equals the inclusion mapping $\mathfrak{h}_2 \hookrightarrow \mathfrak{g}_2$. Denote*

$$\mathfrak{h}_1 = \mathbf{L}(f)^{-1}(\mathfrak{h}_2) \quad \text{and} \quad H_1 = f^{-1}(H_2).$$

Then H_1 in turn has a structure of Banach-Lie group with respect to the topology inherited from G_1 such that the inclusion mapping $\iota_1\colon H_1 \hookrightarrow G_1$ is smooth, $\mathbf{L}(H_1) = \mathfrak{h}_1$ and $\mathbf{L}(\iota_1)$ equals the inclusion mapping $\mathfrak{h}_1 \hookrightarrow \mathfrak{g}_1$, where $\mathfrak{g}_1 = \mathbf{L}(G_1)$.

PROOF It follows by Proposition 4.4 that there exist the open neighborhoods U_2 of $0 \in \mathfrak{g}_2$ and V_2 of $\mathbf{1} \in G_2$ such that $\exp_{G_2}|_{U_2}\colon U_2 \to V_2$ is a homeomorphism and $\exp_{G_2}(\mathfrak{h}_2 \cap U_2) = H_2 \cap V_2$.

Now recall from Remark 2.34 the commutative diagram

$$\begin{array}{ccc} \mathfrak{g}_1 & \xrightarrow{\mathbf{L}(f)} & \mathfrak{g}_2 \\ {\scriptstyle \exp_{G_1}}\big\downarrow & & \big\downarrow{\scriptstyle \exp_{G_2}} \\ G_1 & \xrightarrow{f} & G_2 \end{array}$$

whence it easily follows that

$$\mathfrak{h}_1 = \{x \in \mathfrak{g}_1 \mid (\forall t \in \mathbb{R}) \quad \exp_{G_1}(tx) \in H_1\}.$$

Since H_1 is clearly a closed subgroup of G_1, it follows by Corollary 3.7 that there exists on H_1 a unique topology τ and a corresponding structure of Banach-Lie group such that $\mathbf{L}(H_1) = \mathfrak{h}_1$, the inclusion mapping $\iota_1\colon H_1 \hookrightarrow G_1$ is smooth, and $\mathbf{L}(\iota_1)$ equals the inclusion mapping $\mathfrak{h}_1 \hookrightarrow \mathfrak{g}_1$. We are going to make use of Proposition 4.4 to prove that the topology τ coincides with the topology which H_1 inherits from G_1.

To this end let U_1 be an open neighborhood of $0 \in \mathfrak{g}_1$ and V_1 an open neighborhood of $\mathbf{1} \in G_1$ such that $U_1 \subseteq \mathbf{L}(f)^{-1}(U_2)$ and $\exp_{G_1}|_{U_1}\colon U_1 \to V_1$ is a homeomorphism. In particular there exists the commutative diagram

$$\begin{array}{ccc} U_1 & \xrightarrow{\mathbf{L}(f)|_{U_1}} & U_2 \\ {\scriptstyle \exp_{G_1}|_{U_1}}\big\downarrow & & \big\downarrow{\scriptstyle \exp_{G_2}|_{U_2}} \\ V_1 & \xrightarrow{f|_{V_2}} & V_2 \end{array}$$

whose vertical arrows are homeomorphisms. We are going to prove that

$$\exp_{G_1}(\mathfrak{h}_1 \cap U_1) = H_1 \cap V_1,$$

and then the desired conclusion will follow by means of Proposition 4.4. The inclusion $\subseteq$ is obvious. To prove the converse inclusion, let $h_1 \in H_1 \cap V_1$

arbitrary. Since $h_1 \in V_1$, there exists a uniquely determined element $x_1 \in U_1$ such that $h_1 = \exp_{G_1} x_1$. On the other hand, since $h_1 \in H_1$, we have $f(h_1) = f(\exp_{G_1} x_1) = \exp_{G_2}(\mathbf{L}(f)x_1) \in \exp_{G_2}(U_2) = V_2$. Thus $f(h_1) \in H_2 \cap V_2 = \exp_{G_2}(\mathfrak{h}_2 \cap U_2)$. Consequently there exists $x_2 \in \mathfrak{h}_2 \cap U_2$ such that $f(h_1) = \exp_{G_2} x_2$. The latter equality implies that $\exp_{G_2} x_2 = f(\exp_{G_1} x_1) = \exp_{G_2}(\mathbf{L}(f)x_1)$. Since $x_2, \mathbf{L}(f)x_1 \in U_2$, we get $x_2 = \mathbf{L}(f)x_1$, so that $x_1 \in \mathbf{L}(f)^{-1}(\mathfrak{h}_2) = \mathfrak{h}_1$. Consequently $x_1 \in \mathfrak{h}_1 \cap U_1$, whence $h_1 = \exp_{G_1} x_1 \in \exp_{G_1}(\mathfrak{h}_1 \cap U_1)$, and we are done. ∎

Algebraic subgroups

DEFINITION 4.9 Let A be a real or complex associative unital Banach algebra, n a positive integer, and G a subgroup of $A^\times$. We say that G is an *algebraic subgroup of* $A^\times$ *of degree* $\leq n$ if we have

$$G = \{a \in A^\times \mid (\forall p \in \mathcal{P}) \quad p(a, a^{-1}) = 0\}$$

for a certain family $\mathcal{P}$ of continuous polynomials on $A \times A$.

EXAMPLE 4.10 Let $\mathfrak{X}$ be a real or complex Banach algebra and consider the associative Banach algebra $A := \mathcal{B}(\mathfrak{X})$ and the group $G := \mathrm{Aut}(\mathfrak{X})$. Then G is an algebraic subgroup of $A^\times$ of degree ≤ 2. Moreover, G is a Banach-Lie group with the topology inherited from A, and the Lie algebra of G is $\mathrm{Der}(\mathfrak{X})$. (Compare Theorem 4.13 below.) In particular, note that for every $\delta \in \mathrm{Der}(\mathfrak{X})$ we have $\exp \delta \in \mathrm{Aut}(\mathfrak{X})$.

To prove one of the basic properties of the notion introduced in Definition 4.9 (see Theorem 4.13 below), we need the following two lemmas.

LEMMA 4.11 *Let $f \colon \mathbb{C} \to \mathbb{C}$ be a holomorphic function satisfying the following conditions:*

(i) *There exist $A, B > 0$ such that $|f(z)| \leq A\exp(B|z|)$ for all $z \in \mathbb{C}$.*

(ii) *We have $\limsup\limits_{t\to\infty}(1/t)\log|f(\pm\mathrm{i}t)| < \pi$.*

(iii) *For each integer $n \geq 0$ we have $f(n) = 0$.*

Then $f = 0$ on $\mathbb{C}$.

PROOF See e.g., [Boa54]. ∎

LEMMA 4.12 *If A be a complex associative unital Banach algebra and $a \in A$ then*

$$\limsup_{t \to \infty} \frac{1}{t} \log \| \exp(ta) \| = \sup_{z \in \sigma(a)} \operatorname{Re} z$$

and

$$\liminf_{t \to -\infty} \frac{1}{t} \log \| \exp(ta) \| = \inf_{z \in \sigma(a)} \operatorname{Re} z,$$

where $\sigma(\cdot)$ stands for the spectrum of an element in A.

PROOF See e.g., [Va82]. ∎

Now we are able to prove the main result on algebraic subgroups of invertible elements in an associative Banach algebra.

THEOREM 4.13 *Let A be a complex associative unital Banach algebra and n a positive integer. Assume that G is an algebraic subgroup of $A^\times$ of degree $\leq n$. Then G has a structure of Banach-Lie group whose underlying topology is the topology inherited from A.*

PROOF Denote

$$\mathcal{U} = \Big\{ a \in A \mid \sigma(a) \subseteq \mathbb{R} + \mathrm{i}(-\frac{\pi}{n}, \frac{\pi}{n}) \Big\}$$

and

$$\mathcal{V} = \Big\{ a \in A^\times \mid (\forall z \in \sigma(a)) \quad |\arg z| < \frac{\pi}{n} \Big\}.$$

Then Proposition A.48 shows that

$$\exp \colon \mathcal{U} \to \mathcal{V}, \quad a \mapsto \exp a = \sum_{j=0}^{\infty} \frac{a^j}{j!}$$

is a complex analytic bijection with the inverse denoted by $\log \colon \mathcal{V} \to \mathcal{U}$.

Furthermore denote

$$\mathfrak{g} = \{ a \in A \mid (\forall t \in \mathbb{R}) \quad \exp(ta) \in G \}.$$

It is clear from Definition 4.9 that G is closed in $A^\times$, hence Corollary 3.7 shows that $\mathfrak{g}$ is a closed Lie subalgebra of A. Moreover, according to Corollary 3.7, what we still have to prove is that the Lie group topology of G coincides with the topology inherited from $A^\times$. To this end, in view of Proposition 4.4, it suffices to prove that

$$\exp(\mathcal{U} \cap \mathfrak{g}) = \mathcal{V} \cap G, \tag{4.1}$$

The inclusion $\subseteq$ is obvious. The converse inclusion is equivalent to

$$\mathcal{U} \cap \mathfrak{g} \supseteq \log(\mathcal{V} \cap G).$$

We already know that $\log(\mathcal{V}) = \mathcal{U}$, so that we still have to check that $\log(\mathcal{V}\cap G)\subseteq\mathfrak{g}$. To this end, let $g\in\mathcal{V}\cap G$ arbitrary.

Denote $B = A\times A$ and $b = (x,-x)\in B$. Let $p\in\mathcal{P}$ arbitrary. What we have to prove is that

$$(\forall t\in\mathbb{R})\quad p(\exp(tb)) = 0.$$

In fact, consider the holomorphic function

$$f\colon\mathbb{C}\to\mathbb{C},\quad f(z) = p(\exp(zb)).$$

Then for each integer $k\ge 0$ we have $f(k) = p(\exp(kb)) = p(g^k, g^{-k}) = 0$ since $g^k\in G$. Moreover the function f clearly satisfies condition (i) in Lemma 4.11.

To check that condition (iii) of that lemma is satisfied as well, first use the fact that p is a polynomial of degree $\le n$ to write $p = p_0 + p_1 + \cdots + p_n$, where each $p_j\colon B\to\mathbb{C}$ is defined by $p_j(y) = \psi_j(y,\dots,y)$ for all $y\in B$, where $\psi_j\colon B\times\cdots\times B\to\mathbb{C}$ is a certain j-linear continuous functional. Then for all $y\in B$ we have

$$|p(y)|\le\|\psi_0\| + \|\psi_1\|\cdot\|y\| + \cdots + \|\psi_n\|\cdot\|y\|^n\le M\cdot\max\{1,\|y\|^n\},$$

whence $\log|p(y)|\le\log M + n\max\{0,\log\|y\|\}$, where $M>0$ is a constant. Consequently

$$\begin{aligned}\limsup_{t\to\infty}\frac{1}{t}\log|p(\exp(\pm itb))| &\le n\limsup_{t\to\infty}\frac{1}{t}\max\{0,\log\|\exp(\pm itb)\|\}\\ &\le n\sup_{w\in\sigma(b)}\operatorname{Im} w<\pi,\end{aligned}$$

where the last but one inequality follows by Lemma 4.12. Now Lemma 4.11 shows that $f(z) = 0$ for all $z\in\mathbb{C}$, and we are done. ∎

Until the end of the proof of Theorem 4.18, we denote by $\mathfrak{B}$ an associative unital complex Banach algebra.

REMARK 4.14 Let $\mathfrak{B}$ be an associative unital complex Banach algebra. The set

$$\mathrm{U}_{\mathfrak{B}} := \{u\in\mathfrak{B}^\times\mid\|u\| = \|u^{-1}\| = 1\}$$

is a closed subgroup of $\mathfrak{B}^\times$, hence it is a real Banach-Lie group with respect to a Hausdorff topology τ uniquely determined by the fact that the Lie algebra of $\mathrm{U}_{\mathfrak{B}}$ is the closed real Lie subalgebra of $\mathrm{U}_{\mathfrak{B}}$ defined by

$$\mathfrak{u}(\mathfrak{B}) = \{b\in\mathfrak{B}\mid(\forall t\in\mathbb{R})\quad e^{tb}\in\mathrm{U}_{\mathfrak{B}}\}$$

and the diagram

$$\begin{array}{ccc}\mathrm{U}_{\mathfrak{B}} & \longrightarrow & \mathfrak{B}^\times\\ {\scriptstyle\exp_{\mathrm{U}_{\mathfrak{B}}}}\uparrow & & \uparrow{\scriptstyle\exp_{\mathfrak{B}^\times}}\\ \mathfrak{u}(\mathfrak{B}) & \longrightarrow & \mathfrak{B}\end{array}$$

is commutative, where the horizontal arrows stand for the inclusion maps.

In general, the topology τ is stronger than the one inherited by $\mathrm{U}_{\mathfrak{B}}$ from $\mathfrak{B}$. Nevertheless, if $\mathfrak{B}$ is a C^*-algebra, then $\mathrm{U}_{\mathfrak{B}}$ is an algebraic subgroup of $\mathfrak{B}^\times$ since

$$\mathrm{U}_{\mathfrak{B}} = \{u \in \mathfrak{B} \mid u^*u = uu^* = 1\},$$

hence Theorem 4.13 can be used to deduce that the topology τ coincides with the norm topology of $\mathrm{U}_{\mathfrak{B}}$.

We always consider $\mathrm{U}_{\mathfrak{B}}$ endowed with the topology τ.

DEFINITION 4.15 An *algebraic subgroup of* $\mathrm{U}_{\mathfrak{B}}$ *of degree* $\leq n$ is a subgroup H of $\mathrm{U}_{\mathfrak{B}}$ such that

$$H = \{u \in \mathrm{U}_{\mathfrak{B}} \mid (\forall p \in \mathcal{P}) \quad p(u, u^{-1}) = 0\}$$

for some set $\mathcal{P}$ of vector-valued continuous polynomial functions on $\mathfrak{A} \times \mathfrak{A}$. That is, for every $p \in \mathcal{P}$ there exist a real Banach space $\mathfrak{V}_p$ and a k-linear map

$$\psi_k \colon \underbrace{(\mathfrak{B} \times \mathfrak{B}) \times \cdots \times (\mathfrak{B} \times \mathfrak{B})}_{k \text{ pairs}} \to \mathfrak{V}_p$$

(for $k = 0, 1, \ldots, n$) such that $p(b) = \psi_n(b, \ldots, b) + \cdots + \psi_1(b) + \psi_0$ for every $b \in \mathfrak{B} \times \mathfrak{B}$ (where $\psi_0 \in \mathfrak{V}_p$).

REMARK 4.16 In the setting of Definition 4.15, an algebraic subgroup of $\mathrm{U}_{\mathfrak{B}}$ of degree $\leq n$ is simply the intersection of $\mathrm{U}_{\mathfrak{B}}$ with an algebraic subgroup of $\mathfrak{B}^\times$.

REMARK 4.17 Let n be a positive integer. It is obvious that every algebraic subgroup H of $\mathrm{U}_{\mathfrak{B}}$ of degree $\leq n$ is closed with respect to the norm topology of $\mathrm{U}_{\mathfrak{B}}$. Since the norm topology is coarser than the topology τ of the *Banach-Lie group* $\mathrm{U}_{\mathfrak{B}}$ (see Remark 4.14), it then follows that H is also a closed subgroup of the Banach-Lie group $\mathfrak{B}^\times$. Then Corollary 3.7 shows that H is in turn a real Banach-Lie group with respect to a Hausdorff topology τ_H which is *a priori* stronger than the restriction of τ to H. Actually, Theorem 4.18 below shows that τ_H always coincides with the restriction of τ to H.

THEOREM 4.18 *If n is a positive integer, H is an algebraic subgroup of $\mathrm{U}_{\mathfrak{B}}$ of degree $\leq n$, and*

$$\mathfrak{h} := \{b \in \mathfrak{B} \mid (\forall t \in \mathbb{R}) \quad e^{tb} \in H\},$$

then $\mathfrak{h}$ is a closed subalgebra of $\mathfrak{u}(\mathfrak{B})$, H is a real Banach-Lie group with the topology inherited from the topology τ of the Banach-Lie group $\mathrm{U}_{\mathfrak{B}}$, and $\mathfrak{h}$ is the Lie algebra of H.

PROOF Let $\widetilde{H}$ be an algebraic subgroup of $\mathfrak{B}^\times$ such that

$$H = \widetilde{H} \cap \mathrm{U}_{\mathfrak{B}}$$

(see Remark 4.16). If $j\colon \mathrm{U}_{\mathfrak{B}} \to \mathfrak{B}^\times$ stands for the inclusion map, then

$$H = j^{-1}(\widetilde{H}),$$

hence the desired conclusion follows by Theorem 4.13 and Proposition 4.8. ∎

Smooth quotient spaces

Here is the central result of the present section.

THEOREM 4.19 *Let G be a Banach-Lie group, H a Banach-Lie subgroup of G and $\pi\colon G \to G/H$ the natural projection. Endow G/H with the quotient topology and consider the natural transitive action*

$$\alpha\colon G \times G/H \to G/H, \quad (g, kH) \mapsto \alpha_g(kH) := gkH.$$

Then G/H has a structure of real analytic manifold such that the following conditions are satisfied:

(i) *The mapping π is real analytic and has real analytic local cross sections near every point of G/H.*

(ii) *For every $g \in G$ the mapping*

$$\alpha_g\colon G/H \to G/H$$

is real analytic.

NOTATION 4.20 We are going to keep the following notation until the proof of Theorem 4.19 will be accomplished:

- $\mathbf{L}(G) = \mathfrak{g}$, $\mathbf{L}(H) = \mathfrak{h}$.
- For all $g \in G$ we define $L_g\colon G \to G$ by $L_g(k) = gk$ for all $k \in G$.
- We pick a closed subspace $\mathfrak{M}$ of $\mathfrak{g}$ such that $\mathfrak{g} = \mathfrak{M} \dotplus \mathfrak{h}$ (see Definition 4.1).
- We define $\psi\colon \mathfrak{g} \to G/H$ by $\psi = \pi \circ \exp_G$.

LEMMA 4.21 *There exists an open neighborhood U of $0 \in \mathfrak{M}$ such that $\psi(U)$ is an open subset of G/H and the mapping*

$$\psi_1 := \psi|_U\colon U \to \psi(U)$$

is a homeomorphism.

PROOF The proof has several stages.

1° Let $\mathcal{W}$ be a basis of neighborhoods of $\mathbf{1} \in G$ as in Remark 4.3. Since H is a Banach-Lie subgroup of G it follows by Proposition 4.4 that there exist $W \in \mathcal{W}$ and an open neighborhood $U_{\mathfrak{g}}$ of $0 \in \mathfrak{g}$ such that $\exp_G|_{U_{\mathfrak{g}}}: U_{\mathfrak{g}} \to W$ is a homeomorphism, $\exp_G(U_{\mathfrak{h},W}) = H \cap W$ and $U_{\mathfrak{h},W} \subseteq U_{\mathfrak{g}} \cap \mathfrak{h}$.

Now let U be an open neighborhood of $0 \in \mathfrak{M}$ such that

$$U \subseteq U_{\mathfrak{M},W} \quad \text{and} \quad \exp_G(-U) \cdot \exp_G(U) \subseteq W.$$

We are going to show that U has the desired properties.

2° At this stage we prove that $\psi|_U$ is injective. In fact, let $x, x' \in U$ such that $\psi(x) = \psi(x')$. Then $\pi(\exp_G x) = \pi(\exp_G x')$, that is, $\exp_G(-x') \cdot \exp_G x \in H \cap W = \exp_G(U_{\mathfrak{h},W})$. Thus there exists $y \in U_{\mathfrak{h},W}$ such that $\exp_G x = \exp_G x' \cdot \exp_G y$. In the notation of Remark 4.3, the latter equality can be written $\Phi_W(x, 0) = \Phi_W(x', y)$, hence $x = x'$ (and $y = 0$). Consequently $\psi|_U$ is injective.

3° Now note that

$$\psi(U) = \pi(\exp_G(U)) = \pi(\exp_G(U) \cdot \exp_G(U_{\mathfrak{h},W})) = \pi(\Phi_W(U \times U_{\mathfrak{h},W})).$$

Since Φ_W is a homeomorphism and $\pi: G \to G/H$ is an open map, it then follows that $\psi(U)$ is open in G/H.

4° Define

$$\sigma: \psi(U) \to G, \quad \sigma = \exp_G \circ (\psi|_U)^{-1}.$$

Then for every open subset D of G we have

$$\sigma^{-1}(D) = \psi|_U(\exp_G^{-1}(D)) = \psi((\exp_G|_U)^{-1}(D)) = \pi(D),$$

which is open in G/H since π is an open map. Thus we see that σ is continuous. Moreover note that

$$\pi \circ \sigma = \mathrm{id}_{\psi(U)}.$$

Also note that the mapping $\pi|_{\exp_G(U)}: \exp_G(U) \to \psi(U)$ is bijective and $(\pi|_{\exp_G(U)})^{-1} = \sigma$ is continuous, hence both mappings $\sigma: \psi(U) \to \exp_G(U)$ and $\pi|_{\exp_G(U)}: \exp_G(U) \to \psi(U) = \pi(\exp_G(U))$ are homeomorphisms.

On the other hand, the mapping $\exp_G|_U: U \to \exp_G(U)$ is in turn a homeomorphism, so that the commutative diagram

$$\begin{array}{ccc} \exp_G(U) & \xrightarrow{\pi|_{\exp_G(U)}} & \psi(U) \\ {\scriptstyle \exp_G} \uparrow & \nearrow {\scriptstyle \psi|_U} & \\ U & & \end{array}$$

shows that $\psi_1 = \psi|_U$ is a homeomorphism. ∎

LEMMA 4.22 *Let $U \subseteq \mathfrak{M}$ and $\psi_{\mathbf{1}}$ as in* Lemma 4.21 *and denote $V := \psi(U) \subseteq G/H$. Assume that $g \in G$ has the property that $V \cap \alpha_g(V) \neq \emptyset$. Then the mapping*

$$\psi_{\mathbf{1}}^{-1} \circ \alpha_g \circ \psi_{\mathbf{1}}|_{\psi^{-1}(V \cap \alpha_{g^{-1}}(V))} \colon \psi_{\mathbf{1}}^{-1}(V \cap \alpha_{g^{-1}}(V)) \to \psi_{\mathbf{1}}^{-1}(V \cap \alpha_g(V))$$

is a real analytic mapping between two open subsets of the Banach space $\mathfrak{M}$.

PROOF Let $\sigma \colon V \to G$ be as in the proof of Lemma 4.22. For all $x \in V \cap \alpha_{g^{-1}}(V)$ we have

$$\pi(L_g(\sigma(x))) = \alpha_g(\pi(\sigma(x))) = \alpha_g(x) \tag{4.2}$$

since $\pi \circ \sigma = \mathrm{id}_V$. Now denote $y = \psi_{\mathbf{1}}^{-1}(x)$, so that $y \in \psi_{\mathbf{1}}^{-1}(V \cap \alpha_{g^{-1}}(V)) \subseteq \psi_{\mathbf{1}}^{-1}(V) = U \subseteq \mathfrak{M}$. Then equality (4.2) implies

$$\alpha_g(\psi(y)) = \pi(L_g(\sigma(\psi(y)))) = \pi(L_g(\sigma(\pi(\exp_G y)))).$$

Note that $\exp_G y \in \exp_G(U)$, hence $\sigma(\pi(\exp_G y)) = \exp_G y$. Consequently the above equality leads to $\alpha_g(\psi(y)) = \pi(L_g(\exp_G y))$, whence

$$(\forall y \in \psi_{\mathbf{1}}^{-1}(V \cap \alpha_{g^{-1}}(V))) \qquad (\psi_{\mathbf{1}}^{-1} \circ \alpha_g \circ \psi_{\mathbf{1}})(y) = \psi_{\mathbf{1}}^{-1}(\pi(L_g(\exp_G y))).$$

On the other hand, in the notation of Remark 4.3, note that we have a commutative diagram

$$\begin{array}{ccc} U \times U_{\mathfrak{h},W} & \xrightarrow{\Phi_W|_{U \times U_{\mathfrak{h},W}}} & \Phi_W(U \times U_{\mathfrak{h},W}) \\ {\scriptstyle pr_1}\downarrow & & \downarrow{\scriptstyle \pi} \\ U & \xrightarrow{\psi|_U} & V \end{array}$$

where W is the one used in the proof of Lemma 4.21, and $\Phi_W(U \times U_{\mathfrak{h},W})$ is open in $\Phi_W(U_{\mathfrak{M},W} \times U_{\mathfrak{h},W}) = W$. Also note that both horizontal arrows in the preceding diagram are bijections. Consequently,

$$\psi_{\mathbf{1}}^{-1} \circ \pi|_{\Phi_W(U \times U_{\mathfrak{h},W})} = pr_1 \circ (\Phi_W|_{U \times U_{\mathfrak{h},W}})^{-1} \colon \Phi_W(U \times U_{\mathfrak{h},W}) \to U.$$

Additionally we note that for each $y \in \psi_{\mathbf{1}}^{-1}(V \cap \alpha_{g^{-1}}(V))$ we have

$$\begin{aligned} \exp_G y \in (\exp_G \circ \psi_{\mathbf{1}}^{-1})(V \cap \alpha_{g^{-1}}(V)) &\subseteq (\pi|_{\exp_G(U)})^{-1}(V \cap \alpha_{g^{-1}}(V)) \\ &\subseteq \exp_G(U) \subseteq \Phi_W(U \times U_{\mathfrak{h},W}) \end{aligned}$$

and similarly $L_g(\exp_G y) \in \Phi_W(U \times U_{\mathfrak{h},W})$. Consequently for $y \in \psi_{\mathbf{1}}^{-1}(V \cap \alpha_{g^{-1}}(V))$ we get

$$(\psi_{\mathbf{1}}^{-1} \circ \alpha_g \circ \psi_{\mathbf{1}})(y) = (pr_1 \circ (\Phi_W|_{U \times U_{\mathfrak{h},W}})^{-1} \circ L_g \circ \exp_G)(y),$$

whence the desired assertion follows at once (see Remark 4.3 and Theorem 2.42). ∎

PROOF (of Theorem 4.19) The proof has several stages.

1° At this stage we construct a real analytic atlas on G/H. Let $\psi_{\mathbf{1}} = \psi|_U : U \to \psi(U) = V$ given by Lemma 4.21 and for all $g \in G$ define

$$\psi_g : U \to \alpha_g(V), \quad \psi_g = \alpha_g \circ \psi_{\mathbf{1}}.$$

Then it is clear that

$$\bigcup_{g\in G} \psi_g(U) = G/H.$$

On the other hand, if $g_1, g_2 \in G$ and $\psi_{g_1}(U) \cap \psi_{g_2}(U) \neq \emptyset$, then the mapping

$$\psi_{g_2}^{-1} \circ \psi_{g_1}|_{\psi_{g_1}^{-1}(\psi_{g_1}(U)\cap\psi_{g_2}(U))} : \psi_{g_1}^{-1}(\psi_{g_1}(U) \cap \psi_{g_2}(U)) \to \psi_{g_2}^{-1}(\psi_{g_1}(U) \cap \psi_{g_2}(U))$$

satisfies

$$(\psi_{g_2}^{-1} \circ \psi_{g_1})(x) = \psi_{\mathbf{1}}^{-1}(\alpha_{g_2^{-1}}(\alpha_{g_1}(\psi(x)))) = (\psi_{\mathbf{1}}^{-1} \circ \alpha_{g_2^{-1}g_1} \circ \psi_{\mathbf{1}})(x)$$

for all $x \in \psi_{g_1}^{-1}(\psi_{g_1}(U) \cap \psi_{g_2}(U))$. Now Lemma 4.22 applied for $g = g_2^{-1}g_1$ shows that the mapping $\psi_{g_2}^{-1} \circ \psi_{g_1}|_{\psi_{g_1}^{-1}(\psi_{g_1}(U)\cap\psi_{g_2}(U))}$ is real analytic. Consequently $\{\psi_g\}_{g\in G}$ is a real analytic atlas on G/H.

2° At this stage we show that for $g_0 \in G$ arbitrary the mapping $\alpha_{g_0} : G/H \to G/H$ is real analytic. To this end let $x \in G/H$ arbitrary. Then there exists $k \in G$ such that $x = kH$. Then $\psi_k : U \to \alpha_k(V)$ is a chart about $x \in G/H$, $\psi_{gk} : U \to \alpha_{gk}(V)$ is a chart about $\alpha_g(x) = gkH \in G/H$, and $\psi_{gk} = \alpha_g \circ \psi_k$, so that the mapping $\psi_{gk}^{-1} \circ \alpha_g \circ \psi_k = \mathrm{id}_U : U \to U$ is real analytic.

3° At this stage we prove the assertion concerning the cross sections. To this end, denote

$$\sigma := \exp_G \circ \psi_{\mathbf{1}}^{-1} : V \to G.$$

We already know that σ is a cross section of π on the neighborhood V of $\pi(\mathbf{1}) \in G/H$. This cross section is real analytic since it is a composition of real analytic mappings.

Now take $k \in G$ arbitrary. Then the mapping

$$\sigma_k := L_k \circ \sigma \circ \alpha_{k^{-1}}|_{\alpha_k(V)} : \alpha_k(V) \to G$$

is a real analytic cross section of π on the neighborhood $\alpha_k(V)$ of $\pi(k) \in G/H$. This follows since the diagram

$$\begin{array}{ccc} G & \xrightarrow{L_k} & G \\ {\scriptstyle\pi}\downarrow & & {\scriptstyle\pi}\downarrow \\ G/H & \xrightarrow{\alpha_k} & G/H \end{array}$$

is commutative. ∎

4.2 Symplectic homogeneous spaces

In this section we introduce symplectic manifolds and symplectic homogeneous spaces of various types. The most important result is a characterization of symplectic homogeneous spaces in Lie algebraic terms (Theorem 4.30). We then discuss two important examples: the orbit symplectic form (see Example 4.31) and the natural symplectic structure of the projective space of a Hilbert space (Example 4.32).

DEFINITION 4.23 Let $\mathfrak{z}$ be a real Banach space. A *weakly symplectic manifold of type* $\mathfrak{z}$ is a pair (M, ω) with the following properties:

(i) The symbol M stands for a Banach manifold.

(ii) We have $\omega \in \Omega^2(M, \mathfrak{z})$ and ω is weakly nondegenerate and closed.

If the 2-form ω is actually strongly nondegenerate (see Definition A.71), then we say that (M, ω) is a *strongly symplectic manifold.*

DEFINITION 4.24 Let G be a Banach-Lie group, H a Banach-Lie subgroup of G and $M := G/H$. Consider the natural transitive action

$$\alpha \colon G \times G/H \to G/H, \quad (g, kH) \mapsto \alpha_g(kH) := gkH.$$

Also let $\mathfrak{z}$ be a real Banach space and $n \geq 1$ an integer.

An *invariant* differential n-form of type $\mathfrak{z}$ on M is a form $\psi \in \Omega^n(M, \mathfrak{z})$ satisfying $(\alpha_g)^*(\psi) = \psi$ for all $g \in G$. We denote by $\Omega^n_G(M, \mathfrak{z})$ the vector space of all these invariant forms.

DEFINITION 4.25 Let $\mathfrak{z}$ be a real Banach space and G a real Banach-Lie group. A *weakly symplectic homogeneous space of* G *of type* $\mathfrak{z}$ is a pair (M, ω) satisfying the following conditions.

(a) There exists a Banach-Lie subgroup H of G with $M = G/H$.

(b) The pair (M, ω) is a weakly symplectic manifold.

(c) We have $\omega \in \Omega^2_G(M, \mathfrak{z})$.

If moreover the 2-form ω is strongly symplectic, then we say that (M, ω) is a *strongly* symplectic homogeneous space of G of type $\mathfrak{z}$.

REMARK 4.26 In Definition 4.25, it follows by the invariance condition (c) that if the operator

$$\alpha_m \colon T_m M \to \mathcal{B}(T_m M, \mathfrak{z}), \quad v \mapsto \omega_m(z, \cdot)$$

is injective (respectively surjective) for some $m \in M$, then it has this property for all $m \in M$.

DEFINITION 4.27 If $\mathfrak{g}$ is a real Banach-Lie algebra, $\mathfrak{h}$ is a closed subalgebra of $\mathfrak{g}$ and $\mathfrak{z}$ is a real Banach space, then $Z_c^2(\mathfrak{g}, \mathfrak{z})$ stands for the space of all *continuous* $\mathfrak{z}$*-valued* 2*-cocycles* of $\mathfrak{g}$. That is, $Z_c^2(\mathfrak{g}; \mathfrak{z})$ is the set of all bilinear skew-symmetric maps

$$\eta\colon \mathfrak{g} \times \mathfrak{g} \to \mathfrak{z}$$

such that

$$(\forall a, b, c \in \mathfrak{g}) \qquad \eta(a, [b, c]) + \eta(b, [c, a]) + \eta(c, [a, b]) = 0.$$

Then $Z_c^2(\mathfrak{g}; \mathfrak{z})$ is a real Banach space under the norm defined by

$$\|\omega\| = \sup\{\|\eta(a, b)\|_{\mathfrak{z}} \mid a, b \in \mathfrak{g} \text{ and } \max\{\|a\|_{\mathfrak{g}}, \|b\|_{\mathfrak{g}}\} \leq 1\}$$

for every $\eta \in Z_c^2(\mathfrak{g}, \mathfrak{z})$. Moreover we need the closed subspace of $Z_c^2(\mathfrak{g}, \mathfrak{z})$ defined by

$$Z_c^2(\mathfrak{g}, \mathfrak{h}; \mathfrak{z}) = \{\eta \in Z_c^2(\mathfrak{g}, \mathfrak{z}) \mid (\forall a \in \mathfrak{h})(\forall b \in \mathfrak{g}) \quad \eta(a, b) = 0\}.$$

Clearly $Z_c^2(\mathfrak{g}, \{0\}; \mathfrak{z}) = Z_c^2(\mathfrak{g}; \mathfrak{z})$.

DEFINITION 4.28 Let G be a Banach-Lie group, H a Banach-Lie subgroup of G and $\pi\colon G \to M := G/H$ the natural projection. Denote $x_0 = \pi(\mathbf{1}) \in M$ and consider the natural transitive action

$$\alpha\colon G \times G/H \to G/H, \quad (g, kH) \mapsto \alpha_g(kH) := gkH.$$

Also let $\mathfrak{z}$ be a real Banach space and $n \geq 1$ an integer. Denote $\mathbf{L}(G) = \mathfrak{g}$, $\mathbf{L}(H) = \mathfrak{h}$ and let $\mathfrak{M}$ be a closed subspace of $\mathfrak{g}$ such that $\mathfrak{M} \dot{+} \mathfrak{h} = \mathfrak{g}$. Now let U be an open neighborhood of $0 \in \mathfrak{M}$ such that $\psi_{\mathbf{1}} := \pi \circ \exp_G|_U \colon U \to \psi_{\mathbf{1}}(U)$ is a local chart of $M = G/H$ (see Lemma 4.21). Then the chart ψ_1 leads as in Definition A.55 to an isomorphism of Banach spaces $\mathfrak{M} \simeq T_{x_0}M$. On the other hand, we clearly have an isomorphism of Banach spaces $\mathfrak{M} \simeq \mathfrak{g}/\mathfrak{h}$, hence eventually

$$T_{x_0}M \simeq \mathfrak{g}/\mathfrak{h}.$$

Now let $\eta \in Z_c^2(\mathfrak{g}, \mathfrak{h}; \mathfrak{z})$ arbitrary. Then in view of the aforementioned description of $T_{x_0}M$ we can define

$$\omega_{x_0}\colon T_{x_0}M \times T_{x_0}M \to \mathfrak{z}, \quad \omega_{x_0}(a + \mathfrak{h}, b + \mathfrak{h}) = \eta(a, b).$$

Next for $g \in G$ arbitrary and $x := \alpha_g(x_0) \in M$ we have $T_x(\alpha_{g^{-1}})\colon T_xM \to T_{x_0}M$ so that we can define

$$\omega_x\colon T_xM \times T_xM \to \mathfrak{z}, \quad \omega_x(v, w) = \omega_{x_0}(T_x(\alpha_{g^{-1}})v, T_x(\alpha_{g^{-1}})w).$$

We thus get a differential 2-form $\omega = \{\omega_x\}_{x\in M} \in \Omega^2_G(M,\mathfrak{z})$. We denote $\omega = \Sigma(\eta)$.

LEMMA 4.29 *In the setting of* Definition 4.28 *the following assertions hold:*

(i) *For every $\eta \in Z^2_c(\mathfrak{g},\mathfrak{h};\mathfrak{z})$ the 2-form $\Sigma(\eta)$ is well defined and indeed belongs to $\Omega^2_G(M,\mathfrak{z})$.*

(ii) *The correspondence*

$$\Sigma\colon Z^2_c(\mathfrak{g},\mathfrak{h};\mathfrak{z}) \to \Omega^2_G(M,\mathfrak{z})$$

is a linear bijection.

PROOF Let $\eta \in Z^2_c(\mathfrak{g},\mathfrak{h};\mathfrak{z})$. If $g_1, g_2 \in G$ and $\alpha_{g_1}(x_0) = \alpha_{g_2}(x_0) =: x \in M$ then $\alpha_{g_1^{-1}g_2}(x_0) = x_0$, hence $h := g_1^{-1}g_2 \in H$. Consequently $\alpha_h = \mathrm{id}_M$, whence $T(\alpha_h) = \mathrm{id}_{TM}$. Thus for all $v, w \in T_xM$ we have

$$\begin{aligned}\omega_{x_0}(T_x(\alpha_{g_1^{-1}})v, T_x(\alpha_{g_1^{-1}})w) &= \omega_{x_0}((T_{x_0}(\alpha_h)T_x(\alpha_{g_2^{-1}}))v, (T_{x_0}(\alpha_h)T_x(\alpha_{g_2^{-1}}))w)\\ &= \omega_{x_0}(T_x(\alpha_{g_2^{-1}})v, T_x(\alpha_{g_2^{-1}})w),\end{aligned}$$

and it follows that the value $\omega_{x_0}(T_x(\alpha_{g^{-1}})v, T_x(\alpha_{g^{-1}})w)$ is independent on the choice of $g \in G$ with $\alpha_g(x_0) = x$. The axiom (ii) in Definition A.71 is clearly satisfied and thus $\omega = \Sigma(\eta) \in \Omega(M,\mathfrak{z})$. Moreover it clearly follows by the construction of ω that it satisfies the invariance property $(\alpha_g)^*(\omega) = \omega$ for all $g \in G$. Consequently $\omega \in \Omega_G(M,\mathfrak{z})$.

That the map Σ is linear is obvious. To see that it is bijective we denote by $\Pi\colon \mathfrak{g} \to \mathfrak{g}/\mathfrak{h} \simeq T_{x_0}M$ the natural projection and define

$$\Delta\colon \Omega^2_G(M,\mathfrak{z}) \to Z^2_c(\mathfrak{g},\mathfrak{h};\mathfrak{z}), \quad \Delta(\omega) = \omega_{x_0} \circ (\Pi \times \Pi),$$

that is, $\Delta(\omega)(a,b) = \omega_{x_0}(\Pi(a),\Pi(b))$ whenever $a, b \in \mathfrak{g}$ and $\omega \in \Omega^2_G(M,\mathfrak{z})$. Now it is easy to see that Δ is an inverse for the map Σ, and we are done. ∎

THEOREM 4.30 *Let G be a Banach-Lie group and H a Banach-Lie subgroup of G, and denote $\mathbf{L}(G) = \mathfrak{g}$ and $\mathbf{L}(H) = \mathfrak{h}$. Also let $\mathfrak{z}$ be a real Banach space and $M = G/H$. Then the map*

$$\Sigma\colon Z^2_c(\mathfrak{g},\mathfrak{h};\mathfrak{z}) \to \Omega^2_G(M,\mathfrak{z})$$

has the following properties:

(i) *For every $\eta \in Z^2_c(\mathfrak{g},\mathfrak{h};\mathfrak{z})$ the differential form $\Sigma(\eta)$ is closed.*

(ii) *If* $\eta \in Z^2_c(\mathfrak{g}, \mathfrak{h}; \mathfrak{z})$ *satisfies*

$$\mathfrak{h} = \{a \in \mathfrak{g} \mid (\forall b \in \mathfrak{g}) \quad \eta(a, b) = 0\}$$

and we denote $\omega = \Sigma(\eta)$, *then* (M, ω) *is a weakly symplectic homogeneous space of* G *of type* $\mathfrak{z}$.

PROOF Assertion (ii) clearly follows from assertion (i) along with Remark 4.26.

The proof of assertion (i) has several stages. Denote

$$\omega = \Sigma(\eta) \in \Omega^2_G(M, \mathfrak{z}).$$

1° At this stage we assume that $H = \{\mathbf{1}\}$, hence $M = G$, and denote by $\widetilde{\omega} \in \Omega^2_G(G, \mathfrak{z})$ the corresponding form, which will turn out to be closed. Let $g_0 \in G$ arbitrary and $v_1, v_2, v_3 \in T_{g_0}G$. According to Definition A.74, we have to prove that $(d\widetilde{\omega})_{g_0}(v_1, v_2, v_3) = 0$.

To this end, denote $z_j = T_{g_0}(\alpha_{g_0^{-1}})(v_j) \in T_{\mathbf{1}}G = \mathfrak{g}$ and then $w_j = \iota(v_j) \in \mathcal{V}^l(G)$ for $j = 1, 2, 3$, where we use the isomorphism of Lie algebras $\iota \colon \mathfrak{g} \to \mathcal{V}^l(G)$ from Lemma 2.17 (see also the proof of Theorem 2.12). By Proposition A.73 we have

$$\begin{aligned}(d\widetilde{\omega})(w_1, w_2, w_3) = &\sum_{i=1}^{3} (-1)^i D_{w_i}(\widetilde{\omega}(w_1, \dots, \widehat{w_i}, \dots, w_3)) \\ &+ \sum_{1 \le i < j \le 3} (-1)^{i+j} \widetilde{\omega}([w_i, w_j], w_1, \dots, \widehat{w_i}, \dots, \widehat{w_j}, \dots, w_3),\end{aligned} \tag{4.3}$$

where $D_{w_j} \colon \mathcal{C}^\infty(G, \mathfrak{z}) \to \mathcal{C}^\infty(G, \mathfrak{z})$ are the linear operators introduced in Definition A.64.

Note that since $\widetilde{\omega} \in \Omega^2_G(G, \mathfrak{z})$ and $w_1, w_2, w_3 \in \mathcal{V}^l(G)$, that is, $\widetilde{\omega}$ and all of the vector fields w_1, w_2, w_3 are invariant with respect to the left translations on G, it follows that the three functions

$$\widetilde{\omega}(w_1, w_2), \widetilde{\omega}(w_1, w_3), \widetilde{\omega}(w_2, w_3) \colon G \to \mathfrak{z}$$

are constant, hence $D_{w_3}(\widetilde{\omega}(w_1, w_2)) = D_{w_2}(\widetilde{\omega}(w_1, w_3)) = D_{w_1}(\widetilde{\omega}(w_2, w_3)) = 0$. Consequently the first sum in the right-hand side of (4.3) vanishes. Thus using the fact that $\widetilde{\omega}$ is a family of skew-symmetric bilinear functionals we get

$$(d\widetilde{\omega})(w_1, w_2, w_3) = -\widetilde{\omega}([w_1, w_2], w_3) - \widetilde{\omega}([w_2, w_3], w_1) - \widetilde{\omega}([w_3, w_1], w_2).$$

Evaluating this equality at $\mathbf{1} \in G$ and using the fact that $\mathcal{V}^l(G) \to \mathfrak{g}$, $w \mapsto w(\mathbf{1})$, is a Lie algebra homomorphism, we get

$$(d\widetilde{\omega})_{\mathbf{1}}(w_1(\mathbf{1}), w_2(\mathbf{1}), w_3(\mathbf{1})) = 0, \tag{4.4}$$

since $\widetilde{\omega}_1 = \psi \in Z^2_c(\mathfrak{g};\mathfrak{z})$. Now recall that $w_j(\mathbf{1}) = z_j = T_{g_0}(\alpha_{g_0^{-1}})(v_j)$ for $j = 1,2,3$. Then equation (4.4) implies that

$$\begin{aligned} 0 &= ((\alpha_{g_0^{-1}})^*(d\widetilde{\omega}))_{g_0}(v_1,v_2,v_3) \\ &= (d((\alpha_{g_0^{-1}})^*(\widetilde{\omega})))_{g_0}(v_1,v_2,v_3) && \text{(by Proposition A.73)} \\ &= (d\widetilde{\omega})_{g_0}(v_1,v_2,v_3) && \text{(since } \widetilde{\omega} \text{ is invariant)} \end{aligned}$$

and the proof is complete in the case $H = \{\mathbf{1}\}$.

2° Now we treat the general case. Define $\widetilde{\omega} \in \Omega^2_G(G,\mathfrak{z})$ as in the case 1°. Then it is easy to see that $\widetilde{\omega} = \pi^*(\omega)$, whence by Proposition A.73 we get $\pi^*(d\omega) = d\pi^*(\omega) = d\widetilde{\omega} = 0$.

On the other hand, since $T_g\pi\colon T_gG \to T_{\pi(g)}M$ is surjective for all $g \in G$, it is easy to see that for all $n \geq 1$ and $\tau \in \Omega^n(M,\mathfrak{z})$ we have $\tau = 0$ if and only if $\pi^*(\tau) = 0$. In particular, since $\pi^*(d\omega) = 0$, we get $d\omega = 0$, and the proof ends. □

EXAMPLE 4.31 Let G be a Banach-Lie group with $\mathbf{L}(G) = \mathfrak{g}$, and $\xi \in \mathfrak{g}^\#$ such that

$$G_\xi := \{g \in G \mid \xi \circ \mathrm{Ad}_G(g) = \xi\}$$

is a Banach-Lie subgroup of G. Then it is easy to see that

$$\mathbf{L}(G_\xi) = \{x \in \mathfrak{g} \mid (\forall y \in \mathfrak{g}) \quad \xi([x,y]) = 0\}$$

and the correspondence $g \mapsto \xi \circ \mathrm{Ad}_G(g)$ defines a bijection

$$G/G_\xi \simeq \mathcal{O}_\xi := \{\xi \circ \mathrm{Ad}_G(g) \mid g \in G\}.$$

Since G_ξ is a Banach-Lie subgroup of G, it follows by Theorem 4.19 that the *coadjoint orbit* $\mathcal{O}_\xi$ of ξ is a smooth homogeneous space of G. Now define

$$\eta\colon \mathfrak{g} \times \mathfrak{g} \to \mathbb{R}, \quad \eta(x,y) = \xi([x,y]).$$

Then $\eta \in Z^2_c(\mathfrak{g},\mathbf{L}(G_\xi);\mathbb{R})$, so that Theorem 4.30 shows that $\omega := \Sigma(\eta) \in \Omega^2_G(\mathcal{O}_\xi,\mathbb{R})$ and $(\mathcal{O}_\xi,\omega)$ is a weakly symplectic homogeneous space of G of type $\mathbb{R}$. The 2-form ω constructed above is known as the *orbit symplectic form* on the coadjoint orbit $\mathcal{O}_\xi$ of $\xi \in \mathfrak{g}^\#$.

EXAMPLE 4.32 We now show that the construction described in Theorem 4.30 leads to a structure of *strongly* symplectic homogeneous space on the projective space of a complex Hilbert space. More precisely, we are going to show that the set of one-dimensional subspaces $\mathbb{P}_\mathcal{H} = \{\mathbb{C}x \mid 0 \neq x \in \mathcal{H}\}$ in a complex Hilbert space $\mathcal{H}$, which is called the *projective space* of $\mathcal{H}$, carries a natural structure of strongly symplectic homogeneous space of the Banach-Lie group of all unitary operators on $\mathcal{H}$.

To this end, denote the scalar product of $\mathcal{H}$ by $(\cdot \mid \cdot)$ and fix $x_0 \in \mathcal{H}$ with $\|x_0\| = 1$. Next consider the Banach-Lie group of all unitary operators on $\mathcal{H}$,

$$G = \{u \in \mathcal{B}(\mathcal{H}) \mid uu^* = u^*u = \mathbf{1}\},$$

with $\mathbf{L}(G) = \mathfrak{g} = \{a \in \mathcal{B}(\mathcal{H}) \mid a^* = -a\}$, and its subgroup

$$H = \{u \in G \mid ux_0 \in \mathbb{C}x_0\} = \{u \in G \mid up = pup\},$$

where $p = (\cdot \mid x_0)x_0 \in \mathcal{B}(\mathcal{H})$ is the orthogonal projection onto the subspace $\mathbb{C}x_0$ of $\mathcal{H}$. It is easy to see that H is an algebraic subgroup of G in the sense of Definition 4.15, hence Theorem 4.18 shows that H is a Banach-Lie group with the topology inherited from G and with

$$\mathbf{L}(H) = \mathfrak{h} = \{a \in \mathcal{B}(\mathcal{H}) \mid a = -a^* \text{ and } ax_0 \in \mathrm{i}\mathbb{R}x_0\} = \{a \in \mathfrak{g} \mid ap = pa\}.$$

Then the bounded linear mapping

$$E\colon \mathfrak{g} \to \mathfrak{g}, \quad a \mapsto E(a) = (\mathbf{1} - p)a(\mathbf{1} - p)$$

satisfies $E^2 = E$ and $\operatorname{Ran} E = \mathfrak{h}$, hence $\mathbf{L}(G) = \mathfrak{M} \dot{+} \mathbf{L}(H)$, where $\mathfrak{M} = \operatorname{Ker} E$. Consequently H is a Banach-Lie subgroup of G, and clearly the mapping

$$G \to \mathbb{P}_{\mathcal{H}}, \quad u \mapsto u(\mathbb{C}x_0),$$

defines a bijection

$$G/H \simeq \mathbb{P}_{\mathcal{H}}.$$

Now Theorem 4.19 shows that $\mathbb{P}_{\mathcal{H}}$ gets a structure of Banach manifold.

To define the symplectic structure on $\mathbb{P}_{\mathcal{H}}$ consider the skew-symmetric bilinear form

$$\eta\colon \mathfrak{g} \times \mathfrak{g} \to \mathbb{R}, \quad \eta(a_1, a_2) = \mathrm{i}\mathrm{Tr}\,(p[a_1, a_2]).$$

It is easy to check that actually $\eta \in Z_c^2(\mathfrak{g}, \mathfrak{h}; \mathbb{R})$. Since the elements of $\mathfrak{g}$ are skew symmetric, it follows that for all $a_1, a_2 \in \mathfrak{g}$ we have the equality $\eta(a_1, a_2) = -\mathrm{i} \cdot \mathrm{Tr}\,((\cdot \mid [a_1, a_2]x_0)x_0) = \mathrm{i}([a_1, a_2]x_0 \mid x_0)$, whence

$$\eta(a_1, a_2) = 2\mathrm{Im}\,(a_1x_0 \mid a_2x_0). \tag{4.5}$$

We clearly have $\{a \in \mathfrak{g} \mid (\forall b \in \mathfrak{g})\ \eta(a, b) = 0\} = \{a \in \mathfrak{g} \mid ax_0 \in \mathrm{i}\mathbb{R}x_0\} = \mathfrak{h}$. Consequently Theorem 4.30 shows that $(\mathbb{P}_{\mathcal{H}}, \omega)$ is a weakly symplectic homogeneous space of G of type $\mathbb{R}$, where $\omega = \Sigma(\eta) \in \Omega_G^2(G/H, \mathbb{R})$. To see that $(\mathbb{P}_{\mathcal{H}}, \omega)$ is actually strongly symplectic, note that the kernel of the mapping

$$\mathfrak{g} \to \operatorname{Ran}\,(\mathbf{1} - p), \quad a \mapsto (\mathbf{1} - p)ax_0$$

equals $\mathfrak{h}$, whence we get the isomorphisms of real Banach spaces

$$T_{x_0}(\mathbb{P}_{\mathcal{H}}) \simeq \mathfrak{g}/\mathfrak{h} \simeq \operatorname{Ran}\,(\mathbf{1} - p), \quad a + \mathfrak{h} \mapsto (\mathbf{1} - p)ax_0.$$

It follows by (4.5) that the bilinear functional $\omega_{x_0}\colon T_{x_0}(\mathbb{P}_{\mathcal{H}}) \times T_{x_0}(\mathbb{P}_{\mathcal{H}}) \to \mathbb{R}$ is taken by the above isomorphisms into the bilinear functional

$$\omega_0\colon \mathrm{Ran}\,(\mathbf{1}-p) \times \mathrm{Ran}\,(\mathbf{1}-p) \to \mathbb{R}, \quad (x_1, x_2) \mapsto \mathrm{Im}\,(x_1 \mid x_2),$$

and it is well known that $x \mapsto \omega_0(x \mid \cdot)$ is an isomorphism of real Banach spaces from $\mathrm{Ran}\,(\mathbf{1}-p)$ onto the topological dual of the *real* Hilbert space $\mathrm{Ran}\,(\mathbf{1}-p)$. Consequently $v \mapsto \omega_{x_0}(v \mid \cdot)$ is an isomorphism of real Banach spaces from $T_{x_0}(\mathbb{P}_{\mathcal{H}})$ onto its own topological dual. Now the conclusion that ω is strongly nondegenerate follows by Remark 4.26.

4.3 Some homogeneous spaces related to operator algebras

This section includes several examples of smooth homogeneous spaces that show up in connection with various problems in the theory of operator algebras. For most of the theorems stated below we sketch only the proofs, or we omit the proofs entirely and refer the reader to the literature. The reason we include here a list of examples of homogeneous spaces is twofold: firstly, these examples illustrate the main results of this chapter. Secondly, we hope to give the reader an idea of the wide range of possible applications of differential geometry techniques in the theory of operator algebras.

Unitary orbits of self-adjoint operators

THEOREM 4.33 *Let $\mathcal{H}$ be a complex separable Hilbert space, $M = \mathcal{B}(\mathcal{H})$, and $a = a^* \in M$. Let*

$$G = \mathrm{U}_M = \{u \in M \mid u^*u = uu^* = \mathbf{1}\}$$

be the Banach-Lie group of all unitary elements of M, with the Lie algebra

$$\mathfrak{g} = \mathfrak{u}_M = \{b \in M \mid b^* = -b\}.$$

Next let

$$H = \{u \in \mathrm{U}_M \mid ua = au\}$$

be the group of unitary elements in M that commute with a. Then H is a Banach-Lie subgroup of G and the mapping $u \mapsto uau^*$ induces a bijection

$$G/H \simeq \mathcal{O}_a := \{uau^* \mid u \in \mathrm{U}_M\}$$

of the smooth homogeneous space G/H onto the unitary orbit $\mathcal{O}_a$ of the self-adjoint operator a.

PROOF It is not hard to see that H is an algebraic subgroup of G, and

$$\mathfrak{h} = \mathbf{L}(H) = \{b \in \mathfrak{u}_M \mid ba = ab\}.$$

On the other hand, let $\mathcal{A}$ be the von Neumann algebra of operators on $\mathcal{H}$ generated by $\mathbf{1} = \mathrm{id}_{\mathcal{H}}$ and a. Since $\mathcal{A}$ is commutative, it follows that it is an injective von Neumann algebra (see e.g., Proposition 6.5 in [Con76]), hence Proposition 3.2(iii) in Chapter XV of [Ta03] shows that the commutant

$$\mathcal{A}' := \{b \in M \mid (\forall c \in \mathcal{A}) \quad bc = cb\}$$

is an injective von Neumann algebra as well. Consequently there exists a conditional expectation $E\colon M \to \mathcal{A}'$. That is, E is a bounded linear operator satisfying $E^2 = E$, $\|E\| = 1$ and $E(b^*) = E(b)^*$ for all $b \in M$. Then $E|_{\mathfrak{u}_M} : \mathfrak{u}_M \to \mathfrak{u}_M \cap \mathcal{A}' = \mathfrak{h}$ is a bounded linear idempotent mapping of $\mathfrak{g}$ onto $\mathfrak{h}$, hence $\mathfrak{g} = \mathfrak{M} \dot{+} \mathfrak{h}$, where $\mathfrak{M} = \mathrm{Ker}\,(E|_{\mathfrak{u}_M})$.

Consequently H is a Banach-Lie subgroup of G, and Theorem 4.19 applies. The other assertions are obvious. □

THEOREM 4.34 *In the setting of* Theorem 4.33, *the following assertions are equivalent:*

(i) *The quotient topology of $G/H \simeq \mathcal{O}_a$ coincides with the topology which $\mathcal{O}_a$ inherits from M.*

(ii) *The unitary orbit $\mathcal{O}_a$ is closed in M.*

(iii) *The unital C^*-subalgebra of M generated by a is finite dimensional.*

(iv) *There exist operators b and c on certain finite dimensional Hilbert spaces such that T is unitarily equivalent to the Hilbert space operator defined by the infinite block-diagonal matrix*

$$\begin{pmatrix} b & & & & & 0 \\ & c & & & & \\ & & c & & & \\ & & & \ddots & & \\ & & & & c & \\ 0 & & & & & \ddots \end{pmatrix}.$$

PROOF See Theorem 1.1 in [DF79], Theorem 4.1 in [AFHV84], and Theorems 1.1 and 1.3 in [AS91]. See also Theorem 3.1 in [BR04] □

Unitary orbits in preduals of W^*-algebras

THEOREM 4.35 *Let M be a W^*-algebra and $\varphi \in M_*$ a functional in the predual of M such that $\varphi(a^*) = \overline{\varphi(a)}$ for all $a \in M$. Denote*

$$G = \mathrm{U}_M = \{u \in M \mid u^*u = uu^* = \mathbf{1}\}$$

and

$$H = \{u \in \mathrm{U}_M \mid (\forall a \in M) \quad \varphi(u^*au) = \varphi(a)\}.$$

Also for all $u \in \mathrm{U}_M$ define $\varphi_u \colon M \to \mathbb{C}$ by $\varphi_u(a) = \varphi(u^*au)$ whenever $a \in M$. Then H is a Banach-Lie subgroup of the Banach-Lie group G and the mapping $u \mapsto \varphi_u$ induces a natural bijection

$$G/H \simeq \mathcal{O}_\varphi := \{\varphi_u \mid u \in \mathrm{U}_M\}$$

of the smooth homogeneous space G/H onto the unitary orbit $\mathcal{O}_\varphi$ of φ.

PROOF See Proposition 2.8 and Corollary 2.9 in [BR04]. ∎

REMARK 4.36 In the setting of Theorem 4.35, one can make use of the construction of Example 4.31 to make the unitary orbit $\mathcal{O}_\varphi$ into a weakly symplectic homogeneous space of G of type $\mathbb{R}$. See [BR04] for details.

THEOREM 4.37 *Let $\mathcal{H}$ be a complex separable Hilbert space and $M = \mathcal{B}(\mathcal{H})$. Pick $a = a^* \in \mathcal{C}_1(\mathcal{H}) = M_*$ and denote*

$$G = \mathrm{U}_M \quad \text{and} \quad H = \{u \in G \mid u^*au = a\}.$$

Then the following assertions are equivalent.

(i) *The quotient topology of $G/H \simeq \mathcal{O}_a$ coincides with the topology which $\mathcal{O}_a$ inherits from M_*.*

(ii) *The unitary orbit $\mathcal{O}_a$ is closed in M_*.*

(iii) *The operator a has finite rank.*

PROOF Some references for this fact are [DF79], [Bon04], and [BR04]. ∎

Unitary orbits of spectral measures

THEOREM 4.38 *Let $\mathcal{H}$ be a complex separable Hilbert space and $M = \mathcal{B}(\mathcal{H})$. Also let $\mathcal{T}$ be a σ-algebra of subsets of a certain set, and $E \colon \mathcal{T} \to \mathcal{B}(\mathcal{H})$*

a spectral measure whose values are self-adjoint projections on $\mathcal{H}$. Denote

$$G = \mathrm{U}_M = \{u \in M \mid u^*u = uu^* = \mathbf{1}\}$$

and

$$H = \{u \in \mathrm{U}_M \mid (\forall T \in \mathcal{T}) \quad uE(T) = E(T)u\}.$$

Then H is a Banach-Lie subgroup of G and the mapping $u \mapsto uE(\cdot)u^*$ induces a bijection

$$G/H \simeq \mathcal{U}_E := \{uE(\cdot)u^* \mid u \in \mathrm{U}_M\}$$

of the smooth homogeneous space G/H onto the unitary orbit $\mathcal{O}_E$ of the spectral measure E.

PROOF See [ARS93]. ∎

Unitary orbits of conditional expectations

THEOREM 4.39 *Let M be a W^*-algebra and $E\colon M \to M$ a faithful normal conditional expectation. Denote*

$$G = \mathrm{U}_M = \{u \in M \mid u^*u = uu^* = \mathbf{1}\}$$

and

$$H = \{u \in \mathrm{U}_M \mid (\forall a \in M) \quad E(u^*au) = u^*E(a)u\}.$$

Also for all $u \in \mathrm{U}_M$ define $E_u\colon M \to M$ by $E_u(a) = uE(u^*au)u^*$ for all $a \in M$. Then H is a Banach-Lie subgroup of G and the mapping $u \mapsto E_u$ induces a bijection

$$G/H \simeq \mathcal{O}_E := \{E_u \mid u \in \mathrm{U}_M\}$$

of the smooth homogeneous space G/H onto the unitary orbit $\mathcal{O}_E$ of the conditional expectation E.

PROOF See [ArS01]. ∎

REMARK 4.40 In the setting of Theorem 4.39, if $\operatorname{Ran} E = \mathbb{C}\mathbf{1}$, then E is actually a faithful normal state of M and we thus get a special case of Theorem 4.35.

Unitary orbits of group representations

THEOREM 4.41 *Let $\mathcal{H}$ be a complex separable Hilbert space and $M = \mathcal{B}(\mathcal{H})$. Moreover let $\rho\colon K \to \mathrm{U}_M$ be a norm-continuous unitary representation* *of the compact group K on $\mathcal{H}$. Denote $G = \mathrm{U}_M$ and*

$$H = \{u \in \mathrm{U}_M \mid (\forall k \in K) \quad u\rho(k) = \rho(k)u\}.$$

Then H is a Banach-Lie subgroup of G and the mapping $u \mapsto u\rho(\cdot)u^*$ induces a bijection

$$G/H \simeq \mathcal{O}_\rho := \{u\rho(\cdot)u^* \mid u \in \mathrm{U}_M\}$$

of the smooth homogeneous space G/H onto the unitary orbit $\mathcal{O}_\rho$ of the representation ρ.

PROOF See [Mr90] and [CG99], and also [BP05]. ∎

Unitary orbits of algebra representations

THEOREM 4.42 *Assume that $\mathcal{H}$ is a complex separable Hilbert space and $\rho\colon M \to \mathcal{B}(\mathcal{H})$ is a normal $*$-representation of an injective W^*-algebra M. Denote*

$$G = \mathrm{U}_{\mathcal{B}(\mathcal{H})} = \{u \in \mathcal{B}(\mathcal{H}) \mid u^*u = uu^* = \mathbf{1}\}$$

and

$$H = \{u \in \mathrm{U}_{\mathcal{B}(\mathcal{H})} \mid (\forall a \in M) \quad u\rho(a) = \rho(a)u\}.$$

Then H is a Banach-Lie subgroup of G and the mapping $u \mapsto u\rho(\cdot)u^*$ induces a bijection

$$G/H \simeq \mathcal{O}_\rho := \{u\rho(\cdot)u^* \mid u \in \mathrm{U}_{\mathcal{B}(\mathcal{H})}\}$$

of the smooth homogeneous space G/H onto the unitary orbit $\mathcal{O}_\rho$ of the representation ρ.

PROOF See [ACS95a]. ∎

Notes

We refer to [Up85] for more details on Banach-Lie subgroups. A very important early reference for homogeneous Banach manifolds is the paper [Rae77].

The proof of Proposition 4.4 follows the lines of the proof of Lemma 2.5 in Chapter II in [Hel62]. The proof of Proposition 4.5 uses the method of proof of Proposition 2.1 in [Ho65]. For another proof of Proposition 4.8 see Lemma IV.11 in [Ne04] or Lemma II.1 in [GN03]. Theorem 4.13 is the main result of the paper [HK77]. The method of proof of Theorem 4.19 follows the proof of the corresponding result for finite-dimensional Lie groups from [Hel62]. Our Remark 4.14 is inspired by Corollary 15.22 in [Up85].

The symplectic manifolds of type $\mathfrak{z}$ with $\mathfrak{z} = \mathbb{R}$ are the usual symplectic manifolds that play a central role in many areas of differential geometry that claim their origins from mechanics; see e.g., [OR04]. See also [CM74] for a discussion of weakly symplectic manifolds. Symplectic vector spaces with vector-valued symplectic form ("of type $\mathfrak{z}$," in the terminology of Definition 4.25) proved to be useful e.g., in the index problems in [Ni97], $\mathfrak{z}$ being there the topological dual of a Clifford algebra.

The orbit symplectic form described in our Example 4.31 plays a central role in many problems of representation theory and differential geometry. See e.g., Section 7 in [OR03] for a discussion of that construction in the setting of Banach-Lie groups. Some of the classical references are [Lie1890], [Ki76], [Ko70], [So66], and [So67]. A slightly different approach to Example 4.32 can be found in Remark 2.11 in [BR04].

There exist numerous papers devoted to the differential geometry of various homogeneous spaces associated with operator algebras. See the reference list at the end of the book. In the present chapter we have included only a few examples of such homogeneous spaces, without going the details of their differential geometry.

We have to mention also the interesting theory of symmetric Banach manifolds, which extends the classical theory of symmetric spaces exposed e.g., in [Hel62]. We do not discuss the symmetric Banach manifolds in the present book since there already exists the excellent monograph [Up85] treating this topic.

Chapter 5

Quasimultiplicative Maps

Abstract. This chapter has a technical character, including several tools with a distribution theoretic flavor. Quasimultiplicative maps are Banach-algebra-valued distributions with compact support on a finite-dimensional real vector space, which satisfy a weak multiplicativity condition. The first part of the chapter concerns some basic properties of the convolution of two distributions. We then prove the central result of this chapter: a theorem on separate parts of the support of a quasimultiplicative operator-valued distribution. Subsequently we collect some useful facts on intertwining properties of certain spectral subspaces associated with quasimultiplicative maps. The chapter concludes by an exposition of some basic examples of quasimultiplicative maps, namely the Weyl functional calculus of hermitian maps and the Borel functional calculus of Borel measures. During that exposition we see that the support of a quasimultiplicative map should be thought of as a sort of spectrum (more precisely a joint spectrum).

Throughout the present chapter, $\mathfrak{z}$ denotes a finite-dimensional real vector space, $\mathcal{C}^\infty(\mathfrak{z})$ stands for the Fréchet space of all complex-valued smooth functions on $\mathfrak{z}$, and $\mathcal{C}_0^\infty(\mathfrak{z})$ is the set of all functions in $\mathcal{C}^\infty(\mathfrak{z})$ having compact support.

5.1 Supports, convolution, and quasimultiplicativity

The goal of this section is to introduce the notion of quasimultiplicative map (Definition 5.9). We use the convolution of distributions to investigate the relationship between quasimultiplicative maps and derivations of Banach algebras (see e.g., Corollary 5.11).

LEMMA 5.1 *The linear subspace spanned by the set* $\{e^{i\gamma} \mid \gamma \in \mathfrak{z}^\#\} \subseteq \mathcal{C}^\infty(\mathfrak{z})$ *is dense in* $\mathcal{C}^\infty(\mathfrak{z})$. *Moreover*

$$\mathcal{C}^\infty(\mathfrak{z})\widehat{\otimes}\mathcal{C}^\infty(\mathfrak{z}) \simeq \mathcal{C}^\infty(\mathfrak{z}\oplus\mathfrak{z})$$

as Fréchet spaces and the comultiplication map

$$\Delta\colon \mathcal{C}^\infty(\mathfrak{z}) \to \mathcal{C}^\infty(\mathfrak{z})\widehat{\otimes}\mathcal{C}^\infty(\mathfrak{z}) = \mathcal{C}^\infty(\mathfrak{z}\oplus\mathfrak{z}),$$
$$(\forall f\in\mathcal{C}^\infty(\mathfrak{z}))(\forall z_1,z_2\in\mathfrak{z})\quad (\Delta f)(z_1,z_2) = f(z_1+z_2)$$

has the property

$$(\forall\gamma\in\mathfrak{z}^\#)\qquad \Delta(e^{i\gamma}) = e^{i\gamma}\otimes e^{i\gamma}\quad (\in\mathcal{C}^\infty(\mathfrak{z})\otimes\mathcal{C}^\infty(\mathfrak{z})).$$

PROOF All of these facts are well known in distribution theory. In view of the Hahn-Banach theorem, the first assertion follows by the fact that the Fourier transform is injective on the space of distributions with compact support (see e.g., in [Tr67], Proposition 29.1, Theorem 25.6 and Example 1 preceding Theorem 25.5).

For the assertion concerning the topological tensor product, see e.g., equality (51.4) in Theorem 51.6 in [Tr67].

Finally, the asserted property of the comultiplication map is obvious. ∎

REMARK 5.2 For a complex Banach space $\mathfrak{B}$, we consider the vector space of "$\mathfrak{B}$-valued distributions with compact support" (see Lemma 5.7(a) below)

$$\mathcal{E}'(\mathfrak{z},\mathfrak{B}) := \{\Psi\colon \mathcal{C}^\infty(\mathfrak{z})\to\mathfrak{B}\mid \Psi \text{ linear and continuous}\},$$

usually endowed with the topology of uniform convergence on the bounded subsets of $\mathcal{C}^\infty(\mathfrak{z})$. Then, denoting by $\widehat{\otimes}$ the projective tensor product, formula (50.18) in [Tr67] shows that

$$\mathcal{E}'(\mathfrak{z},\mathfrak{B}) \simeq \mathcal{E}'(\mathfrak{z},\mathbb{C})\widehat{\otimes}\mathfrak{B},$$

because $\mathcal{C}^\infty(\mathfrak{z})$ is a nuclear space by the Corollary to Theorem 51.5 in [Tr67]. Now, by using formula (5.18) in [Tr67], we get the following isomorphisms of topological vector spaces:

$$\begin{aligned}\mathcal{E}'(\mathfrak{z},\mathfrak{B})\widehat{\otimes}\mathcal{E}'(\mathfrak{z},\mathfrak{B}) &\simeq \big(\mathcal{E}'(\mathfrak{z},\mathbb{C})\widehat{\otimes}\mathfrak{B}\big)\widehat{\otimes}\big(\mathcal{E}'(\mathfrak{z},\mathbb{C})\widehat{\otimes}\mathfrak{B}\big)\\ &\simeq \big(\mathcal{E}'(\mathfrak{z},\mathbb{C})\widehat{\otimes}\mathcal{E}'(\mathfrak{z},\mathbb{C})\big)\widehat{\otimes}(\mathfrak{B}\widehat{\otimes}\mathfrak{B})\\ &\simeq \mathcal{E}'(\mathfrak{z}\oplus\mathfrak{z},\mathbb{C})\widehat{\otimes}(\mathfrak{B}\widehat{\otimes}\mathfrak{B})\\ &\simeq \mathcal{E}'(\mathfrak{z}\oplus\mathfrak{z},\mathfrak{B}\widehat{\otimes}\mathfrak{B}),\end{aligned}$$

hence we have a natural (jointly) continuous bilinear map

$$\mathcal{E}'(\mathfrak{z},\mathfrak{B})\times\mathcal{E}'(\mathfrak{z},\mathfrak{B})\to\mathcal{E}'(\mathfrak{z}\oplus\mathfrak{z},\mathfrak{B}\widehat{\otimes}\mathfrak{B}),\quad (\Psi_1,\Psi_2)\mapsto\Psi_1\otimes\Psi_2,$$

satisfying $(\Psi_1\otimes\Psi_2)(f_1\otimes f_2) = \Psi_1(f_1)\otimes\Psi_2(f_2)$ whenever $\Psi_j\in\mathcal{E}'(\mathfrak{z}_j,\mathfrak{B})$, $f_j\in\mathcal{C}^\infty(\mathfrak{z}_j)$, $j=1,2$. (See also Corollary 2 after Lemma 41.1 in [Tr67].)

DEFINITION 5.3 Let $\mathfrak{B}$ be a complex Banach algebra, and denote by

$$m\colon \mathfrak{B}\widehat{\otimes}\mathfrak{B} \to \mathfrak{B}$$

the map induced by the multiplication of $\mathfrak{B}$. If $\Delta\colon \mathcal{C}^\infty(\mathfrak{z}) \to \mathcal{C}^\infty(\mathfrak{z}\oplus\mathfrak{z})$ is as in Lemma 5.1, then for all $\Psi_1, \Psi_2 \in \mathcal{E}'(\mathfrak{z}, \mathfrak{B})$ we define their *convolution* $\Psi_1 * \Psi_2 \in \mathcal{E}'(\mathfrak{z}, \mathfrak{B})$ by $\Psi_1 * \Psi_2 = m \circ (\Psi_1 \otimes \Psi_2) \circ \Delta$.

REMARK 5.4 It easily follows by Remark 5.2 that, in the framework of Definition 5.3, the map

$$\mathcal{E}'(\mathfrak{z}, \mathfrak{B}) \times \mathcal{E}'(\mathfrak{z}, \mathfrak{B}) \to \mathcal{E}'(\mathfrak{z}, \mathfrak{B}), \quad (\Psi_1, \Psi_2) \mapsto \Psi_1 * \Psi_2$$

is bilinear and jointly continuous.

DEFINITION 5.5 If $\mathfrak{B}$ is a complex Banach space and $\Psi \in \mathcal{E}'(\mathfrak{z}, \mathfrak{B})$, then the *support* of Ψ, denoted $\operatorname{supp}\Psi$, is the intersection of all closed subsets K of $\mathfrak{z}$ such that $\Psi(f) = 0$ whenever f vanishes on $\mathfrak{z} \setminus K$.

If moreover $\mathfrak{B} = \mathcal{B}(\mathfrak{Y})$ for some complex Banach space $\mathfrak{Y}$, then for all $y \in \mathfrak{Y}$ we have $\Psi(\cdot)y \in \mathcal{E}'(\mathfrak{z}, \mathfrak{Y})$ and thus for each closed subset F of $\mathfrak{z}$ it makes sense to define

$$\mathfrak{Y}_\Psi(F) = \{y \in \mathfrak{Y} \mid \operatorname{supp}\Psi(\cdot)y \subseteq F\}.$$

This is a closed linear subspace of $\mathfrak{Y}$ as a consequence of Lemma 5.7(b) below.

REMARK 5.6 If $\mathfrak{B}$ is a complex Banach space, then $\mathcal{E}'(\mathfrak{z}, \mathfrak{B})$ has a natural structure of $\mathcal{C}^\infty(\mathfrak{z})$-module, in the sense that there exists a separately continuous bilinear map

$$\mathcal{C}^\infty(\mathfrak{z}) \times \mathcal{E}'(\mathfrak{z}, \mathfrak{B}) \to \mathcal{E}'(\mathfrak{z}, \mathfrak{B}), \quad (g, \Psi) \mapsto g\Psi,$$

where, for $\Psi \in \mathcal{E}'(\mathfrak{z}, \mathfrak{B})$ and $f, g \in \mathcal{C}^\infty(\mathfrak{z})$ we define $(g\Psi)(f) = \Psi(gf)$.

We also note that, if $g \in \mathcal{C}^\infty(\mathfrak{z})$ and $g \equiv 1$ on some open neighborhood of $\sup\Psi$, then $g\Psi = \Psi$. Also, if $g \equiv 0$ on some open neighborhood of $\operatorname{supp}\Psi$, then $g\Psi = 0$.

LEMMA 5.7 *Let $\mathfrak{B}$ be a complex Banach space.*

(a) *If $\Psi \in \mathcal{E}'(\mathfrak{z}, \mathfrak{B})$, then $\operatorname{supp}\Psi$ is a compact subset of $\mathfrak{z}$. Moreover, we have $\operatorname{supp}\Psi = \emptyset$ if and only if $\Psi = 0$.*

(b) *If F is a closed subset of $\mathfrak{z}$, then*

$$\{\Psi \in \mathcal{E}'(\mathfrak{z}, \mathfrak{B}) \mid \operatorname{supp}\Psi \subseteq F\}$$

is a closed linear subspace of $\mathcal{E}'(\mathfrak{z}, \mathfrak{B})$.

(c) *If* $\mathfrak{B}$ *is moreover a complex Banach algebra, then*

$$\mathrm{supp}\,(\Psi_1 \otimes \Psi_2) = (\mathrm{supp}\,\Psi_1) \times (\mathrm{supp}\,\Psi_2) \quad (\subseteq \mathfrak{z} \oplus \mathfrak{z})$$

and

$$\mathrm{supp}\,(\Psi_1 * \Psi_2) \subseteq \mathrm{supp}\,\Psi_1 + \mathrm{supp}\,\Psi_2$$

for all $\Psi_1, \Psi_2 \in \mathcal{E}'(\mathfrak{z}, \mathfrak{B})$.

PROOF The assertions in (a) can be proved just as in the case when $\mathfrak{B} = \mathbb{C}$ (see e.g., Theorem 2.3.1 and Theorem 2.2.1 in [Ho90]).

To prove (b), one only has to use the fact that $\mathcal{E}'(\mathfrak{z}, \mathfrak{B})$ is endowed with the topology of uniform convergence on the bounded subsets of $\mathcal{C}^\infty(\mathfrak{z})$.

To prove (c), first note that the Hahn-Banach theorem implies that for every subset $\widetilde{\mathfrak{B}}$ of $\mathfrak{B}^\#$ generating a w^*-dense linear subspace we have

$$\mathrm{Ker}\,\Psi = \bigcap_{\lambda \in \widetilde{\mathfrak{B}}} \mathrm{Ker}\,(\lambda \circ \Psi),$$

whence

$$\mathrm{supp}\,\Psi = \bigcup_{\lambda \in \widetilde{\mathfrak{B}}} \mathrm{supp}\,(\lambda \circ \Psi). \tag{5.1}$$

Furthermore, for all $\lambda_1, \lambda_2 \in \mathfrak{B}^\#$ we have

$$(\lambda_1 \circ \Psi_1) \otimes (\lambda_2 \circ \Psi_2) = (\lambda_1 \otimes \lambda_2) \circ (\Psi_1 \otimes \Psi_2),$$

hence

$$\mathrm{supp}\,((\lambda_1 \otimes \lambda_2) \circ (\Psi_1 \otimes \Psi_2)) = \mathrm{supp}\,(\lambda_1 \circ \Psi_1) \times \mathrm{supp}\,(\lambda_2 \circ \Psi_2)$$

by formula (5.1.3) in [Ho90]. Consequently, by (5.1) we have

$$\begin{aligned}\bigcup_{\lambda_1, \lambda_2 \in \mathfrak{B}^\#} & \mathrm{supp}\,((\lambda_1 \otimes \lambda_2) \circ (\Psi_1 \otimes \Psi_2)) \\ &= \Big(\bigcup_{\lambda_1 \in \mathfrak{B}^\#} \mathrm{supp}\,(\lambda_1 \circ \Psi_1) \Big) \times \Big(\bigcup_{\lambda_2 \in \mathfrak{B}^\#} \mathrm{supp}\,(\lambda_2 \circ \Psi_2) \Big) \\ &= \mathrm{supp}\,\Psi_1 \times \mathrm{sup}\,\Psi_2.\end{aligned}$$

But

$$\bigcup_{\lambda_1, \lambda_2 \in \mathfrak{B}^\#} \mathrm{supp}\,((\lambda_1 \otimes \lambda_2) \circ (\Psi_1 \otimes \Psi_2)) = \mathrm{supp}\,(\Psi_1 \otimes \Psi_2)$$

as a consequence of (5.1) and of the fact that $\mathfrak{B}^\# \otimes \mathfrak{B}^\#$ is a w^*-dense linear subspace of $(\mathfrak{B} \widehat{\otimes} \mathfrak{B})^\#$.

The last assertion of (c) follows by the previous one in a standard way: defining $S\colon \mathfrak{z} \oplus \mathfrak{z} \to \mathfrak{z}$, $(z_1, z_2) \mapsto z_1 + z_2$, we have

$$(\forall \Psi \in \mathcal{E}'(\mathfrak{z} \oplus \mathfrak{z}, \mathfrak{B})) \quad \operatorname{supp}(\Psi \circ \Delta) = S(\operatorname{supp} \Psi).$$

Thus

$$\begin{aligned}
\operatorname{supp}(\Psi_1 * \Psi_2) &= \operatorname{supp}(m \circ (\Psi_1 \otimes \Psi_2) \circ \Delta) \\
&\subseteq \operatorname{supp}((\Psi_1 \otimes \Psi_2) \circ \Delta) \\
&= S(\operatorname{supp}(\Psi_1 \otimes \Psi_2)) \\
&= S(\operatorname{supp} \Psi_1 \times \operatorname{supp} \Psi_2) \\
&= \operatorname{supp} \Psi_1 + \operatorname{supp} \Psi_2,
\end{aligned}$$

and the proof ends. ∎

PROPOSITION 5.8 *Let $\mathfrak{A}$ be a complex Banach algebra and assume that $\Psi \in \mathcal{E}'(\mathfrak{z}, \mathcal{B}(\mathfrak{A}))$ satisfies*

$$(\forall \gamma \in \mathfrak{z}^{\#}) \qquad \Psi(e^{\mathrm{i}\gamma}) \in \operatorname{Hom}(\mathfrak{A}, \mathfrak{A}).$$

Then

$$\Psi(\cdot)a_1 * \Psi(\cdot)a_2 = \Psi(\cdot)(a_1 a_2) \qquad (\in \mathcal{E}'(\mathfrak{z}, \mathfrak{A}))$$

for all $a_1, a_2 \in \mathfrak{A}$.

PROOF According to Lemma 5.1, it suffices to check that the sides of the desired relation take equal values on each function of the form $e^{i\gamma}$ $(\in \mathcal{C}^\infty(\mathfrak{z}))$ with $\gamma \in \mathfrak{z}^{\#}$. By Definition 5.3 we get

$$\begin{aligned}
(\Psi(\cdot)a_1 * \Psi(\cdot)a_2)(e^{\mathrm{i}\gamma}) &= (m \circ (\Psi(\cdot)a_1 \otimes \Psi(\cdot)a_2) \circ \Delta)(e^{\mathrm{i}\gamma}) \\
&= (m \circ (\Psi(\cdot)a_1 \otimes \Psi(\cdot)a_2))(e^{\mathrm{i}\gamma} \otimes e^{\mathrm{i}\gamma}) \\
&= m(\Psi(e^{\mathrm{i}\gamma})a_1 \otimes \Psi(e^{\mathrm{i}\gamma})a_2) \\
&= (\Psi(e^{\mathrm{i}\gamma})a_1) \cdot (\Psi(e^{\mathrm{i}\gamma})a_2) \\
&= \Psi(e^{\mathrm{i}\gamma})(a_1 a_2),
\end{aligned}$$

where the second equality follows by Lemma 5.1, while the last equality follows by the hypothesis. The proof is finished. ∎

The following definition helps us to describe situations where Proposition 5.8 applies. (See Corollary 5.10 below.)

DEFINITION 5.9 Assume that $\mathfrak{B}$ is a complex Banach algebra, and $\Psi \in \mathcal{E}'(\mathfrak{z}, \mathfrak{B})$. We say that Ψ is *quasimultiplicative* if for every $\gamma \in \mathfrak{z}^{\#}$ the map

$$\mathcal{C}^\infty(\mathbb{R}) \to \mathfrak{B}, \quad g \mapsto \Psi(g \circ \gamma)$$

is an algebra homomorphism (i.e., it is multiplicative).

COROLLARY 5.10 *Let $\mathfrak{A}$ be a complex Banach algebra and assume that $\Psi \in \mathcal{E}'(\mathfrak{z}, \mathcal{B}(\mathfrak{A}))$ is quasimultiplicative and satisfies $\Psi(\mathfrak{z}^{\#}) \subseteq \mathrm{Der}(\mathfrak{A})$. Then for all $a_1, a_2 \in \mathfrak{A}$ we have*

$$\Psi(\cdot)a_1 * \Psi(\cdot)a_2 = \Psi(\cdot)(a_1 a_2).$$

PROOF Let $\gamma \in \mathfrak{z}^{\#}$. The fact that the map

$$\mathcal{C}^\infty(\mathbb{R}) \to \mathcal{B}(\mathfrak{A}), \quad g \mapsto \Psi(g \circ \gamma)$$

is a homomorphism of topological algebras implies that $\Psi(e^{\mathrm{i}\gamma}) = e^{\mathrm{i}\Psi(\gamma)}$. Since $\Psi(\gamma) \in \mathrm{Der}(\mathfrak{A})$ (see Example 4.10), it follows that $\Psi(e^{\mathrm{i}\gamma}) \in \mathrm{Aut}(\mathfrak{A})$, and then Proposition 5.8 applies. ∎

COROLLARY 5.11 *If $\mathfrak{A}$ is a complex Banach algebra and $\Psi \in \mathcal{E}'(\mathfrak{z}, \mathcal{B}(\mathfrak{A}))$ is quasimultiplicative with $\Psi(\mathfrak{z}^{\#}) \subseteq \mathrm{Der}(\mathfrak{A})$, then the following assertions hold.*

(a) *For all $a_1, a_2 \in \mathfrak{A}$ we have*

$$\mathrm{supp}\, \Psi(\cdot)(a_1 a_2) \subseteq \mathrm{supp}\, \Psi(\cdot)a_1 + \mathrm{supp}\, \Psi(\cdot)a_2.$$

(b) *If S is a (closed) subsemigroup of $(\mathfrak{z}, +)$, then*

$$\mathfrak{A}_\Psi(S) = \{a \in \mathfrak{A} \mid \mathrm{supp}\, \Psi(\cdot)a \subseteq S\}$$

is a (closed) subalgebra of $\mathfrak{A}$.

PROOF The assertion (a) follows by Corollary 5.10 and the second assertion of Lemma 5.7(c). Then the assertion (b) easily follows by the already proved assertion (a) and Lemma 5.7(b). ∎

5.2 Separate parts of supports

In this section we prove that each separate part (that is, union of some connected components) of the support of a quasimultiplicative map $\Psi \in \mathcal{E}'(\mathfrak{z}, \mathcal{B}(\mathfrak{A}))$ is naturally associated with an idempotent operator $P \in \mathcal{B}(\mathfrak{A})$ (see Theorem 5.14 for details).

LEMMA 5.12 *Let $\mathfrak{A}$ be a complex Banach space and $\Psi \in \mathcal{E}'(\mathfrak{z}, \mathcal{B}(\mathfrak{A}))$ quasimultiplicative. Denote*

$$\Psi(1) =: T \quad (\in \mathcal{B}(\mathfrak{A}))$$

and define

$$T\delta \in \mathcal{E}'(\mathfrak{z}, \mathcal{B}(\mathfrak{A})), \quad f \mapsto f(0)T.$$

Next define for every $t \in \mathbb{R}$

$$H^t \colon \mathcal{C}^\infty(\mathfrak{z}) \to \mathcal{C}^\infty(\mathfrak{z}), \quad (\forall f \in \mathcal{C}^\infty(\mathfrak{z}))(\forall z \in \mathfrak{z}) \quad (H^t f)(z) = f(tz),$$

and

$$\Psi^t := \Psi \circ H^t \quad (\in \mathcal{E}'(\mathfrak{z}, \mathcal{B}(\mathfrak{A})).$$

Then

$$(\forall t, s \in \mathbb{R}) \qquad \Psi^t * \Psi^s = \Psi^{t+s}$$

and $\lim_{t \to 0} \Psi^t = T\delta$ in $\mathcal{E}'(\mathfrak{z}, \mathcal{B}(\mathfrak{A}))$.

PROOF The proof of the assertion concerning the limit is straightforward, and we omit it.

To prove the other assertion, fix $t, s \in \mathbb{R}$. By Lemma 5.1, it suffices to compare the values the sides of the desired equality take on $e^{\mathrm{i}\gamma}$ for $\gamma \in \mathfrak{z}^{\#}$. By Definition 5.3 and then by Lemma 5.1 we get

$$\begin{aligned}
(\Psi^t * \Psi^s)(e^{\mathrm{i}\gamma}) &= (m \circ (\Psi^t \otimes \Psi^s) \circ \Delta)(e^{\mathrm{i}\gamma}) \\
&= (m \circ (\Psi^t \otimes \Psi^s))(e^{\mathrm{i}\gamma} \otimes e^{\mathrm{i}\gamma}) \\
&= \Psi^t(e^{\mathrm{i}\gamma}) \cdot \Psi^s(e^{\mathrm{i}\gamma}) \\
&= \Psi(e^{\mathrm{i}t\gamma}) \cdot \Psi(e^{\mathrm{i}s\gamma}) \\
&= \Psi(e^{\mathrm{i}(t+s)\gamma}) \\
&= \Psi^{t+s}(e^{\mathrm{i}\gamma}),
\end{aligned}$$

where the last but one equality follows since Ψ is quasimultiplicative. The proof is finished. $\square$

LEMMA 5.13 *Let* $\mathfrak{B}$ *be a complex Banach space and* $\Psi_0, \Psi_1, U_0, U_1 \in \mathcal{E}'(\mathfrak{z}, \mathfrak{B})$ *such that* $\Psi_0 + \Psi_1 = U_0 + U_1$ *and* $(\operatorname{supp} \Psi_0 \cup \operatorname{supp} U_0) \cap (\operatorname{supp} \Psi_1 \cup \operatorname{supp} U_1) = \emptyset$. *Then* $\Psi_0 = U_0$ *and* $\Psi_1 = U_1$.

PROOF Clearly $\operatorname{supp}(\pm(\Psi_j - U_j)) \subseteq \operatorname{supp} \Psi_j \cup \operatorname{supp} U_j$ for $j = 0, 1$. Hence for $\Psi := \Psi_0 - U_0 = -(\Psi_1 - U_1)$ we have

$$\operatorname{supp} \Psi \subseteq (\operatorname{supp} \Psi_0 \cup \operatorname{supp} U_0) \cap (\operatorname{supp} \Psi_1 \cup \operatorname{supp} U_1) = \emptyset$$

according to the hypothesis. Then Lemma 5.7(a) implies that $\Psi = 0$, whence $\Psi_0 = U_0$ and $\Psi_1 = U_1$. $\square$

We now come to one of the main results of the present chapter.

THEOREM 5.14 *Let $\mathfrak{A}$ be a complex Banach space, $\Psi \in \mathcal{E}'(\mathfrak{z}, \mathcal{B}(\mathfrak{A}))$ quasimultiplicative with*

$$\bigcap_{f \in \mathcal{C}^\infty(\mathfrak{z})} \operatorname{Ker} \Psi(f) = \{0\},$$

and F_0, F_1 closed subsets of $\mathfrak{z}$ such that

$$\operatorname{supp} \Psi = F_0 \cup F_1 \text{ and } F_0 \cap F_1 = \emptyset.$$

Pick $g \in \mathcal{C}^\infty(\mathfrak{z})$ such that $g \equiv 1$ on some neighborhood of F_0 and $g \equiv 0$ on some neighborhood of F_1 and denote

$$P := \Psi(g) \quad (\in \mathcal{B}(\mathfrak{A})).$$

Then the following assertions hold.

(i) *The operator P is idempotent, i.e.,*

$$P^2 = P.$$

(ii) *For all $f \in \mathcal{C}^\infty(\mathfrak{z})$ we have*

$$P\Psi(f) = \Psi(f)P.$$

(iii) *If $\Psi_0, \Psi_1 \in \mathcal{E}'(\mathfrak{z}, \mathcal{B}(\mathfrak{A}))$ are defined by $\Psi_0 = g\Psi$, $\Psi_1 = (1-g)\Psi$, then*

$$\operatorname{supp} \Psi_j = F_j \quad \text{for} \quad j = 0, 1$$

and for all $f \in \mathcal{C}^\infty(\mathfrak{z})$ we have

$$\Psi_0(f) = \Psi(f) \quad \text{and} \quad \Psi_1(f) = (\mathrm{id}_{\mathfrak{A}} - P)\Psi(f).$$

(iv) *We have*

$$\operatorname{Ran} P = \mathfrak{A}_\Psi(F_0) \quad \text{and} \quad \operatorname{Ran}(\mathrm{id}_{\mathfrak{A}} - P) = \mathfrak{A}_\Psi(F_1).$$

PROOF Obviously $\Psi = \Psi_0 + \Psi_1$, hence Lemma 5.12 implies that

$$(\forall t > 0) \qquad \Psi^t = \Psi_0^t + \Psi_1^t,$$

and then

$$(\forall t, s > 0) \qquad \Psi_0^{t+s} + \Psi_1^{t+s} (= \Psi^{t+s}) = \Psi_0^t * \Psi^s + \Psi_1^t * \Psi^s. \tag{5.2}$$

Next, for $j = 0, 1$, we easily get that $\sup \Psi_j = F_j$ (compare Remark 5.6), whence

$$\sup \Psi_j^{t+s} \subseteq (t+s)F_j \quad \text{and} \quad \operatorname{supp}(\Psi_j^t * \Psi^s) \subseteq tF_j + sF. \tag{5.3}$$

(The latter inclusion follows by the second assertion of Lemma 5.7(c).)

If $t > 0$ is fixed for the moment then for $s > 0$ small enough, we have

$$(t+s)F_j \cap (tF_{1-j} + sF) = \emptyset \quad \text{for} \quad j = 0, 1.$$

Hence (5.2), (5.3), and Lemma 5.13 imply that for fixed t and s small enough, we have $\Psi_j^{t+s} = \Psi_j * \Psi^s$ for $j = 0, 1$. By iterating the latter equality for s small enough, we get by Lemma 7.12 that

$$(\forall t, s > 0) \qquad \Psi_j^{t+s} = \Psi_j * \Psi^s \quad \text{for} \quad j = 0, 1.$$

Since $P = \Psi(g) = \Psi_0(1)$, we then deduce by Lemma 5.12 and Remark 5.4 first that

$$(\forall t > 0) \qquad \Psi_j^t = \Psi_j^t * P\delta \quad \text{for} \quad j = 0, 1,$$

and then that

$$P\delta = P\delta * P\delta,$$

where

$$P\delta \in \mathcal{E}'(\mathfrak{z}, \mathcal{B}(\mathfrak{A})), \quad f \mapsto f(0)P.$$

Now, evaluating both sides of the equality $P\delta = P\delta * P\delta$ on the constant function $1 \in \mathcal{C}^\infty(\mathfrak{z})$, we easily get $P = P^2$.

Next, relation (5.3) implies for $j = 0$, $s = 1$, and $t \to 0$ that $\Psi_0 = P\delta * \Psi$, hence for $f \in \mathcal{C}^\infty(\mathfrak{z})$ we have by Definition 5.3 that

$$\Psi_0(f) = (P\delta * \Psi)(f) = (m \circ (P\delta \otimes \Psi) \circ \Delta)(f) = P \cdot \Psi(f).$$

Since $\Psi_0 + \Psi_1 = \Psi$, it also follows that $\Psi_1(f) = (\mathrm{id}_{\mathfrak{A}} - P)\Psi(f)$.

We can similarly prove that $\Psi_0(f) = \Psi(f) \cdot P$, hence the desired assertion (ii) follows.

It only remains to prove the assertion (iv). Let $a \in \operatorname{Ran} P$. Then $a = Pa$, hence

$$\operatorname{supp} \Psi(\cdot)a = \operatorname{supp} \Psi(\cdot)Pa = \operatorname{supp} \Psi_0(\cdot)a \subseteq \operatorname{supp} \Psi_0 = F_0,$$

hence $a \in \mathfrak{A}_\Psi(F_0)$ (see Definition 5.5). Conversely, let $a \in \mathfrak{A}_\Psi(F_0)$, that is, $\operatorname{supp} \Psi(\cdot)a \subseteq F_0$. We have already seen that $\operatorname{Ran} P \subseteq \mathfrak{A}_\Psi(F_0)$, hence $\operatorname{supp} \Psi(\cdot)Pa \subseteq F_0$. Then

$$\sup \Psi(\cdot)(\mathrm{id}_{\mathfrak{A}} - P)a = \operatorname{supp}(\Psi(\cdot)a - \Psi(\cdot)Pa) \subseteq \operatorname{supp} \Psi(\cdot)a \cup \operatorname{supp} \Psi(\cdot)Pa \subseteq F_0.$$

But we have as above $\operatorname{Ran}(\mathrm{id}_{\mathfrak{A}} - P) \subseteq \mathfrak{A}_\Psi(F_1)$, whence $\operatorname{supp} \Psi(\cdot)(\mathrm{id}_{\mathfrak{A}} - P)a \subseteq F_1$. Thus

$$\sup \Psi(\cdot)(\mathrm{id}_{\mathfrak{A}} - P)a \subseteq F_0 \cap F_1 = \emptyset,$$

and then Lemma 5.7(a) implies that $\Psi(\cdot)(\mathrm{id}_{\mathfrak{A}} - P)a = 0$. Now the hypothesis

$$\bigcap_{f \in \mathcal{C}^\infty(\mathfrak{z})} \operatorname{Ker} \Psi(f) = \{0\}$$

shows that $(\mathrm{id}_{\mathfrak{A}} - P)a = 0$, whence $a \in \operatorname{Ran} P$, and the proof is finished. ∎

COROLLARY 5.15 *Let $\mathfrak{A}$ be a complex Banach algebra and assume that $\Psi \in \mathcal{E}'(\mathfrak{z}, \mathcal{B}(\mathfrak{A}))$ satisfies*

$$\Psi(\mathfrak{z}^{\#}) \subseteq \mathrm{Der}(\mathfrak{A})$$

and

$$\bigcap_{f \in \mathcal{C}^{\infty}(\mathfrak{z})} \operatorname{Ker} \Psi(f) = \{0\}.$$

Also assume that there exists a closed subsemigroup S of $(\mathfrak{z}, +)$ such that $S \cap (-S) = \emptyset$ and

$$\operatorname{supp} \Psi \subseteq (-S) \cup \{0\} \cup S.$$

Then $\mathfrak{A}$ has the following decomposition into a direct sum of closed subalgebras

$$\mathfrak{A} = \mathfrak{A}_{\Psi}(-S) \dotplus \mathfrak{A}_{\Psi}(\{0\}) \dotplus \mathfrak{A}_{\Psi}(S),$$

and $\mathfrak{A}_{\Psi}(\{0\}) \cdot \mathfrak{A}_{\psi}(\pm S) + \mathfrak{A}_{\psi}(\pm S) \cdot \mathfrak{A}_{\Psi}(\{0\}) \subseteq \mathfrak{A}_{\psi}(\pm S)$.

PROOF Use Theorem 5.14 and Corollary 5.11(b). ∎

Some intertwining properties of quasimultiplicative maps

In the following we collect several intertwining properties of quasimultiplicative maps, which will prove to be very useful later on. It will be convenient to use the following remark.

REMARK 5.16 Denote by $\mathcal{P}(\mathfrak{z})$ the algebra of complex-valued polynomial functions on $\mathfrak{z}$. In other words, $\mathcal{P}(\mathfrak{z})$ is the complex subalgebra of $\mathcal{C}^{\infty}(\mathfrak{z})$ generated by $\mathfrak{z}^{\#} \cup \{1\}$. Then for every $p \in \mathcal{P}(\mathfrak{z})$ we have

$$p = t_N(\gamma_N)^N + \cdots + t_1\gamma_1 + t_0$$

for certain $N \in \mathbb{Z}_+$, $t_0, \ldots, t_N \in \mathbb{C}$ and $\gamma_1, \ldots, \gamma_N \in \mathfrak{z}^{\#}$.

PROPOSITION 5.17 *Let $\mathfrak{A}$ be a complex Banach space and assume that $\Psi \in \mathcal{E}'(\mathfrak{z}, \mathcal{B}(\mathfrak{A}))$ is quasimultiplicative. Also assume that $\mathfrak{A}_0$ is a closed subspace of $\mathfrak{A}$ such that*

$$\Psi(\gamma)\mathfrak{A}_0 \subseteq \mathfrak{A}_0 \text{ whenever } \gamma \in \mathfrak{z}^{\#} \cup \{1\} \quad (\subseteq \mathcal{C}^{\infty}(\mathfrak{z})).$$

Then $\Psi(f)\mathfrak{A}_0 \subseteq \mathfrak{A}_0$ for all $f \in \mathcal{C}^{\infty}(\mathfrak{z})$, and

$$\Psi(\cdot)|_{\mathfrak{A}_0} \in \mathcal{E}'(\mathfrak{z}, \mathcal{B}(\mathfrak{A}_0)).$$

PROOF The only nontrivial assertion is that $\mathfrak{A}_0$ is invariant to $\Psi(f)$ for all $f \in \mathcal{C}^\infty(\mathfrak{z})$. With the notation of Remark 5.16, since $\mathcal{P}(\mathfrak{z})$ is dense in $\mathcal{C}^\infty(\mathfrak{z})$, it will be enough to show that for $p \in \mathcal{P}(\mathfrak{z})$ arbitrary we have $\Psi(p)\mathfrak{A}_0 \subseteq \mathfrak{A}_0$.

To this end, write $p = t_N(\gamma_N)^N + \cdots + t_1\gamma_1 + t_0$ as in Remark 5.16. Since Ψ is quasimultiplicative we have $\Psi((\gamma_j)^j) = \Psi(\gamma_j)^j$ for $j = 1, \dots, N$, whence

$$\Psi(p) = t_N\Psi(\gamma_N)^N + \cdots + t_1\Psi(\gamma_1) + t_0\Psi(1).$$

Since $\mathfrak{A}_0$ is invariant to each of the operators $\Psi(\gamma_N), \dots, \Psi(\gamma_1), \Psi(1)$, it then follows that $\mathfrak{A}_0$ is invariant to $\Psi(p)$ as well, and the proof is finished. ∎

PROPOSITION 5.18 *For $j = 1, 2$ let $\mathfrak{A}_j$ be a complex Banach space and $\Psi_j \in \mathcal{E}'(\mathfrak{z}, \mathcal{B}(\mathfrak{A}_j))$ quasimultiplicative. Consider $\varphi \in \mathcal{B}(\mathfrak{A}_1, \mathfrak{A}_2)$ such that, for each $\gamma \in \mathfrak{z}^\# \cup \{1\}$ ($\subseteq \mathcal{C}^\infty(\mathfrak{z})$), the diagram*

$$\begin{array}{ccc} \mathfrak{A}_1 & \xrightarrow{\varphi} & \mathfrak{A}_2 \\ {\scriptstyle \Psi_1(\gamma)}\downarrow & & \downarrow{\scriptstyle \Psi_2(\gamma)} \\ \mathfrak{A}_1 & \xrightarrow{\varphi} & \mathfrak{A}_2 \end{array}$$

is commutative. Then the following assertions hold.

(a) *If* $\operatorname{Ran}\varphi$ *is dense in* $\mathfrak{A}_2$, *then* $\operatorname{supp}\Psi_2 \subseteq \operatorname{supp}\Psi_1$.

(b) *If* $\operatorname{Ker}\varphi = \{0\}$, *then* $\operatorname{supp}\Psi_1 \subseteq \sup\Psi_2$.

(c) *If* $\operatorname{Ran}\varphi \not\subseteq \bigcap\limits_{f \in \mathcal{C}^\infty(\mathfrak{z})} \operatorname{Ker}\Psi_2(f)$, *then* $\operatorname{supp}\Psi_1 \cap \operatorname{supp}\Psi_2 \neq \emptyset$.

PROOF Using Remark 5.16, write an arbitrary $p \in \mathcal{P}(\mathfrak{z})$ under the form $p = t_N(\gamma_N)^N + \cdots + t_1\gamma_1 + t_0$. Now, since Ψ_j is quasimultiplicative, we get $\Psi_j(p) = \sum\limits_{k=0}^{N} t_k\Psi_j(\gamma_k)^k$ as in the proof of Proposition 5.17. Then the hypothesis implies that $\Psi_2(p) \circ \varphi = \varphi \circ \Psi_1(p)$. But $\mathcal{P}(\mathfrak{z})$ is dense in $\mathcal{C}^\infty(\mathfrak{z})$, hence

$$(\forall f \in \mathcal{C}^\infty(\mathfrak{z})) \qquad \Psi_2(f) \circ \varphi = \varphi \circ \Psi_1(f).$$

This fact easily implies that if $\operatorname{Ran}\varphi$ is dense in $\mathfrak{A}_2$, then $\operatorname{Ker}\Psi_1 \subseteq \operatorname{Ker}\Psi_2$. The latter inclusion in turn implies that $\operatorname{supp}\Psi_2 \subseteq \operatorname{supp}\Psi_1$, thus proving the assertion (a).

Similarly, if $\operatorname{Ker}\varphi = \{0\}$, then we obtain by that $\operatorname{Ker}\Psi_2 \subseteq \operatorname{Ker}\Psi_1$, which shows that $\operatorname{supp}\Psi_1 \subseteq \sup\Psi_2$, and (b) is proved.

In order to prove the assertion (c), pick $a_1 \in \mathfrak{A}_1$ with

$$\varphi(a_1) \notin \bigcap_{f \in \mathcal{C}^\infty(\mathfrak{z})} \operatorname{Ker}\Psi_2(f).$$

Then Lemma 5.7(a) shows that

$$\operatorname{supp} \Psi_2(\cdot)(\varphi(a_1)) \neq \emptyset.$$

We have obviously

$$\operatorname{supp} \Psi_2(\cdot)(\varphi(a_1)) \subseteq \operatorname{supp} \Psi_2.$$

On the other hand, for every $f \in \mathcal{C}^\infty(\mathfrak{z})$ with $\operatorname{supp} f \subseteq \mathfrak{z} \setminus \sup \Psi_1(\cdot)a_1$, we have $\Psi_1(f)a_1 = 0$, whence $0 = \varphi(\Psi_1(f)a_1) = \Psi_2(f)(\varphi(a_1))$. Consequently we obtain $(\operatorname{supp} \Psi_2(\cdot)(\varphi(a_1))) \cap (\mathfrak{z} \setminus \operatorname{supp} \Psi_1(\cdot)a_1) = \emptyset$, hence

$$\operatorname{supp} \Psi_2(\cdot)(\varphi(a_1)) \subseteq \operatorname{supp} \Psi_1(\cdot)a_1 \subseteq \operatorname{supp} \Psi_1.$$

Thus

$$\emptyset \neq \operatorname{supp} \Psi_2(\cdot)(\varphi(a_1)) \subseteq \operatorname{supp} \Psi_1 \cap \operatorname{supp} \Psi_2$$

and the proof is finished. □

REMARK 5.19 For later use we now state a by-product of the proof of Proposition 5.18(c). *For $j = 1, 2$, let $\mathfrak{A}_j$ be a complex Banach space and $\Psi_j \in \mathcal{E}'(\mathfrak{z}, \mathcal{B}(\mathfrak{A}_j))$ quasimultiplicative. If $\varphi \in \mathcal{B}(\mathfrak{A}_1, \mathfrak{A}_2)$ and $\varphi \circ \Psi_1(\gamma) = \Psi_2(\gamma) \circ \varphi$ for all $\gamma \in \mathfrak{z}^\# \cup \{1\}$, then for each closed subset F of $\mathfrak{z}$ we have*

$$\varphi\big((\mathfrak{A}_1)_{\Psi_1}(F)\big) \subseteq (\mathfrak{A}_2)_{\Psi_2}(F).$$

(See Definition 5.5.)

To prove this fact, just recall from the proof of Proposition 5.18(c) that

$$\operatorname{supp} \Psi_2(\cdot)(\varphi(a_1)) \subseteq \sup \Psi_1(\cdot)a_1$$

for each $a_1 \in \mathfrak{A}_1$.

LEMMA 5.20 *Let $\mathfrak{A}$ be a complex Banach space, $\Psi \in \mathcal{E}'(\mathfrak{z}, \mathcal{B}(\mathfrak{A}))$ quasimultiplicative with $\Psi(1) = \mathrm{id}_{\mathfrak{A}}$, and F_0, F_1 closed subsets of $\mathfrak{z}$ such that*

$$\operatorname{supp} \Psi = F_0 \cup F_1 \quad \text{and} \quad F_0 \cap F_1 = \emptyset.$$

Let $\mathfrak{B}$ be another Banach space, denote $\widetilde{\mathfrak{A}} = \mathcal{B}(\mathfrak{A}, \mathfrak{B})$ and define

$$\widetilde{\Psi} := M_{\Psi(\cdot)} \in \mathcal{E}'(\mathfrak{z}, \mathcal{B}(\widetilde{\mathfrak{A}})),$$

where

$$(\forall f \in \mathcal{C}^\infty(\mathfrak{z})) \qquad M_{\Psi(f)} \colon \widetilde{\mathfrak{A}} \to \widetilde{\mathfrak{A}}, \quad T \mapsto T\Psi(f).$$

Then $\widetilde{\Psi}$ is quasimultiplicative, $\widetilde{\Psi}(1) = \mathrm{id}_{\widetilde{\mathfrak{A}}}$ and

$$\widetilde{\mathfrak{A}}_{\widetilde{\Psi}}(F_j)\big(\mathfrak{A}_\Psi(F_k)\big) = \{0\}$$

for $j, k \in \{0,1\}$ with $j \neq k$.

PROOF It is clear that $\widetilde{\Psi}$ is quasimultiplicative and $\widetilde{\Psi}(1) = \mathrm{id}_{\widetilde{\mathfrak{A}}}$, and it is also easy to check that

$$\operatorname{supp} \widetilde{\Psi} = \operatorname{supp} \Psi = F_0 \cup F_1.$$

Then we can apply Theorem 5.14 to both Ψ and $\widetilde{\Psi}$. We thus obtain a decomposition of $\mathfrak{A}$ into a direct sum of closed subspaces,

$$\mathfrak{A} = \mathfrak{A}_0 \dotplus \mathfrak{A}_1 \qquad (\text{where } \mathfrak{A}_j = \mathfrak{A}_\Psi(F_j) \text{ for } j = 0, 1),$$

and a similar decomposition of $\widetilde{\mathfrak{A}}$,

$$\widetilde{\mathfrak{A}} = \widetilde{\mathfrak{A}}_0 \dotplus \widetilde{\mathfrak{A}}_1 \qquad (\text{with } \widetilde{\mathfrak{A}}_j = \widetilde{\mathfrak{A}}_{\widetilde{\Psi}(F_j)} \text{ for } j = 0, 1).$$

Moreover, for $j = 0, 1$, $\mathfrak{A}_j$ is invariant to $\Psi(f)$ for all $f \in \mathcal{C}^\infty(\mathfrak{z})$, and we define

$$\Psi_j := \Psi(\cdot)|_{\mathfrak{A}_j} \in \mathcal{E}'(\mathfrak{z}, \mathcal{B}(\mathfrak{A}_j)).$$

We similarly define

$$\widetilde{\Psi}_j := \widetilde{\Psi}(\cdot)|_{\widetilde{\mathfrak{A}}_j} \in \mathcal{E}'(\mathfrak{z}, \mathcal{B}(\widetilde{\mathfrak{A}}_j)).$$

It follows at once by Theorem 5.14 that

$$\operatorname{supp} \Psi_j = \operatorname{supp} \widetilde{\Psi}_j = F_j \quad \text{for} \quad j = 0, 1.$$

Now fix $j \in \{0,1\}$ and define

$$\varphi \colon \widetilde{\mathfrak{A}}_j \to \mathcal{B}(\mathfrak{A}_{1-j}, \mathfrak{B}), \quad T \mapsto T|_{\mathfrak{A}_{1-j}}.$$

What we have to prove is that $\varphi \equiv 0$. To this end, note that

$$(\forall f \in \mathcal{C}^\infty(\mathfrak{z})) \qquad M_{\Psi_{1-j}(f)} \circ \varphi = \varphi \circ \widetilde{\Psi}_j(f),$$

(where $M_{\Psi_{1-j}(\cdot)} \in \mathcal{E}'\big(\mathfrak{z}, \mathcal{B}(\mathcal{B}(\mathfrak{A}_{1-j}, \mathfrak{B}))\big)$ is defined similarly to $M_{\Psi(\cdot)}$). Since

$$\operatorname{supp} M_{\Psi_{1-j}(\cdot)} = \operatorname{supp} \Psi_{1-j} = F_{1-j}$$

and $\operatorname{supp} \widetilde{\Psi}_j = F_j$, it then follows by Proposition 5.18(c) that

$$\operatorname{Ran} \varphi \subseteq \bigcap_{f \in \mathcal{C}^\infty(\mathfrak{z})} \operatorname{Ker} M_{\Psi_{1-j}(f)}.$$

But $M_{\Psi_{1-j}(1)} = \mathrm{id}_{\mathcal{B}(\mathfrak{A}_{1-j}, \mathfrak{B})}$, hence the above inclusion implies that $\operatorname{Ran} \varphi = \{0\}$, i.e., $\varphi \equiv 0$. The proof is finished. ∎

PROPOSITION 5.21 *Let $\mathfrak{A}$ and $\mathfrak{V}$ be complex Banach spaces and*

$$\omega\colon \mathfrak{A}\times\mathfrak{A}\to\mathfrak{V}$$

a bounded skew-symmetric bilinear map. Assume that $\Psi\in\mathcal{E}'(\mathfrak{z},\mathcal{B}(\mathfrak{A}))$ is quasimultiplicative, $\Psi(1)=\mathrm{id}_{\mathfrak{A}}$, and we have

$$\omega\big(\mathfrak{A}_\Psi(\{0\}),\mathfrak{A}_\Psi(\{0\})\big)=\{0\}$$

and

$$(\forall\gamma\in\mathfrak{z}^{\#})(\forall a,b\in\mathfrak{A})\qquad \omega(\Psi(\gamma)a,b)=-\omega(a,\Psi(\gamma)b).$$

If S is a closed subset of $\mathfrak{z}$ such that $S\cap(-S)=\emptyset$ and

$$\mathrm{supp}\,\Psi\subseteq(-S)\cup\{0\}\cup S$$

and we denote

$$\mathfrak{p}=\mathfrak{A}_\Psi((-S)\cup\{0\}),$$

then

$$\omega(\mathfrak{p},\mathfrak{p})=\{0\}.$$

PROOF As in Lemma 5.20, denote $\widetilde{\mathfrak{A}}=\mathcal{B}(\mathfrak{A},\mathfrak{V})$ and define $\widetilde{\Psi}\in\mathcal{E}'(\mathfrak{z},\mathcal{B}(\widetilde{\mathfrak{A}})$ by

$$\widetilde{\Psi}(f)\colon\widetilde{\mathfrak{A}}\to\widetilde{\mathfrak{A}},\quad T\mapsto T\Psi(f),$$

for all $f\in\mathcal{C}^\infty(\mathfrak{z})$. Moreover, define $\widetilde{\Psi}N\in\mathcal{E}'(\mathfrak{z},\mathcal{B}(\widetilde{\mathfrak{A}}))$ by

$$(\forall f\in\mathcal{C}^\infty(\mathfrak{z}))\qquad \widetilde{\Psi}N(f)=\Psi(f\circ N),$$

where

$$N\colon\mathfrak{z}\to\mathfrak{z},\quad z\mapsto -z.$$

Then

$$\mathrm{supp}\,\widetilde{\Psi}=\mathrm{supp}\,\Psi\quad\text{and}\quad\mathrm{supp}\,(\widetilde{\Psi}N)=-\mathrm{supp}\,\Psi.$$

Further define

$$\theta\colon\mathfrak{A}\to\widetilde{\mathfrak{A}},\quad a\mapsto\omega(a,\cdot).$$

Then the hypothesis implies that, for each $\gamma\in\mathfrak{z}^{\#}\cup\{1\}$, the diagram

$$\begin{array}{ccc}\mathfrak{A} & \xrightarrow{\theta} & \widetilde{\mathfrak{A}}\\ {\scriptstyle\Psi(\gamma)}\downarrow & & \downarrow{\scriptstyle(\widetilde{\Psi}N)(\gamma)}\\ \mathfrak{A} & \xrightarrow{\theta} & \widetilde{\mathfrak{A}}\end{array}$$

is commutative, hence Remark 5.19 shows that

$$\theta(\mathfrak{A}_\Psi(-S))\subseteq\widetilde{\mathfrak{A}}_{\widetilde{\Psi}N}(-S)=\widetilde{\mathfrak{A}}_{\widetilde{\Psi}}(S)\tag{5.4}$$

and

$$\theta(\mathfrak{A}_\Psi(\{0\})) \subseteq \widetilde{\mathfrak{A}}_{\widetilde{\Psi}N}(\{0\}) = \widetilde{\mathfrak{A}}_{\widetilde{\Psi}}(\{0\}). \tag{5.5}$$

On the other hand, Theorem 5.14 implies that

$$\mathfrak{A} = \mathfrak{A}_\Psi(-S) \dot{+} \mathfrak{A}_\Psi(\{0\}) \dot{+} \mathfrak{A}_\Psi(S).$$

In particular, we have

$$\mathfrak{p} = \mathfrak{A}_\Psi(-S) \dot{+} \mathfrak{A}_\Psi(\{0\}),$$

hence

$$\begin{aligned}\omega(\mathfrak{p},\mathfrak{p}) &= \theta(\mathfrak{p})\mathfrak{p}\\ &= \big(\theta(\mathfrak{A}_\Psi(-S)) + \theta(\mathfrak{A}_\Psi(\{0\}))\big)\big(\mathfrak{A}_\Psi(-S) + \mathfrak{A}_\Psi(\{0\})\big)\\ &\subseteq \widetilde{\mathfrak{A}}_{\widetilde{\Psi}}(S)(\mathfrak{A}_\Psi(-S)) + \widetilde{\mathfrak{A}}_{\widetilde{\Psi}}(S)(\mathfrak{A}_\Psi(\{0\})) + \widetilde{\mathfrak{A}}_{\widetilde{\Psi}}(\{0\})(\mathfrak{A}_\Psi(-S))\\ &\quad + \theta(\mathfrak{A}_\Psi(\{0\}))\mathfrak{A}_\Psi(\{0\})\end{aligned}$$

by (5.4) along with (5.5). In view of Lemma 5.20 we then get

$$\omega(\mathfrak{p},\mathfrak{p}) \subseteq \theta(\mathfrak{A}_\Psi(\{0\})\mathfrak{A}_\Psi(\{0\}) = \omega(\mathfrak{A}_\Psi(\{0\}), \mathfrak{A}_\psi(\{0\})) = \{0\},$$

where the latter equality follows by the hypothesis. The proof ends. ∎

PROPOSITION 5.22 *For $j = 1, 2$, let $\mathfrak{A}_j$ be a complex Banach space and $\Psi_j \in \mathcal{E}'(\mathfrak{z}, \mathcal{B}(\mathfrak{A}_j))$ quasimultiplicative with $\Psi_j(1) = \mathrm{id}_{\mathfrak{A}_j}$. Consider*

$$\kappa\colon \mathfrak{A}_1 \to \mathfrak{A}_2$$

continuous and conjugate-linear such that for every $\gamma \in \mathfrak{z}^\#$ the diagram

$$\begin{array}{ccc}\mathfrak{A}_1 & \xrightarrow{\kappa} & \mathfrak{A}_2\\ {\scriptstyle -\Psi_1(\gamma)}\big\downarrow & & \big\downarrow{\scriptstyle \Psi_2(\gamma)}\\ \mathfrak{A}_1 & \xrightarrow{\kappa} & \mathfrak{A}_2\end{array}$$

is commutative. Then

$$\kappa((\mathfrak{A}_1)_{\Psi_1}(F)) \subseteq (\mathfrak{A}_2)_{\Psi_2}(-F)$$

for each closed subset F of $\mathfrak{z}$.

PROOF The proof has two stages.

1° We first prove that if $N\colon \mathfrak{z} \to \mathfrak{z}$ is given by $N(z) = -z$ for all $z \in \mathfrak{z}$, then

$$(\forall f \in \mathcal{C}^\infty(\mathfrak{z})) \qquad \kappa \circ \Psi_1(\bar{f} \circ N) = \Psi_2(f) \circ \kappa. \tag{5.6}$$

Since the complex-valued polynomial functions are dense in $\mathcal{C}^\infty(\mathfrak{z})$, it suffices to prove (5.6) in the case when $f = p$ is a polynomial function. To this end, write it under the form

$$p = \sum_{j=1}^{m} t_j(\gamma_j)^j + t_0 \cdot 1 \tag{5.7}$$

as in Remark 5.16, with $m \in \mathbb{Z}_+$, $t_0, \dots, t_m \in \mathbb{C}$ and $\gamma_1, \dots, \gamma_m \in \mathfrak{z}^\#$. Since Ψ_2 is quasimultiplicative and $\Psi_2(1) = \mathrm{id}_{\mathfrak{A}_2}$, we get as in the proof of Proposition 5.17 that

$$\Psi_2(p) = \sum_{j=1}^{m} t_j \Psi(\gamma_j)^j + t_0 \cdot \mathrm{id}_{\mathfrak{A}_2},$$

whence

$$\Psi_2(p) \circ \kappa = \sum_{j=1}^{m} t_j \Psi(\gamma_j)^j \circ \kappa + t_0 \cdot \kappa. \tag{5.8}$$

On the other hand, we get by (5.7)

$$\bar{p} \circ N = \sum_{j=1}^{m} \bar{t}_j(-\gamma_j)^j + \bar{t}_0 \cdot 1,$$

hence the fact that Ψ_1 is quasimultiplicative and κ is conjugate-linear implies that

$$\kappa \circ \Psi_1(\bar{p} \circ N) = \sum_{j=1}^{m} t_j \kappa \circ \Psi_1(-\gamma_j)^j + t_0 \cdot \kappa = \sum_{j=1}^{m} t_j \Psi_2(\gamma_j)^j \circ \kappa + t_0 \cdot \kappa. \tag{5.9}$$

Now (5.8) and (5.9) show that (5.6) indeed holds for $f = p$.

2° We now come back to the proof and consider an arbitrary closed subset F of $\mathfrak{z}$. Take $a_1 \in (\mathfrak{A}_1)_{\Psi_1}(F)$ arbitrary. We have to prove that $\kappa(a_1) \in (\mathfrak{A}_2)_{\Psi_2}(-F)$, that is,

$$\operatorname{supp} \Psi_2(\cdot)(\kappa(a_2)) \subseteq -F.$$

In order to prove this fact, let $f \in \mathcal{C}^\infty(\mathfrak{z})$ arbitrary with $(\operatorname{supp} f) \cap (-F) = \emptyset$. We have by (5.6)

$$\Psi_2(f)(\kappa(a_1)) = \kappa(\Psi_1(\bar{f} \circ N)a_1). \tag{5.10}$$

But $\operatorname{supp}(\bar{f} \circ N) = -\sup f \subseteq \mathfrak{z} \setminus F$. Since $a_1 \in (\mathfrak{A}_1)_{\Psi_1}(F)$, it then follows that $\Psi_1(\bar{f} \circ N)a_1 = 0$, hence $\Psi_2(f)(\kappa(a_1)) = 0$ by (5.10). In view of the way f was chosen, we get $\operatorname{supp} \Psi_2(\cdot)(\kappa(a_1)) \subseteq -F$, and the proof is finished. $\square$

5.3 Hermitian maps

Our aim in this section is to discuss some basic examples of quasimultiplicative maps: the Weyl functional calculus of hermitian maps (Example 5.25) and, as a special case, the Borel functional calculus of spectral measures (Example 5.26).

DEFINITION 5.23 If $\mathfrak{A}$ is a complex Banach space and $\Psi_0 \in \mathcal{B}(\mathfrak{z}^{\#}, \mathcal{B}(\mathfrak{A}))$ has the property that

$$(\forall \gamma \in \mathfrak{z}^{\#}) \qquad \|e^{\mathrm{i}\Psi_0(\gamma)}\| \leq 1,$$

then we say that Ψ_0 is *hermitian.* An operator $A \in \mathcal{B}(\mathfrak{A})$ is said to be *hermitian* if the map $\mathbb{R} \to \mathcal{B}(\mathfrak{A})$, $t \mapsto tA$, is hermitian.

REMARK 5.24 In Definition 5.23, if $n = \dim \mathfrak{z}$ and $\gamma_1, \dots, \gamma_n$ is a basis of $\mathfrak{z}^{\#}$, then the condition that Ψ_0 is hermitian is equivalent to the fact that, denoting $A_k = \Psi_0(\gamma_k) \in \mathcal{B}(\mathfrak{A})$, we have

$$(\forall k \in \{1, \dots, n\})(\forall t \in \mathbb{R}) \qquad \|e^{\mathrm{i}tA_k}\| = 1,$$

by Lemma 2.1 in [An69] (see also Remarks 4/5 in §14 in [BS01]). When $\mathfrak{A}$ is a complex Hilbert space, this is further equivalent to the fact that $A_1, \dots, A_n$ are self-adjoint operators.

EXAMPLE 5.25 As an example of quasimultiplicative map, we now describe the Weyl functional calculus developed in [An69].

Let $\mathfrak{A}$ be a complex Banach space and $\Psi_0 \in \mathcal{B}(\mathfrak{z}, \mathcal{B}(\mathfrak{A}))$ hermitian in the sense of Definition 5.23. Denote $\dim \mathfrak{z} = n$.

In order to extend Ψ_0 from $\mathfrak{z}^{\#}$ to $\mathcal{C}^\infty(\mathfrak{z})$, first recall the *Fourier transform* $\hat{f}$ of a continuous absolutely integrable function $f \colon \mathfrak{z} \to \mathcal{B}(\mathfrak{A})$

$$\hat{f} \colon \mathfrak{z}^{\#} \to \mathcal{B}(\mathfrak{A}), \qquad \hat{f}(\gamma) = \frac{1}{(2\pi)^{n/2}} \int\limits_{\mathfrak{z}} e^{-i\langle \gamma, z \rangle} f(z) dz. \tag{5.11}$$

If $\hat{f}$ is also absolutely integrable, then the *Fourier inversion formula*

$$(\forall z \in \mathfrak{z}) \qquad f(z) = \frac{1}{(2\pi)^{n/2}} \int\limits_{\mathfrak{z}^{\#}} e^{\mathrm{i}\langle \gamma, z \rangle} \hat{f}(\gamma) d\gamma \tag{5.12}$$

holds (see e.g., Lemma 8.2 in Chapter IV in [Va82]) and suggests to define the linear map $\Psi \colon \mathcal{C}_0^\infty(\mathfrak{z}) \to \mathcal{B}(\mathfrak{A})$ by

$$(\forall f \in \mathcal{C}_0^\infty(\mathfrak{z})) \qquad \Psi(f) = \frac{1}{(2\pi)^{n/2}} \int\limits_{\mathfrak{z}^{\#}} e^{\mathrm{i}\Psi_0(\gamma)} \hat{f}(\gamma) d\gamma. \tag{5.13}$$

Then Lemma 2.3 in [An69] shows that we have actually $\Psi \in \mathcal{E}'(\mathfrak{z}, \mathcal{B}(\mathfrak{A}))$ (i.e., Ψ extends from $\mathcal{C}_0^\infty(\mathfrak{z})$ to $\mathcal{C}^\infty(\mathfrak{z})$). Moreover, Lemma 3.1 in [An69] implies that Ψ is quasimultiplicative.

The compact subset $\operatorname{supp}\Psi$ of $\mathfrak{z}$ (see Lemma 5.7(a)) will be called the *Weyl spectrum* of Ψ_0 and will be denoted by

$$\sigma_W(\Psi_0).$$

It clearly agrees with the notation of Weyl joint spectrum $\sigma_W(\cdot)$ introduced in [An69]. Moreover, if $[\operatorname{Ran}\Psi_0, \operatorname{Ran}\Psi_0] = \{0\}$, then $\Psi\colon \mathcal{C}^\infty(\mathfrak{z}) \to \mathcal{B}(\mathfrak{A})$ is multiplicative (see e.g., Theorem 2.4(c) in [An69] or Satz 3.3 in [Al74]), hence $\sigma_W(\Psi_0)$ agrees with the Taylor joint spectrum of commuting tuples of operators (see e.g., Folgerung 2.16 in [Al74] or Proposition 7.7 (1)) in Chapter IV in [Va82]).

We also note that for the constant function $1 \in \mathcal{C}^\infty(\mathfrak{z})$, we have $\Psi(1) = \mathrm{id}_{\mathfrak{A}}$. (See e.g., Theorem 2.9 (b) in [An69].) In particular, it follows that

$$\bigcap_{f \in \mathcal{C}^\infty(\mathfrak{z})} \operatorname{Ker}\big(\Psi(f)\big) = \{0\}.$$

Finally, denoting

$$(\forall F \subseteq \mathfrak{z}) \qquad \mathfrak{A}_{\Psi_0}(F) := \mathfrak{A}_{\Psi}(F)$$

(cf. Definition 5.5), we have

$$\mathfrak{A}_{\Psi_0}(\{0\}) = \bigcap_{\gamma \in \mathfrak{z}^\#} \operatorname{Ker}\big(\Psi_0(\gamma)\big). \tag{5.14}$$

Indeed, if $\Psi_0(\gamma)a = 0$ for all $\gamma \in \mathfrak{z}^\#$, then it easily follows by (5.12) and (5.13) that

$$(\forall f \in \mathcal{C}^\infty(\mathfrak{z})) \qquad \Psi(f)a = f(0)a,$$

hence $\operatorname{supp}\Psi(\cdot)a = \{0\}$. Conversely, if the latter equality holds, then for each $\gamma \in \mathfrak{z}^\#$ fixed we have $\operatorname{supp}\Psi_\gamma(\cdot)a = \{0\}$, where

$$\Psi_\gamma \in \mathcal{E}'(\mathbb{R}, \mathcal{B}(\mathfrak{A})), \quad \Psi_\gamma(f) = \Psi(f \circ \gamma).$$

Then the local spectrum of a with respect to the operator $\Psi_\gamma(\mathrm{id}_{\mathbb{R}})$ $(= \Psi(\gamma))$ is $\{0\}$ by Proposition 7.7 (2) in Chapter IV in [Va82]. Then Lemma 4.4.4 in [CF68] and Proposition 1 in [Al78] (or Corollary 4 in §14 in [BS01]) imply that $\Psi(\gamma)a = 0$, and (5.14) is completely proved.

EXAMPLE 5.26 Let $\mathfrak{A}$ be a complex Banach space. Consider a compact subset K of $\mathfrak{z}$ and denote by $\mathrm{Bor}(K)$ the σ-algebra of all Borel subsets of K. Let

$$E\colon \mathrm{Bor}(K) \to \mathcal{B}(\mathfrak{A})$$

be a *spectral measure on* $\mathfrak{z}$, i.e., $E(\cdot)$ has the following properties:

(a) $E(\emptyset) = 0$, $E(K) = \mathrm{id}_{\mathfrak{A}}$;

(b) $E(\delta_1 \cap \delta_2) = E(\delta_1)E(\delta_2)$ whenever $\delta_1, \delta_2 \in \mathrm{Bor}(K)$;

(c) $\sup\{\|E(\delta)\| \mid \delta \in \mathrm{Bor}(K)\} =: M < \infty$; and

(d) if $\{\delta_j\}_{j\geq 1}$ is a sequence of pairwise disjoint elements of $\mathrm{Bor}(K)$, then

$$E(\bigcup_{j=1}^{\infty} \delta_j)a = \sum_{j=1}^{\infty} E(\delta_j)a$$

for each $a \in \mathfrak{A}$.

Then E defines an element Ψ_E of $\mathcal{E}'(\mathfrak{z}, \mathcal{B}(\mathfrak{A}))$ by

$$\Psi_E \colon \mathcal{C}^\infty(\mathfrak{z}) \to \mathcal{B}(\mathfrak{A}), \quad \Psi_E(f) = \int_{\mathfrak{z}} f(z)dE(z).$$

(See e.g., Theorem 7.21 in Chapter IV in [Va82] for the details of construction of the above integral.) We denote

$$\mathrm{supp}\, E := \mathrm{supp}\, \Psi_E$$

and note that $\mathrm{supp}\, E$ is the smallest closed subset K_0 of K with $E(K_0) = \mathrm{id}_{\mathfrak{A}}$.

The property (b) of the spectral measure E implies that Ψ_E is *multiplicative*:

$$(\forall f, g \in \mathcal{C}^\infty(\mathfrak{z})) \qquad \Psi_E(fg) = \Psi_E(f)\Psi_E(g). \tag{5.15}$$

We then have for every $\gamma \in \mathfrak{z}^\# \; (\subseteq \mathcal{C}^\infty(\mathfrak{z}))$

$$\|e^{\mathrm{i}\Psi_E(\gamma)}\| = \|\Psi_E(e^{\mathrm{i}\gamma})\| = \|\int_{\mathfrak{z}} e^{\mathrm{i}\langle\gamma,z\rangle} dE(z)\| \leq 4M$$

(see estimate (7.10) in Chapter IV in [Va82]). Since $\{e^{\mathrm{i}\Psi_E(\gamma)} \mid \gamma \in \mathfrak{z}^\#\}$ is a group of invertible operators on $\mathfrak{A}$ (by (5.15)), it then follows that the function $\|\cdot\|_1 \colon \mathfrak{A} \to \mathbb{R}_+$ defined by

$$(\forall a \in \mathfrak{A}) \qquad \|a\|_1 = \sup\{\|e^{\mathrm{i}\Psi_E(\gamma)}\| \mid \gamma \in \mathfrak{z}^\#\}$$

is an equivalent norm on $\mathfrak{A}$ such that, denoting also by $\|\cdot\|_1$ the corresponding operator norm on $\mathcal{B}(\mathfrak{A})$, we have

$$(\forall \gamma \in \mathfrak{z}^\#) \qquad \|e^{\mathrm{i}\Psi_E(\gamma)}\|_1 \leq 1,$$

hence the map $\Psi_E|_{\mathfrak{z}^\#} \colon \mathfrak{z}^\# \to \mathcal{B}(\mathfrak{A})$ is hermitian in the sense of Definition 5.23, provided we consider $\mathfrak{A}$ endowed with the equivalent norm $\|\cdot\|_1$.

We now prove a version of Proposition 5.17 in the framework of spectral measures.

PROPOSITION 5.27 *Let $\mathfrak{A}$ be a complex Banach space, K a compact subset of $\mathfrak{z}$, and $E\colon \mathrm{Bor}(K) \to \mathcal{B}(\mathfrak{A})$ a spectral measure. If $\mathfrak{A}_0$ is a closed subspace of $\mathfrak{A}$ that is invariant to $\Psi_E(\gamma)$ for all $\gamma \in \mathfrak{z}^{\#}$, then $\mathfrak{A}_0$ is invariant to $E(\delta)$ for all $\delta \in \mathrm{Bor}(K)$, and*

$$E(\cdot)|_{\mathfrak{A}_0} : \mathrm{Bor}(K) \to \mathcal{B}(\mathfrak{A}_0)$$

is again a spectral measure.

PROOF Take $a_0 \in \mathfrak{A}_0$ and $\beta \in \mathfrak{A}^{\#}$ arbitrary with $\beta|_{\mathfrak{A}_0} = 0$, and observe that $\beta(E(\cdot)a_0)$ is a complex Radon measure on K, vanishing on the polynomials by Proposition 5.17. Hence $\beta(E(\delta)a_0) = 0$ for all $\delta \in \mathrm{Bor}(K)$. Since $\beta \in \mathfrak{A}^{\#}$ was arbitrary with $\beta|_{\mathfrak{A}_0} = 0$, we get $E(\delta)a_0 \in \mathfrak{A}_0$ as an easy consequence of the Hahn-Banach Theorem. Consequently, $\mathfrak{A}_0$ is invariant to $E(\delta)$ for all $\delta \in \mathrm{Bor}(K)$.

The other assertion in the statement is obvious. ∎

REMARK 5.28 For later use, let us note a more general fact that can be obtained by the argument used in the proof of Proposition 5.27. *For $j = 1, 2$, let $\mathfrak{A}_j$ be a complex Banach space and*

$$E_j\colon \mathrm{Bor}(K) \to \mathcal{B}(\mathfrak{A}_j)$$

a spectral measure, where K is a compact subset of $\mathfrak{z}$. If $\varphi \in \mathcal{B}(\mathfrak{A}_1, \mathfrak{A}_2)$ and

$$(\forall \gamma \in \mathfrak{z}^{\#}) \qquad \Psi_{E_2}(\gamma) \circ \varphi = \varphi \circ \Psi_{E_1}(\gamma),$$

then

$$(\forall \delta \in \mathrm{Bor}(K)) \qquad E_2(\delta) \circ \varphi = \varphi \circ E_1(\delta).$$

REMARK 5.29 We now state a version of Proposition 5.22 in the framework of spectral measures. *Let K be a compact subset of $\mathfrak{z}$; for $j = 1, 2$, $\mathfrak{A}_j$ a complex Banach space; and*

$$E_j\colon \mathrm{Bor}(K) \to \mathcal{B}(\mathfrak{A}_j)$$

a spectral measure. Let $\kappa\colon \mathfrak{A}_1 \to \mathfrak{A}_2$ be continuous and conjugate-linear such that

$$(\forall \gamma \in \mathfrak{z}^{\#}) \qquad \Psi_{E_2}(\gamma) \circ \kappa = -\kappa \circ \Psi_{E_1}(\gamma).$$

Then

$$(\forall \delta \in \mathrm{Bor}(K)) \qquad E_2(\delta) \circ \kappa = \kappa \circ E_1(-\delta).$$

To prove this fact, just use relation (5.6) in the proof of Proposition 5.22, and then the argument in the proof of Proposition 5.27.

Notes

In this chapter we obtain versions of results belonging to the spectral theory of commuting tuples of operators. These versions are suitable for the applications we are going to make in the next chapters. Instead of making use of Koszul complexes as e.g., in [Va82] and [BS01], we use appropriate functional calculi. And these functional calculi are nothing more than operator-valued distributions on finite-dimensional real vector spaces. The support of such a distribution plays the role of a joint spectrum (see Example 5.25).

The present chapter owes much to the papers [An69] and [An70]. In particular in Theorem 5.14 we used the method of proof of Theorem 1 in [An70]. Nevertheless we need (and draw) conclusions which are significantly sharper than the statement of Theorem 1 in [An70]. In particular, nothing like the assertions (iii) and (iv) in the following theorem seems to be noted in [An70].

Definition 5.3 is suggested by e.g., Theorem 40.5 in [Tr67]. Our Remark 5.4 in the special case of scalar valued distributions can be found e.g., as Corollary 1 to Lemma 41.1 in [Tr67]. For our Remark 5.16 see the beginning of the proof of Theorem 2.4 in [An69].

We also note that Corollary 5.11 extends the results of Section 1 in [Be05a]. Also, Proposition 5.21 in the present chapter is a generalization of Proposition 2.1 in [Be05a].

Chapter 6

Complex Structures on Homogeneous Spaces

Abstract. We expose a general method to construct invariant complex structures on homogeneous spaces of Banach-Lie groups. The approach relies on the properties of quasimultiplicative maps. Afterwards we introduce the various types of (pseudo-) Kähler manifolds, which are helpful in order to understand the interaction between complex geometry and symplectic geometry. The (pseudo-) Kähler homogeneous spaces of a Lie group G are described in Lie algebraic terms by means of the complex polarizations, which are particular subalgebras of the complexified Lie algebra of G. The chapter concludes by an application of the aforementioned method to constructions of invariant complex structures on flag manifolds associated with Banach algebras. In the special case when the Banach algebra under consideration is a matrix algebra $M_n(\mathbb{C})$ we get precisely the usual flag manifolds dealt with in the classical complex geometry.

6.1 General results

According to Theorem 4.19, if H is a Banach-Lie subgroup of the real Banach-Lie group G, then G/H carries the structure of a real Banach manifold such that the canonical projection $\pi\colon G \to G/H$ is a real analytic submersion and the natural transitive action

$$G \times G/H \to G/H, \qquad (g, kH) \mapsto gkH$$

is real analytic. In this setting, it is important to know when G/H possesses a *G-invariant complex structure*. That is, when does G/H possess a structure of complex Banach manifold such that for each $g \in G$, the map

$$G/H \to G/H, \qquad kH \mapsto gkH,$$

is holomorphic?

The following theorem contains an answer to that question. The main point here is that the complex G-invariant structure of the homogeneous space G/H is studied only in terms of Lie algebras.

THEOREM 6.1 *Let G be a real Banach-Lie group with the Lie algebra $\mathfrak{g}$. Denote by $a \mapsto \bar{a}$ the involution of $\mathfrak{g}_{\mathbb{C}}$ whose fixed points are just the elements of $\mathfrak{g}$. Assume that H is a connected Banach-Lie subgroup of G with the Lie algebra $\mathfrak{h}$ and fix a closed subspace $\mathfrak{M}$ of $\mathfrak{g}$ with $\mathfrak{g} = \mathfrak{M} \dot{+} \mathfrak{h}$. Moreover denote by*

$$\alpha \colon G \times G/H \to G/H, \quad (g, kH) \mapsto \alpha_g(kH) = gkH,$$

the transitive action of G on the smooth homogeneous space G/H. Then there exists a bijective correspondence between, on one hand, the complex Banach manifold structures on $M = G/H$ satisfying the conditions:

(a) *the smooth structure underlying the complex structure is just the smooth manifold structure of the homogeneous space G/H, and*

(b) *the mapping $\alpha_g \colon G/H \to G/H$ is holomorphic for all $g \in G$,*

and, on the other hand, the closed complex subalgebras $\mathfrak{k}$ of $\mathfrak{g}_{\mathbb{C}}$ having the properties

(i) $[\mathfrak{h}, \mathfrak{k}] \subseteq \mathfrak{k}$,

(ii) $\mathfrak{k} \cap \bar{\mathfrak{k}} = \mathfrak{h} + \mathrm{i}\mathfrak{h}$, *and*

(iii) $\mathfrak{k} + \bar{\mathfrak{k}} = \mathfrak{g}_{\mathbb{C}}$.

The same conclusion holds when H is not necessarily connected, provided (i) *holds under the stronger form*

(i') $\mathrm{Ad}_G(h)\mathfrak{k} \subseteq \mathfrak{k}$ *for all $h \in H$.*

PROOF See [Be05b] for a complete and self-contained discussion. □

Theorem 6.1 shows the importance of the following question:

> *Given a real Banach-Lie algebra $\mathfrak{g}$ with the complexification $\mathfrak{A}$, and a closed subalgebra $\mathfrak{h}$ of $\mathfrak{g}$, describe sufficient conditions for the existence of a closed complex subalgebra $\mathfrak{k}$ of $\mathfrak{A}$ with the properties* (i)–(iii) *as in* Theorem 6.1 *above.*

It is just our next aim to approach this question using the tools developed in Chapter 5.

PROPOSITION 6.2 *Let $\mathfrak{g}$ be a canonically involutive Banach-Lie algebra, $\mathfrak{z}$ a finite-dimensional real vector space, and $\Psi_1 \in \mathcal{B}(\mathfrak{z}^{\#}, \mathrm{Der}(\mathfrak{g}))$ and denote by $\mathfrak{A}$ the complexification of $\mathfrak{g}$. Assume that the map*

$$\mathrm{i}\Psi_1 \colon \mathfrak{z}^{\#} \to \mathrm{Der}(\mathfrak{A})$$

can be extended to a quasimultiplicative map $\Psi \in \mathcal{E}'(\mathfrak{z}, \mathcal{B}(\mathfrak{A}))$ with $\Psi(1) = \mathrm{id}_{\mathfrak{A}}$.

If there exists a closed subsemigroup S of $(\mathfrak{z}, +)$ such that

$$(-S) \cap S = \emptyset$$

and

$$\mathrm{supp}\, \Psi \subseteq (-S) \cup \{0\} \cup S,$$

then

$$\mathfrak{h} := \mathfrak{g} \cap \mathfrak{A}_{\Psi}(\{0\})$$

is a closed subalgebra of $\mathfrak{g}$,

$$\mathfrak{k} := \mathfrak{A}_{\Psi}((-S) \cup \{0\})$$

is a closed complex subalgebra of $\mathfrak{A}$, and the conditions (i)–(iii) *in* Theorem 6.1 *are fulfilled.*

PROOF As in the statement of Theorem 8.4, denote by $a \mapsto \bar{a}$ the involution of $\mathfrak{A}$ whose fixed points are the elements of $\mathfrak{g}$. We have

$$(\forall \gamma \in \mathfrak{z}^{\#})(\forall a \in \mathfrak{A}) \qquad \Psi(\gamma)\bar{a} = -\overline{\Psi(\gamma)a},$$

hence Proposition 5.22 implies that

$$\overline{\mathfrak{A}_{\Psi}(F)} = \mathfrak{A}_{\Psi}(-F) \quad \text{for every closed subset } F \text{ of } \mathfrak{z}. \tag{6.1}$$

In particular, $\overline{\mathfrak{A}_{\Psi}(\{0\})} = \mathfrak{A}_{\Psi}(\{0\})$, whence

$$\mathfrak{h} + \mathrm{i}\mathfrak{h} = \mathfrak{A}_{\Psi}(\{0\}).$$

On the other hand,

$$\mathfrak{k} \cap \bar{\mathfrak{k}} = \mathfrak{A}_{\Psi}(S \cup \{0\}) \cap \mathfrak{A}_{\Psi}((-S) \cup \{0\}) = \mathfrak{A}_{\Psi}(\{0\}),$$

hence

$$\mathfrak{k} \cap \bar{\mathfrak{k}} = \mathfrak{h} + \mathrm{i}\mathfrak{h},$$

which is just condition (ii) in Theorem 6.1. The other conditions (i) and (iii) in Theorem 6.1 follow by Corollary 5.15, using the fact that

$$\mathfrak{A}_{\Psi}(\pm S \cup \{0\}) = \mathfrak{A}_{\Psi}(\pm S) \dotplus \mathfrak{A}_{\Psi}(\{0\}),$$

and the proof ends. ∎

We now apply the previous Proposition 5.5 using Definition 5.23 and Example 5.25.

COROLLARY 6.3 *Let $\mathfrak{g}$ be a canonically involutive Banach-Lie algebra with the complexification $\mathfrak{A}$, $\mathfrak{z}$ a finite-dimensional real vector space, and $\Psi_1 \in \mathcal{B}(\mathfrak{z}^{\#}, \mathrm{Der}(\mathfrak{g}))$. Assume that the map*

$$\Psi_0 := \mathrm{i}\Psi_1 \quad (\in \mathcal{B}(\mathfrak{z}^{\#}, \mathcal{B}(\mathfrak{A})))$$

is hermitian and there exists a closed subsemigroup S of $(\mathfrak{z}, +)$ such that

$$(-S) \cap S = \emptyset$$

and

$$\sigma_W(\Psi_0) \subseteq (-S) \cup \{0\} \cup S.$$

Then

$$\mathfrak{h} := \bigcap_{\gamma \in \mathfrak{z}^{\#}} \mathrm{Ker}\, \Psi_1(\gamma)$$

is a closed subalgebra of $\mathfrak{g}$,

$$\mathfrak{k} := \mathfrak{A}_{\Psi_0}((-S) \cup \{0\})$$

is a closed complex subalgebra of $\mathfrak{A}$, and the conditions (i)–(iii) *in* Theorem 6.1 *are fulfilled.*

PROOF Since the map $\Psi_0 \colon \mathfrak{z}^{\#} \to \mathcal{B}(\mathfrak{A})$ is hermitian, Example 5.25 shows that it can be extended to a quasimultiplicative map $\Psi \in \mathcal{E}'(\mathfrak{z}, \mathcal{B}(\mathfrak{A}))$ with $\Psi(1) = \mathrm{id}_{\mathfrak{A}}$ and

$$\mathfrak{A}_{\Psi}(\{0\}) = \bigcap_{\gamma \in \mathfrak{z}^{\#}} \mathrm{Ker}\, \Psi(\gamma).$$

Now use Proposition 6.2. ∎

The next result is a version of the above Corollary 6.3 involving spectral measures (see Example 5.26). The point here is that we strengthen the condition upon the quasimultiplicative map Ψ in exchange for weakening the condition on the subsemigroup S of $(\mathfrak{z}, +)$.

PROPOSITION 6.4 *Let $\mathfrak{g}$ be a canonically involutive Banach-Lie algebra whose complexification is denoted by $\mathfrak{A}$, $\mathfrak{z}$ a finite-dimensional real vector* *space, and* $\Psi \in \mathcal{B}(\mathfrak{z}^{\#}, \mathrm{Der}(\mathfrak{g}))$. *Assume that there exists a spectral measure* E *on* $\mathfrak{z}$ *with values in* $\mathcal{B}(\mathfrak{A})$ *such that*

$$(\forall \gamma \in \mathfrak{z}^{\#}) \qquad i\Psi(\gamma) = \int\limits_{\mathfrak{z}} \gamma(z) dE(z)$$

and there exists a closed subsemigroup S *of* $(\mathfrak{z}, +)$ *with*

$$(-S) \cap S = \{0\}$$

and

$$\sup E \subseteq (-S) \cup S.$$

Then

$$\mathfrak{h} := \bigcap_{\gamma \in \mathfrak{z}^{\#}} \operatorname{Ker} \Psi(\gamma)$$

is a closed subalgebra of $\mathfrak{g}$,

$$\mathfrak{k} := \operatorname{Ran} E(-S)$$

is a closed complex subalgebra of $\mathfrak{A}$, *and the conditions* (i)–(iii) *in* Theorem 6.1 *are fulfilled.*

PROOF We first note that in the statement and in what follows, we define as usual

$$E(\delta) := E(\delta \cap \sup E)$$

for every Borel subset δ of $\mathfrak{z}$.

Since $\operatorname{supp} E \subseteq (-S) \cup S$, we have $\mathrm{id}_{\mathfrak{A}} = E(-S) + E(S \setminus \{0\})$ by the additivity property of the spectral measure E, hence

$$\mathfrak{A} = \operatorname{Ran} E(-S) + \operatorname{Ran} E(S) \tag{6.2}$$

(because $\operatorname{Ran} E(S \setminus \{0\}) \subseteq \operatorname{Ran} E(S)$).

On the other hand, for each closed subset δ of $\mathfrak{z}$ we have by both Corollary 7.22 and Proposition 7.7 in Chapter IV in [Va82] that

$$\operatorname{Ran} E(\delta) = \mathfrak{A}_{\Psi_E}(\delta), \tag{6.3}$$

where $\Psi_E \in \mathcal{E}'(\mathfrak{z}, \mathcal{B}(\mathfrak{A}))$ is the multiplicative map associated to the spectral measure E as in Example 5.26. Now, we can use (6.1) from the proof of Proposition 6.2 to deduce that

$$\operatorname{Ran} E(S) = \overline{\operatorname{Ran} E(-S)},$$

whence by (6.2)

$$\mathfrak{A} = \mathfrak{k} + \bar{\mathfrak{k}}.$$

Moreover,

$$\mathfrak{k} \cap \bar{\mathfrak{k}} = \operatorname{Ran} E(-S) \cap \operatorname{Ran} E(S) = \operatorname{Ran} E((-S) \cap S) = \operatorname{Ran} E(\{0\}) = \mathfrak{h} + i\mathfrak{h},$$

where the latter equality follows just as in the proof of Proposition 6.2 above.

The fact that $\mathfrak{k}$ is a closed subalgebra of $\mathfrak{A}$ follows by (6.3) and Corollary 5.11(b). Since $\mathfrak{h} \subseteq \mathfrak{k}$, it then follows that $[\mathfrak{h}, \mathfrak{k}] \subseteq \mathfrak{k}$. Finally, since $E(-S)^2 = E(-S)$, it follows that $\mathfrak{k}$ $(= \operatorname{Ran} E(-S))$ is complemented in $\mathfrak{A}$ by $\operatorname{Ker} E(-S)$. ∎

6.2 Pseudo-Kähler manifolds

This section is closely related to Section 4.2 inasmuch as the pseudo-Kähler structures which we discuss below are nothing else than symplectic structures that are compatible in some sense with a complex structure. The main result (Theorem 6.12) is a characterization of homogeneous (pseudo-)Kähler structures in terms of complex polarizations.

DEFINITION 6.5 Let $\mathfrak{z}$ be a real Banach space. A *weakly pseudo-Kähler manifold of type* $\mathfrak{z}$ is a pair (M, ω) satisfying the following conditions.

(i) The pair (M, ω) is a weakly symplectic manifold of type $\mathfrak{z}$.

(ii) The Banach manifold M is actually a complex Banach manifold.

(iii) For all $x \in M$ and $v, w \in T_x M$ we have

$$\omega_x(I_x v, I_x w) = \omega_x(v, w),$$

where $I_x \colon T_x M \to T_x M$ is the multiplication-by-i operator.

If moreover (M, ω) is a strongly symplectic manifold of type $\mathfrak{z}$, then we say that (M, ω) is a *strongly pseudo-Kähler manifold of type* $\mathfrak{z}$.

The pair (M, ω) is said to be a *weakly* (respectively, *strongly*) *Kähler manifold of type* $\mathfrak{z}$ if it is a weakly (respectively, strongly) pseudo-Kähler manifold of type $\mathfrak{z}$ that satisfies the following positivity condition:

(iv) There exists a closed convex cone Γ in $\mathfrak{z}$ such that $\Gamma \cap (-\Gamma) = \{0\}$ and

$$(\forall v \in T_x M \setminus \{0\}) \quad \omega_x(v, I_x v) \in \Gamma \setminus \{0\}$$

for all $x \in M$.

REMARK 6.6 It is easy to see that condition (iv) in Definition 6.5 is equivalent to the following condition: If we extend $\omega_x \colon T_xM \times T_xM \to \mathfrak{z}$ to a $\mathbb{C}$-bilinear map $\omega_x \colon (T_xM)_{\mathbb{C}} \times (T_xM)_{\mathbb{C}} \to \mathfrak{z}_{\mathbb{C}}$ and $I_x \colon T_xM \to T_xM$ to a $\mathbb{C}$-linear operator $I_x \colon (T_xM)_{\mathbb{C}} \to (T_xM)_{\mathbb{C}}$ (which satisfies $(I_x)^2 = -\mathrm{id}_{(T_xM)_{\mathbb{C}}}$) then

$$(\forall a \in V_x^+ \setminus \{0\}) \quad -\mathrm{i}\omega_x(a, \overline{a}) \in \Gamma \setminus \{0\},$$

where $a \mapsto \overline{a}$ denotes the conjugation on $(T_xM)_{\mathbb{C}}$, whose set of fixed points is T_xM, and $V_x^+ := \mathrm{Ker}\,(I_x - \mathrm{i} \cdot \mathrm{id}_{(T_xM)_{\mathbb{C}}})$.

DEFINITION 6.7 Let $\mathfrak{z}$ be a real Banach space and G a Banach-Lie group.

A *weakly* (respectively, *strongly*) *pseudo-Kähler homogeneous space of* G *of type* $\mathfrak{z}$ is a weakly (respectively, strongly) pseudo-Kähler manifold (M, ω) of type $\mathfrak{z}$ that satisfies the following conditions:

(i) The pair (M, ω) is a weakly (respectively, strongly) symplectic homogeneous space of G of type $\mathfrak{z}$.

(ii) For every element $g \in G$ the mapping $\alpha_g \colon M \to M$ is holomorphic, where $\alpha \colon G \times M \to M$, $(g, x) \mapsto \alpha_g(x)$, is the transitive action of G on its homogeneous space M.

If (M, ω) is actually a weakly (respectively, strongly) Kähler manifold of type $\mathfrak{z}$, then we say that the pair (M, ω) is a *weakly* (respectively, *strongly*) *Kähler homogeneous space of* G *of type* $\mathfrak{z}$.

The next definition will help us to get a characterization of pseudo-Kähler homogeneous spaces in Lie algebraic terms (see Theorem 6.12 below).

DEFINITION 6.8 Let $\mathfrak{g}$ be a real Banach-Lie algebra, $\mathfrak{z}$ a real Banach space, and $\eta \in Z_c^2(\mathfrak{g}, \mathfrak{z})$. Denote by $a \mapsto \bar{a}$ the involution of $\mathfrak{g}_{\mathbb{C}}$ whose fixed points are the elements of $\mathfrak{g}$. Also denote

$$\mathfrak{h} := \{a \in \mathfrak{g} \mid \omega(a, \mathfrak{g}) = \{0\}\}.$$

Then a *complex polarization* of $\mathfrak{g}$ in ω is a closed complex subspace $\mathfrak{p}$ of $\mathfrak{g}_{\mathbb{C}}$ such that $\mathfrak{p}$ is complemented in $\mathfrak{g}_{\mathbb{C}}$ and moreover has the following properties:

(C1) $\mathfrak{p}$ is a subalgebra of $\mathfrak{g}_{\mathbb{C}}$ and $[\mathfrak{h}, \mathfrak{p}] \subseteq \mathfrak{p}$,

(C2) $\mathfrak{p} \cap \overline{\mathfrak{p}} = \mathfrak{h} + \mathrm{i}\mathfrak{h}$,

(C3) $\mathfrak{p} + \overline{\mathfrak{p}} = \mathfrak{g}_{\mathbb{C}}$, and

(C4) $\omega(\mathfrak{p} \times \mathfrak{p}) = \{0\}$.

REMARK 6.9

(a) In Definition 6.8, the condition $[\mathfrak{h}, \mathfrak{p}] \subseteq \mathfrak{p}$ actually follows by the other properties of $\mathfrak{p}$, because $\mathfrak{p}$ is a Lie algebra and $\mathfrak{h} \subseteq \mathfrak{p}$ by (C2).

(b) It will turn out in stage 1° of the proof of Theorem 6.12 below that in the setting of Definition 6.8, the properties (C2) and (C3) imply that the inclusion map $\mathfrak{g} \hookrightarrow \mathfrak{g}_{\mathbb{C}}$ induces an invertible operator

$$\mathfrak{g}/\mathfrak{h} \to \mathfrak{g}_{\mathbb{C}}/p.$$

(c) We note that a complex polarization in particular satisfies conditions (i)–(iii) in Theorem 6.1

DEFINITION 6.10 Let $\mathfrak{g}$ be a real Banach-Lie algebra, $\mathfrak{z}$ a real Banach space, and $\omega \in Z^2_c(\mathfrak{g}, \mathfrak{z})$. Denote by $a \mapsto \bar{a}$ the involution of $\mathfrak{g}_{\mathbb{C}}$ whose set of fixed points is just $\mathfrak{g}$. Also denote

$$\mathfrak{h} = \{a \in \mathfrak{g} \mid \omega(a, \mathfrak{g}) = \{0\}\}.$$

We say that a complex polarization $\mathfrak{p}$ of $\mathfrak{g}$ in ω is a *weakly Kähler polarization* if, in addition to conditions (C1)–(C4) in Definition 6.8, it also satisfies the following positivity condition:

(C5) There exists a closed convex cone Γ in $\mathfrak{z}$ such that $\Gamma \cap (-\Gamma) = \{0\}$ and

$$(\forall a \in \mathfrak{p} \setminus \mathfrak{h}_{\mathbb{C}}) \qquad -\mathrm{i}\omega(a, \bar{a}) \in \Gamma \setminus \{0\}.$$

If moreover the injective map

$$\mathfrak{g}/\mathfrak{h} \to \mathcal{B}(\mathfrak{g}/\mathfrak{h}, \mathfrak{z}), \quad a + \mathfrak{h} \mapsto \omega(a, \cdot)$$

is invertible, we say that $\mathfrak{p}$ is a *strongly Kähler polarization* of $\mathfrak{g}$ in ω.

REMARK 6.11

(a) In the condition (C5) in Definition 6.10, one actually considers the complexification $\mathfrak{w}$ of $\mathfrak{z}$ and the complex-bilinear extension of ω to

$$\omega \colon \mathfrak{g}_{\mathbb{C}} \times \mathfrak{g}_{\mathbb{C}} \to \mathfrak{w}.$$

One thinks of Γ as a subset of $\mathfrak{w} = \mathfrak{z} + \mathrm{i}\mathfrak{z}$ as usual:

$$\Gamma \subseteq \mathfrak{z} = \mathfrak{z} + \mathrm{i} \cdot 0 \subseteq \mathfrak{w}.$$

(b) In Definition 6.10, in the case when $\mathfrak{z} = \mathbb{R}$, one obtains the classical notions of weakly and strongly Kähler polarizations, respectively.

THEOREM 6.12 *Let $\mathfrak{z}$ be a real Banach space, G a Banach-Lie group with $\mathbf{L}(G) = \mathfrak{g}$, H a Banach-Lie subgroup of G with $\mathbf{L}(H) = \mathfrak{h}$, and $\mathfrak{M}$ a closed subspace of $\mathfrak{g}$ such that $\mathfrak{g} = \mathfrak{M} \dotplus \mathfrak{h}$. Denote $M = G/H$ and let $\pi\colon G \to M$ be the natural projection.*

Assume that $\eta \in Z_c^2(\mathfrak{g}, \mathfrak{h}; \mathfrak{z})$ satisfies

$$\mathfrak{h} = \{a \in \mathfrak{g} \mid (\forall b \in \mathfrak{g}) \quad \eta(a,b) = 0\},$$

and denote $\omega = \Sigma(\eta) \in \Omega_G^2(M, \mathfrak{z})$. On the other hand, extend η to a $\mathbb{C}$-bilinear map $\eta\colon \mathfrak{g}_{\mathbb{C}} \times \mathfrak{g}_{\mathbb{C}} \to \mathfrak{z}_{\mathbb{C}}$.

Moreover assume that $\mathfrak{p}$ is a closed complex subalgebra of $\mathfrak{g}_{\mathbb{C}}$ with the properties

$$\mathfrak{p} + \overline{\mathfrak{p}} = \mathfrak{g}_{\mathbb{C}}, \quad \mathfrak{p} \cap \overline{\mathfrak{p}} = \mathfrak{h}_{\mathbb{C}}, \quad \text{and} \quad \mathrm{Ad}_G(h)\mathfrak{p} \subseteq \mathfrak{p} \quad \text{for all} \quad h \in H,$$

and assume that M has the structure of a complex manifold given by $\mathfrak{p}$ such that $\alpha_g\colon M \to M$ is a holomorphic map for all $g \in G$.

Then the following assertions hold.

(i) *The pair (M, ω) is a weakly pseudo-Kähler homogeneous space of G of type $\mathfrak{z}$ if and only if $\mathfrak{p}$ is a complex polarization of $\mathfrak{g}$ in η.*

(ii) *The pair (M, ω) is a weakly Kähler homogeneous space of G of type $\mathfrak{z}$ if and only if $\mathfrak{p}$ is a weakly Kähler polarization of $\mathfrak{g}$ in η.*

PROOF The proof has several stages, and only the last of them will concern assertion (ii).

1° To begin with, note that the mapping

$$\theta\colon \mathfrak{g}/\mathfrak{h} \to \mathfrak{g}_{\mathbb{C}}/\mathfrak{p}, \quad x + \mathfrak{h} \mapsto x + \mathfrak{p},$$

is an isomorphism of real Banach spaces. In fact, since $\mathfrak{g}$ is the fixed point set of the conjugation $a \mapsto \overline{a}$, it follows that $\mathfrak{p} \cap \mathfrak{g} = \mathfrak{p} \cap \overline{\mathfrak{p}} \cap \mathfrak{g} = \mathfrak{h}_{\mathbb{C}} \cap \mathfrak{g} = \mathfrak{h}$, hence θ is injective. To see that θ is surjective as well, let $z \in \mathfrak{g}_{\mathbb{C}}$ arbitrary. Since $\mathfrak{p} + \overline{\mathfrak{p}} = \mathfrak{g}_{\mathbb{C}}$, there exist $z_1, z_2 \in \overline{\mathfrak{p}}$ such that $z = z_1 + \overline{z}_2$. Then $z = (z_1 + \overline{z}_1) + (\overline{z}_2 - \overline{z}_1)$ with $z_1 + \overline{z}_1 \in \mathfrak{g}$ and $\overline{z}_2 - \overline{z}_1 \in \mathfrak{p}$, hence $z + \mathfrak{p} \in \operatorname{Ran}\theta$.

For later use we also note the following fact:

$$(\forall x \in \mathfrak{g})(\exists z \in \overline{\mathfrak{p}}) \quad x = z + \overline{z}. \tag{6.4}$$

In fact, let $x \in \mathfrak{g} \subseteq \mathfrak{g}_{\mathbb{C}}$ and write $x = z_1 + \overline{z}_2$ as above, with $z_1, z_2 \in \overline{\mathfrak{p}}$. Then $\overline{z}_1 + z_2 = \overline{x} = x = z_1 + \overline{z}_2$, hence $x = z + \overline{z}$, where $z = (z_1 + z_2)/2 \in \overline{\mathfrak{p}}$.

2° Since θ is an isomorphism of the real Banach space $\mathfrak{g}/\mathfrak{h} = T_{\pi(\mathbf{1})}$ onto the complex Banach space $\mathfrak{g}_{\mathbb{C}}/\mathfrak{p}$, we get a linear operator $I_{\pi(\mathbf{1})}\colon T_{\pi(\mathbf{1})}M \to T_{\pi(\mathbf{1})}M$ which is similar (by means of θ) to the multiplication-by-i operator on $\mathfrak{g}_{\mathbb{C}}/\mathfrak{p}$. We now want to describe in more detail the action of $I_{\pi(\mathbf{1})}$. To this end, let

$x \in \mathfrak{g}$ arbitrary. According to (6.4) there exists $z \in \overline{\mathfrak{p}}$ such that $x = z + \overline{z}$. Then

$$\theta(I_{\pi(\mathbf{1})}(x + \mathfrak{h})) = \mathrm{i}x + \mathfrak{p} = \mathrm{i}z + \mathrm{i}\overline{z} + \mathfrak{p} = \mathrm{i}(z - \overline{z}) + \mathfrak{p} = \theta(\mathrm{i}(z - \overline{z}) + \mathfrak{h}),$$

where $\mathrm{i}(z-\overline{z}) \in \mathfrak{g}$ since $\overline{\mathrm{i}(z - \overline{z})} = \mathrm{i}(z-\overline{z})$. Thus the action of $I_{\pi(1)} \colon \mathfrak{g}/\mathfrak{h} \to \mathfrak{g}/\mathfrak{h}$ can be described in the following way:

$$(\forall z \in \overline{\mathfrak{p}}) \quad I_{\pi(\mathbf{1})}(z + \overline{z} + \mathfrak{h}) = \mathrm{i}(z - \overline{z}) + \mathfrak{h}. \tag{6.5}$$

3° It follows by (6.5) that if $z_1, z_2 \in \overline{\mathfrak{p}}$, then

$$\begin{aligned}\omega_{\pi(\mathbf{1})}(I_{\pi(\mathbf{1})}(z_1 &+ \overline{z}_1 + \mathfrak{h}), I_{\pi(\mathbf{1})}(z_2 + \overline{z}_2 + \mathfrak{h}))\\ &= \eta(\mathrm{i}(z_1 - \overline{z}_1), \mathrm{i}(z_2 - \overline{z}_2))\\ &= \eta(z_1, \overline{z}_2) + \eta(\overline{z}_1, z_2) - (\eta(z_1, z_2) + \overline{\eta(z_1, z_2)}),\end{aligned} \tag{6.6}$$

and

$$\begin{aligned}\omega_{\pi(\mathbf{1})}(z_1 &+ \overline{z}_1 + \mathfrak{h}, z_2 + \overline{z}_2 + \mathfrak{h})\\ &= \eta(z_1 + \overline{z}_1, z_2 + \overline{z}_2)\\ &= \eta(z_1, \overline{z}_2) + \eta(\overline{z}_1, z_2) + (\eta(z_1, z_2) + \overline{\eta(z_1, z_2)}).\end{aligned} \tag{6.7}$$

With these formulas at hand, we are able to prove assertion (i) in the statement of the theorem.

First assume that $\eta(\mathfrak{p} \times \mathfrak{p}) = \{0\}$. Then $\eta(\overline{\mathfrak{p}} \times \overline{\mathfrak{p}}) = \{0\}$, hence (6.6) and (6.7) along with (6.4) show that $\omega_x(I_x v, I_x w) = \omega_x(v, w)$ for $x = \pi(\mathbf{1}) \in M$ and all $v, w \in T_x M$. Since both the 2-form ω and the complex structure I are invariant on M, it then follows that M is a weakly pseudo-Kähler homogeneous space of G of type $\mathfrak{z}$.

Conversely, if M is a weakly pseudo-Kähler homogeneous space of G of type $\mathfrak{z}$, then (6.6) and(6.7) imply that

$$(\forall z_1, z_2 \in \overline{\mathfrak{p}}) \quad \eta(z_1, z_2) + \overline{\eta(z_1, z_2)} = 0.$$

Since $\overline{\mathfrak{p}}$ is a complex vector space, we may replace z_2 by $\mathrm{i}z_2$ in the above equation to get $\eta(z_1, z_2) - \overline{\eta(z_1, z_2)} = 0$ for all $z_1, z_2 \in \overline{\mathfrak{p}}$, whence eventually $\eta(z_1, z_2) = 0$ for all $z_1, z_2 \in \overline{\mathfrak{p}}$.

4° At this stage we prove assertion (ii) in the statement of the theorem. Let $x \in \mathfrak{g}$ arbitrary. According to (6.4), there exists $z \in \overline{\mathfrak{p}}$ such that $x = z + \overline{z}$. By (6.5) we get

$$\omega_{\pi(\mathbf{1})}(z + \overline{z} + \mathfrak{h}, I_{\pi(\mathbf{1})}(z + \overline{z} + \mathfrak{h})) = \eta(z + \overline{z}, \mathrm{i}(z - \overline{z})) = 2\mathrm{i}\eta(z, \overline{z}) = -2\mathrm{i}\eta(z, \overline{z}).$$

Now, note that for $x = z+\overline{z}$ as above we have $x \in \mathfrak{g} \setminus \mathfrak{h}$ if and only if $z \in \overline{\mathfrak{p}} \setminus \mathfrak{h}_{\mathbb{C}}$. Consequently

$$(\forall v \in T_{\pi(\mathbf{1})}M \setminus \{0\}) \ \omega_{\pi(\mathbf{1})}(v, I_{\pi(\mathbf{1})}v) \in \Gamma \setminus \{0\} \Leftrightarrow (\forall a \in \overline{\mathfrak{p}} \setminus \mathfrak{h}_{\mathbb{C}}) \ -\mathrm{i}\eta(a, \overline{a}) \in \Gamma \setminus \{0\},$$

and the desired conclusion follows since ω is invariant under the transitive action of G on M. ∎

6.3 Flag manifolds in Banach algebras

Now we are going to use the previously proved results in order to construct invariant complex structures on flag manifolds in Banach algebras (see Example 6.19 below).

Until the end of the present section, we denote by $\mathfrak{B}$ an associative unital complex Banach algebra, and by $G(\mathfrak{B})$ the complex Banach-Lie group of invertible elements in $\mathfrak{B}$.

PROPOSITION 6.13 *Let n be a positive integer and $c_1, \dots, c_n \in \mathfrak{B}$. Denote*

$$H = \{u \in U(\mathfrak{B}) \mid uc_i = c_i u \text{ for } i = 1, \dots, n\}$$

and

$$\mathfrak{h} = \{b \in u(\mathfrak{B}) \mid bc_i = c_i b \text{ for } i = 1, \dots, n\}.$$

Then $\mathfrak{h}$ is a closed subalgebra of $\mathfrak{u}(\mathfrak{B})$, H is a real Banach-Lie group with the topology inherited from the topology τ of the Banach-Lie group $U(\mathfrak{B})$, and $\mathfrak{h}$ is the Lie algebra of H.

PROOF Note that H is an algebraic subgroup of $U(\mathfrak{B})$ of degree ≤ 1. Then Theorem 4.18 can be used because for $b \in \mathfrak{B}$ we have $bc_i = c_i b$ if and only if $e^{tb}c_i = c_i e^{tb}$ for all $t \in \mathbb{R}$. ∎

LEMMA 6.14 *In the setting of* Proposition 6.13, *assume that $c_1, \dots c_n \in \mathfrak{u}(\mathfrak{B})$ and the hermitian map*

$$\Psi_0 \colon (\mathbb{R}^n)^{\#} \simeq \mathbb{R}^n \to \mathcal{B}(\mathfrak{B}), \qquad (t_1, \dots, t_n) \mapsto \mathrm{ad}_{\mathfrak{B}}(t_1 c_1 + \dots + t_n c_n),$$

has the property that 0 is an isolated point of $\sigma_W(\Psi_0)$. Then H is a Banach-Lie subgroup of $U(\mathfrak{B})$.

PROOF We first note that in the above statement, the identification $(\mathbb{R}^n)^{\#} \simeq \mathbb{R}^n$ is performed by means of the usual structure of real Hilbert space of $\mathbb{R}^n$.

Since $c_1, \dots, c_n \in \mathfrak{u}(\mathfrak{B})$, it follows that the closed subspace $\mathfrak{J}(\mathfrak{B}) := \mathfrak{u}(\mathfrak{B}) + i\mathfrak{u}(\mathfrak{B})$ of $\mathfrak{B}$ is invariant to $\Psi_0(\gamma)$ for all $\gamma \in \mathbb{R}^n$, hence

$$\sigma_W(\Psi_0(\cdot)|_{\mathfrak{J}(\mathfrak{B})}) \subseteq \sigma_W(\Psi_0)$$

by Proposition 5.18(b) and the definition of $\sigma_W(\Psi_0)$ in Example 5.25. It then follows that 0 is an isolated point of $\sigma_W(\Psi_0(\cdot)|_{\mathfrak{J}(\mathfrak{B})})$ as well. Since $\Psi_0(\cdot)|_{\mathfrak{J}(\mathfrak{B})}$

is in turn a hermitian map, we deduce by Example 5.25 and Theorem 5.14 that

$$\mathfrak{J}(\mathfrak{B}) = \mathfrak{J}(\mathfrak{B})_{\Psi_0}(\{0\}) \dotplus \mathfrak{J}(\mathfrak{B})_{\Psi_0}(\sigma_W(\Psi_0(\cdot)|_{\mathfrak{J}(\mathfrak{B})}) \setminus \{0\}). \tag{6.8}$$

Moreover,

$$\mathfrak{J}(\mathfrak{B})_{\Psi_0}(\{0\}) = \mathfrak{J}(\mathfrak{B}) \cap \bigcap_{\gamma \in \mathbb{R}^n} \operatorname{Ker} \Psi_0(\gamma) \tag{6.9}$$

by (5.14) in Example 5.25.

On the other hand, since $c_1, \ldots, c_n \in \mathfrak{u}(\mathfrak{B})$, we have

$$(\forall \gamma \in \mathbb{R}^n)(\forall b \in \mathfrak{J}(\mathfrak{B})) \quad \Psi_0(\gamma)\bar{b} = \overline{\Psi_0(\gamma)b},$$

where $b \mapsto \bar{b}$ is the conjugate-linear map of $\mathfrak{J}(\mathfrak{B})$ onto itself defined by

$$a_1 + \mathrm{i}a_2 \mapsto a_1 - \mathrm{i}a_2$$

for all $a_1, a_2 \in \mathfrak{u}(\mathfrak{B})$. Then Proposition 5.22 implies that

$$(\forall F \text{ closed subset of } \mathbb{R}^n) \qquad \overline{\mathfrak{J}(\mathfrak{B})_{\Psi_0}(F)} = \mathfrak{J}(\mathfrak{B})_{\Psi_0}(F).$$

In particular, each of the spaces involved in (6.8) is invariant under the map $b \mapsto \bar{b}$, and then

$$\mathfrak{u}(\mathfrak{B}) = \Big(\mathfrak{u}(\mathfrak{B}) \cap \bigcap_{\gamma \in \mathbb{R}^n} \operatorname{Ker} \Psi_0(\gamma)\Big) \dotplus \Big(\mathfrak{u}(\mathfrak{B}) \cap \mathfrak{J}(\mathfrak{B})_{\Psi_0}(\sigma_W(\Psi_0(\cdot)|_{\mathfrak{J}(\mathfrak{B})}) \setminus \{0\})\Big)$$

by (6.9) and by the fact that

$$\mathfrak{u}(\mathfrak{B}) = \{b \in \mathfrak{J}(\mathfrak{B}) \mid b = \bar{b}\}.$$

Now the definition of $\mathfrak{h}$ in Proposition 6.13 shows that

$$\mathfrak{u}(\mathfrak{B}) = \mathfrak{h} \dotplus \Big(\mathfrak{u}(\mathfrak{B}) \cap \mathfrak{J}(\mathfrak{B})_{\Psi_0}(\sigma_W(\Psi_0(\cdot)|_{\mathfrak{J}(\mathfrak{B})}) \setminus \{0\})\Big),$$

hence the Lie algebra $\mathfrak{h}$ of H is complemented in the Lie algebra $\mathfrak{u}(\mathfrak{B})$ of $U(\mathfrak{B})$. Taking into account Proposition 6.13, it then follows that H is a Banach-Lie subgroup of $U(\mathfrak{B})$. ∎

COROLLARY 6.15 *In the setting of* Proposition 6.13, *let us assume that $c_1, \ldots, c_n \in \mathfrak{u}(\mathfrak{B})$, $[c_i, c_j] = 0$ for all $i, j \in \{1 \ldots, n\}$, and the spectrum of c_i is finite for $i = 1, \ldots, n$. Then H is a Banach-Lie subgroup of $U(\mathfrak{B})$, having the Lie algebra $\mathfrak{h}$.*

PROOF Let $\Psi_0 \colon \mathbb{R}^n \to \mathcal{B}(\mathfrak{B})$ be the hermitian map in Lemma 6.14. In view of that lemma, it suffices to prove that 0 is an isolated point of $\sigma_W(\Psi_0)$.

To this end, recall that

$$\sigma(\mathrm{ad}_{\mathfrak{B}} c_i) \subseteq \sigma(c_i) - \sigma(c_i) \qquad \text{for } i = 1, \dots, n$$

(see e.g., Corollary 4.8 in Chapter III in [Va82]). In particular, $\sigma(\mathrm{ad}_{\mathfrak{B}} c_i)$ is a finite set for $i = 1, \dots, n$.

On the other hand, the operators $\mathrm{ad}_{\mathfrak{B}} c_1, \dots, \mathrm{ad}_{\mathfrak{B}} c_n \in \mathcal{B}(\mathfrak{B})$ mutually commute according to the hypothesis, hence $\sigma_W(\Psi_0)$ agrees with the Taylor joint spectrum of the commuting n-tuple

$$(\mathrm{ad}_{\mathfrak{B}} c_1, \dots, \mathrm{ad}_{\mathfrak{B}} c_n) \in \mathcal{B}(\mathfrak{B})^n$$

(see Example 5.25). But that Taylor spectrum is contained in

$$\sigma(\mathrm{ad}_{\mathfrak{B}} c_1) \times \cdots \times \sigma(\mathrm{ad}_{\mathfrak{B}} c_n)$$

by the projection property (see e.g., Corollary 2 in §26 in [BS01]), hence it is finite since we have already seen that $\sigma(\mathrm{ad}_{\mathfrak{B}} c_i)$ is a finite set for $i = 1, \dots, n$. Consequently, $\sigma_W(\Psi_0)$ is a finite set.

It follows in particular that 0 is an isolated point of $\sigma_W(\Psi_0)$, hence we can apply Lemma 6.14. ∎

LEMMA 6.16 *Let m be a positive integer. Let $e_1, \dots, e_m \in \mathfrak{u}(\mathfrak{B}) \setminus \{0\}$ such that*

$$e_j e_k = 0 \text{ whenever } j \neq k, \tag{6.10}$$

and

$$e_1 + \cdots + e_m = \mathrm{i} \cdot 1. \tag{6.11}$$

Also consider the m-dimensional real vector space $\mathfrak{z} = (\mathrm{sp}_{\mathbb{R}}\{e_1, \dots, e_m\})^{\#}$, and let $z_1, \dots, z_m$ be the basis of $\mathfrak{z}$ which is dual to the basis $e_1, \dots, e_m$ of $\mathrm{sp}_{\mathbb{R}}\{e_1, \dots, e_m\}$. Then the map

$$\Psi_0 \colon \mathfrak{z}^{\#} \simeq \mathrm{sp}_{\mathbb{R}}\{e_1, \dots, e_m\} \to \mathcal{B}(\mathfrak{B}), \quad \gamma \mapsto \mathrm{ad}_{\mathfrak{B}}(\mathrm{i}\gamma),$$

is hermitian and $\sigma_W(\Psi_0) \subseteq \{z_j - z_k \mid 1 \leq j, k \leq m\}$.

PROOF First note that by multiplying by e_j the equality (6.11) and by taking into account hypothesis (6.10), we get

$$e_j^2 = \mathrm{i} e_j \text{ for } j = 1, \dots, m. \tag{6.12}$$

This latter fact then implies by (6.10), in a straightforward manner, that $e_1, \dots, e_m$ are linearly independent, thus justifying the assertion that $\mathfrak{z}$ is m-dimensional.

The fact that the map Ψ_0 is hermitian follows because for every $\gamma \in \mathfrak{z}^{\#} = \mathrm{sp}_{\mathbb{R}}\{e_1, \dots, e_m\} \subseteq \mathfrak{u}(\mathfrak{B})$ we have $e^{\gamma} \in U(\mathfrak{B})$, whence

$$(\forall b \in \mathfrak{B}) \quad \|e^{\mathrm{i}\Psi_0(\gamma)} b\| = \|e^{-\mathrm{ad}_{\mathfrak{B}}\gamma} b\| = \|e^{-\gamma} b e^{\gamma}\| \leq \|e^{-\gamma}\| \cdot \|b\| \cdot \|e^{\gamma}\| \leq \|b\|,$$

that is, $\|e^{i\Psi_0(\gamma)}\| \le 1$.

In order to prove the inclusion concerning $\sigma_W(\Psi_0)$, note that for all $j, k, l \in \{1, \dots, m\}$ and $b \in \mathfrak{B}$ we have by (6.10) and (6.12) that

$$\big(\Psi_0(e_j)\big)(e_k b e_l) = \mathrm{i}[e_j, e_k b e_l] = -\langle z_k - z_l, e_j\rangle e_k b e_l = \langle e_j, z_l - z_k\rangle e_k b e_l.$$

Consequently, for all $k, l \in \{1, \dots, m\}$ we have

$$(\forall \gamma \in \mathfrak{z}^{\#})(\forall b \in e_k \mathfrak{B} e_l) \qquad \Psi_0(\gamma) b = \langle \gamma, z_l - z_k\rangle b.$$

Denoting by

$$\Psi \in \mathcal{E}'(\mathfrak{z}, \mathcal{B}(\mathfrak{B}))$$

the (quasi)multiplicative extension of Ψ_0 to $\mathcal{C}^\infty(\mathfrak{z})$ given by Example 5.25, it then easily follows that

$$(\forall f \in \mathcal{C}^\infty(\mathfrak{z}))\,(\forall b \in e_k \mathfrak{B} e_l) \qquad \Psi(f) b = f(z_l - z_k) b. \tag{6.13}$$

Next, for every $b \in \mathfrak{B}$ we have by (6.11) that

$$b = -(e_1 + \cdots + e_m) b (e_1 + \cdots + e_m) = \sum_{k,l=1}^{m} (-e_k b e_l),$$

hence we get by (6.13)

$$(\forall f \in \mathcal{C}^\infty(\mathfrak{z})) \quad \Psi(f) b = \sum_{k,l=1}^{m} (-f(z_l - z_k) b) = -\Big(\sum_{k,l=1}^{m} f(z_l - z_k)\Big) b.$$

It then easily follows that $\sup \Psi \subseteq \{z_l - z_k \mid 1 \le l, k \le m\}$, and the desired conclusion follows since $\operatorname{supp} \Psi = \sigma_W(\Psi_0)$ by the very definition of $\sigma_W(\Psi_0)$ (see Example 5.25). ∎

REMARK 6.17

(a) The inclusion in Lemma 6.16 might be strict, as the situation when $\mathfrak{B}$ is an abelian algebra shows.

(b) In the framework of Lemma 6.16 and its proof, it easily follows by (6.13) that for every subset F of (the finite set) $\sigma_W(\Psi_0)$, we have

$$\mathfrak{B}_{\Psi_0}(F) = \bigoplus_{\substack{1 \le k, l \le m \\ z_l - z_k \in F}} e_k \mathfrak{B} e_l$$

as a direct sum of Banach spaces.

THEOREM 6.18 *Let m be a positive integer, and $e_1, \dots, e_n \in \mathfrak{u}(\mathfrak{B}) \setminus \{0\}$ such that*

$$e_j e_k = 0 \text{ whenever } j \neq k, \quad \text{and} \quad e_1 + \dots + e_m = \mathrm{i} \cdot \mathbf{1}.$$

If $U(\mathfrak{B})_0$ stands for the identity component of the real Banach-Lie group $U(\mathfrak{B})$, and

$$H = \{u \in u(\mathfrak{B}) \mid ue_j = e_j u \text{ for } j = 1, \dots, m\},$$

then $U(\mathfrak{B})_0 \cap H$ is a Banach-Lie subgroup of $U(\mathfrak{B})_0$, and there exists a complex $U(\mathfrak{B})_0$-invariant structure on the homogeneous space $U(\mathfrak{B})_0/(U(\mathfrak{B})_0 \cap H)$.

PROOF We use of the notation of Lemma 6.16 and also denote

$$S = \big(\mathbb{Z}_+ \cdot (z_1 - z_2) + \mathbb{Z}_+ \cdot (z_2 - z_3) + \dots + \mathbb{Z}_+ \cdot (z_m - z_{m-1})\big) \setminus \{0\} \quad (\subseteq \mathfrak{z}).$$

Then S is a closed subsemigroup of $\mathfrak{z}$ with $0 \notin S$. Moreover, Lemma 6.16 implies that

$$\sigma_W(\Psi_0) \subseteq (-S) \cup \{0\} \cup S.$$

Since

$$\mathfrak{J}(\mathfrak{B}) = \mathfrak{u}(\mathfrak{B}) + \mathrm{i}\mathfrak{u}(\mathfrak{B})$$

is the complexification of $\mathfrak{u}(\mathfrak{B})$, Corollary 6.3 then shows that

$$\mathfrak{k} := \mathfrak{J}(\mathfrak{B})_{\Psi_0}((-S) \cup \{0\}) \tag{6.14}$$

is a closed complex subalgebra of the Lie algebra $\mathfrak{J}(\mathfrak{B})$,

$$\mathfrak{h} := \bigcap_{\gamma \in \mathfrak{z}^\#} \operatorname{Ker}(\mathrm{i}\Psi_0(\gamma)) = \{a \in \mathfrak{u}(\mathfrak{B}) \mid ae_j = e_j a \text{ for } j = 1, \dots, m\}$$

is a closed subalgebra of $\mathfrak{u}(\mathfrak{B})$, and the conditions (i)–(iii) in Theorem 6.1 are satisfied.

On the other hand, the spectrum of e_j is contained in $\{0, \mathrm{i}\}$ by (6.12) (in the proof of Lemma 6.16), hence we can use Corollary 6.15 to deduce that H is a Banach-Lie subgroup of $U(\mathfrak{B})$, having the Lie algebra $\mathfrak{h}$. It then follows that $U(\mathfrak{B})_0 \cap H$ is a closed Banach-Lie subgroup of $U(\mathfrak{B})_0$, and the Lie algebra of $U(\mathfrak{B})_0 \cap H$ is $\mathfrak{h}$. Consequently, since we have already seen that the conditions (i)–(iii) in Theorem 6.1 are fulfilled, it only remains to check the condition (i') from that theorem.

To this end, recall that

$$(\forall u \in U(\mathfrak{B}))(\forall b \in \mathfrak{J}(\mathfrak{B})) \qquad (\mathrm{Ad}_{U(\mathfrak{B})} u)b = ubu^{-1}.$$

It then follows that for all $u \in U(\mathfrak{B})_0 \cap H$, $b \in \mathfrak{J}(\mathfrak{B})$ and $\gamma \in \mathfrak{z}^\# = \mathrm{sp}_{\mathbb{R}}\{e_1, \dots, e_m\}$ we have

$$((\mathrm{Ad}_{U(\mathfrak{B})} u) \circ \Psi_0(\gamma))b = u[\mathrm{i}\gamma, b]u^{-1} = [\mathrm{i}\gamma, ubu^{-1}] = (\Psi_0(\gamma) \circ \mathrm{Ad}_{U(\mathfrak{B})} u)b,$$

where the last but one equality follows because $u\gamma = \gamma u$ according to the definition of H and to the range where γ runs. It then follows by (6.14) and by Remark 5.19 that

$$(\forall u \in U(\mathfrak{B})_0 \cap H) \qquad (\mathrm{Ad}_{U(\mathfrak{B})} u)\mathfrak{k} \subseteq \mathfrak{k},$$

which is just the condition (i') in Theorem 6.1, whence the desired conclusion follows. ∎

EXAMPLE 6.19 Consider the set

$$\mathcal{E}_{\mathfrak{B}} = \{p \in \mathfrak{B} \mid p = p^2\}$$

of all idempotent elements of $\mathfrak{B}$, endowed with the usual order relation defined by

$$p \leq q \iff pq = qp = p.$$

Next let

$$\mathcal{P}_{\mathfrak{B}} = \mathcal{E}_{\mathfrak{B}} \cap \mathrm{iu}(\mathfrak{B})$$

be the set of all hermitian idempotents in $\mathfrak{B}$. For an arbitrary integer number $n \geq 1$, consider the *flag manifold*

$$\mathrm{Fl}_{\mathfrak{B},n} = \{(p_1, \dots, p_n) \in (\mathcal{P}_{\mathfrak{B}})^n \mid p_1 \leq p_2 \leq \dots \leq p_n\}.$$

Now, it is easy to check that

$$\alpha\colon U(\mathfrak{B}) \times \mathrm{Fl}_{\mathfrak{B},n} \to \mathrm{Fl}_{\mathfrak{B},n}, \quad (u, (p_1, \dots, p_n)) \mapsto (up_1u^{-1}, \dots, up_nu^{-1}) \tag{6.15}$$

is a well-defined action of $U(\mathfrak{B})$ on $\mathrm{Fl}_{\mathfrak{B},n}$. Then *the orbits of the restricted action*

$$\alpha_0 := \alpha|_{U(\mathfrak{B})_0 \times \mathrm{Fl}_{\mathfrak{B},n}} : U(\mathfrak{B})_0 \times \mathrm{Fl}_{\mathfrak{B},n} \to \mathrm{Fl}_{\mathfrak{B},n}$$

have complex $U(\mathfrak{B})_0$-invariant structures, as a consequence of Theorem 6.18.

Indeed, let $(p_1, \dots, p_n) \in \mathrm{Fl}_{\mathfrak{B},n}$ and denote

$$e_j = \mathrm{i}(p_j - p_{j-1}) \text{ for } j = 1, \dots, n+1,$$

where $p_0 := 0$ and $p_{n+1} := 1$. Then

$$e_j e_k = 0 \text{ whenever } j \neq k, \text{ and } e_1 + \dots + e_{n+1} = \mathrm{i} \cdot \mathbf{1}.$$

It is clear that the isotropy group of $(p_1, \dots, p_n)$ under the action α_0 is

$$\begin{aligned} H_0 :=& \{u \in U(\mathfrak{B})_0 \mid up_j = p_ju \text{ for } j = 1, \dots, n\} \\ =& \{u \in U(\mathfrak{B})_0 \mid ue_j = e_ju \text{ for } j = 1, \dots, n+1\}, \end{aligned}$$

while the orbit of $(p_1, \dots, p_n)$ under the action α_0 can be identified with $U(\mathfrak{B})_0/H_0$, hence the desired assertion follows by Theorem 6.18.

EXAMPLE 6.20 We now specialize Example 6.19 to the important case when $\mathfrak{B}$ is a unital C^*-algebra. Then

$$\mathcal{P}_{\mathfrak{B}} = \{p \in \mathfrak{B} \mid p = p^2 = p^*\}.$$

Moreover, it is well known that

$$U(\mathfrak{B}) = \{u \in \mathfrak{B} \mid uu^* = u^*u = \mathbf{1}\}$$

(and, if $\mathfrak{B}$ is even a W^*-algebra, then $U(\mathfrak{B})$ is connected, that is to say, $U(\mathfrak{B}) = U(\mathfrak{B})_0$).

In this special case, the conclusion of Example 6.19 says that the orbits of the natural action (6.15) of $U(\mathfrak{B})_0$ upon the flag manifold $\mathrm{Fl}_{\mathfrak{B},n}$ have complex $U(\mathfrak{B})$-invariant structures.

Notes

We refer to §XV.1 in [Ne00] for a discussion of complex structures on homogeneous spaces of a Lie group G in the case when G is finite dimensional, and to Section VI in [Ne04] for the case when G is a Banach-Lie group. An indication of another proof for Theorem 6.1 can be found in [Ne04]. The idea is that the needed reasoning from the proof of Theorem 1 in Section 13.4 in [Ki76] works in the infinite-dimensional case as well.

For Example 6.20, see [MS98] and Section 2.4 in [MS97]. (See also Section 3 in [Be02b].)

The concept of complex polarization we introduced in Definition 6.8 actually extends the classical concept to vector-valued cocycles (see e.g., the last part of the Introduction of [Be05a]). The proof of our Theorem 6.12 is inspired by the proof of Proposition VI.9 and the comment following it in [Ne04].

We refer to [We80] for a nice introduction to finite-dimensional Kähler manifolds. The compact pseudo-Kähler homogeneous spaces were completely classified in [DG92]. See also the celebrated paper [DN88] for general finite-dimensional homogeneous Kähler manifolds. For further progress in the classification of compact complex homogeneous spaces, see [Gu02] and [Gu04]. See [Bea83], [Gu94], [Gu95a], [Gu95b], and [Bog96] for examples of compact complex symplectic non-Kähler manifolds.

Equivariant monotone operators and Kähler structures

Chapter 7

Equivariant Monotone Operators

Abstract. We firstly introduce the equivariant monotone operators and discuss some of their basic properties. For instance we show how such operators arise in the setting of abelian Banach-Lie algebras and of Heisenberg algebras. We then review a few facts on H^*-algebras and we introduce the closely related notion of H^*-ideal of an involutive Banach algebra. A special attention is paid to the special case of L^*-algebras and L^*-ideals, respectively. We then focus on a key feature of equivariant monotone operators: namely, we look at them from the point of view of abstract reproducing kernels. That discussion will eventually lead us to the deepest properties of equivariant monotone operators, which we shall investigate in the next chapters. In order to get a better understanding of H^*-ideals, we give a description of H^*-ideals of H^*-algebras. This is achieved by using some ideas from the theory of operator ranges. The chapter concludes by a discussion of some elementary properties of H^*-ideals.

7.1 Definition of equivariant monotone operators

In this section we introduce the equivariant monotone operators associated with an involutive (in general non-associative) Banach algebra. These operators will play a critical role for all that follows in the present book. We shall see in Remark 7.5 how such operators can be built in the simple situation of abelian separable Banach-Lie algebras. Another simple case (the Heisenberg algebras) will be dealt with in Remark 7.6.

DEFINITION 7.1 If $\mathfrak{g}$ is an involutive Banach algebra over $\mathbb{K} \in \{\mathbb{R}, \mathbb{C}\}$, then for each $f \in \mathfrak{g}^{\#}$ we define $f^* \in \mathfrak{g}^{\#}$ by $f^*(a) := \overline{f(a^*)}$ for every $a \in \mathfrak{g}$. We then denote by $\mathrm{Hom}^*_{\mathfrak{g}}(\mathfrak{g}^{\#}, \mathfrak{g})$ the set of all continuous linear maps $\iota\colon \mathfrak{g}^{\#} \to \mathfrak{g}$ such that for every $f \in \mathfrak{g}^{\#}$ we have

$$\iota(f^*) = \iota(f)^*$$

and

$$(\forall a \in \mathfrak{g}) \qquad L_a \circ \iota = \iota \circ (L_{a^*})^{\#} \text{ and } R_a \circ \iota = \iota \circ (R_{a^*})^{\#}.$$

The latter two equalities will be referred to as *equivariance* properties of ι. (We recall that the operators $L_a, R_a\colon \mathfrak{g} \to \mathfrak{g}$ are defined by $L_a b = ab$ and $R_a b = ba$ whenever $a, b \in \mathfrak{g}$.)

In the case $\mathbb{K} = \mathbb{R}$, we denote by $\mathcal{C}^+(\mathfrak{g})$ the set of all $\iota \in \mathrm{Hom}^*_{\mathfrak{g}}(\mathfrak{g}^{\#}, \mathfrak{g})$ such that $\iota^{\#} = \iota$ and ι is *monotone* in the sense that

$$(\forall f \in \mathfrak{g}^{\#}) \qquad \langle f, \iota(f)\rangle \geq 0.$$

The elements of $\mathcal{C}^+(\mathfrak{g})$ will be referred to as *equivariant monotone operators.*

We also denote

$$\mathcal{C}^+_0(\mathfrak{g}) = \{\iota \in \mathcal{C}^+(\mathfrak{g}) \mid \mathrm{Ker}\,\iota = \{0\}\}.$$

REMARK 7.2

(a) In the above definition, the property that $\langle f, \iota(f)\rangle \geq 0$ for all $f \in \mathfrak{g}^{\#}$ is equivalent by the linearity of ι to the fact that the inverse of ι is a (multivalent) monotone map in the usual sense in the nonlinear analysis.

(b) Let $\mathfrak{X}$ be a Banach space over $\mathbb{K} \in \{\mathbb{R}, \mathbb{C}\}$ endowed with a continuous map $\mathfrak{X} \to \mathfrak{X}$, $x \mapsto x^*$, such that for every $x, y \in \mathfrak{X}$ and $\lambda \in \mathbb{K}$ we have

$$(x^*)^* = x, \quad (x+y)^* = x^* + y^* \quad \text{and} \quad (\lambda x)^* = \bar{\lambda} x^*.$$

Now consider another Banach space $\mathfrak{Y}$ over $\mathbb{K}$, endowed with a continuous map $\mathfrak{Y} \to \mathfrak{Y}$, denoted also $y \mapsto y^*$, and with similar properties. Then there exists a natural continuous map

$$\mathcal{B}(\mathfrak{X}, \mathfrak{Y}) \to \mathcal{B}(\mathfrak{X}, \mathfrak{Y}), \quad T \mapsto T^*,$$

with similar properties; this last map is defined by

$$(\forall x \in \mathfrak{X}) \qquad T^*(x) = \big(T(x^*)\big)^*$$

for all $T \in \mathcal{B}(\mathfrak{X}, \mathfrak{Y})$. With this notation, in Definition 7.1 we have $\iota^* = \iota$ for all $\iota \in \mathrm{Hom}^*_{\mathfrak{g}}(\mathfrak{g}^{\#}, \mathfrak{g})$.

REMARK 7.3 Let $\mathfrak{X}$ be a real Banach space and $\iota \in \mathcal{B}(\mathfrak{X}^{\#}, \mathfrak{X})$ such that $\iota^{\#} = \iota$ and $\langle f, \iota(f)\rangle \geq 0$ for all $f \in \mathfrak{X}^{\#}$. Then

$$\mathfrak{X}^{\#} \times \mathfrak{X}^{\#} \to \mathbb{R}, \quad (f, g) \mapsto \langle f, \iota(g)\rangle,$$

is a symmetric positively semi-definite bilinear form on $\mathfrak{X}^{\#}$, hence Schwartz' inequality

$$(\forall f, g \in \mathfrak{X}^{\#}) \qquad \langle f, \iota(g)\rangle^2 \leq \langle f, \iota(f)\rangle \cdot \langle g, \iota(g)\rangle$$

holds. This easily implies by the Hahn-Banach theorem that

$$\operatorname{Ker}\iota = \{g \in \mathfrak{X}^{\#} \mid \langle g, \iota(g)\rangle = 0\}.$$

In particular, we have $\operatorname{Ker}\iota = \{0\}$ if and only if $\langle f, \iota(f)\rangle > 0$ whenever $f \in \mathfrak{X}^{\#} \setminus \{0\}$.

LEMMA 7.4 *If $\mathfrak{g}$ is a real Banach-Lie algebra with involution, then $\mathcal{C}^+(\mathfrak{g})$ is a closed convex cone in the real Banach space $\mathcal{B}(\mathfrak{g}^{\#}, \mathfrak{g})$, and $\mathcal{C}_0^+(\mathfrak{g})$ is a convex cone which is a semigroup ideal of $\mathcal{C}^+(\mathfrak{g})$, i.e., $\mathcal{C}_0^+(\mathfrak{g}) + \mathcal{C}^+(\mathfrak{g}) \subseteq \mathcal{C}_0^+(\mathfrak{g})$.*

PROOF The only fact which is not a direct consequence of the definitions is that $\mathcal{C}_0^+(\mathfrak{g}) + \mathcal{C}^+(\mathfrak{g}) \subseteq \mathcal{C}_0^+(\mathfrak{g})$, but this follows by Remark 7.3. ∎

The following remark says that the cones $\mathcal{C}^+(\mathfrak{g})$ and $\mathcal{C}_0^+(\mathfrak{g})$ are non-empty in the simplest case of an abelian canonically involutive separable Banach-Lie algebra $\mathfrak{g}$.

REMARK 7.5 We have $0 \in \mathcal{C}^+(\mathfrak{g})$ by the very definition of $\mathcal{C}^+(\mathfrak{g})$.

On the other hand, if $\mathfrak{g}$ is an abelian canonically involutive Banach-Lie algebra whose underlying Banach space is separable, then we can construct an element of $\mathcal{C}_0^+(\mathfrak{g})$ in the following way. Let $\{a_j \mid j \geq 1\}$ be a countable dense subset of the unit ball of $\mathfrak{g}$ and define

$$\iota\colon \mathfrak{g}^{\#} \to \mathfrak{g}, \qquad \iota(f) = \sum_{j=1}^{\infty} \frac{1}{j^2} f(a_j)a_j \quad \text{for all} \quad f \in \mathfrak{g}^{\#}.$$

(Note that for every $f \in \mathfrak{g}^{\#}$ we have $\|f(a_j)a_j\| \leq \|f\| \cdot \|a_j\|^2 \leq \|f\|$ for each $j \geq 1$, hence the series defining $\iota(f)$ is indeed convergent in $\mathfrak{g}$.) Then for every $f, g \in \mathfrak{g}^{\#}$ we have

$$\langle g, \iota(f)\rangle = \sum_{j=1}^{\infty} \frac{1}{j^2} f(a_j)g(a_j) = \langle f, \iota(g)\rangle,$$

hence $\iota^{\#} = \iota$. The same computation shows that $\langle f, \iota(f)\rangle \geq 0$, and $\langle f, \iota(f)\rangle = 0$ if and only if $f(a_j) = 0$ for all $j \geq 1$. But the latter condition is equivalent to $f = 0$ because the set $\{a_j \mid j \geq 1\}$ is dense in the unit ball of $\mathfrak{g}$.

The following remark shows that if $\mathfrak{g}$ is a Heisenberg algebra, then $\mathcal{C}_0^+(\mathfrak{g}) = \emptyset$ and $\mathcal{C}^+(\mathfrak{g}) = \mathbb{R}_+ \cdot \iota_0$ for some $\iota_0 \neq 0$. Other concrete computations of equivariant monotone operators will be done later. For example, we show in Example 8.23 an algebra $\mathfrak{g}$ with $\mathcal{C}_0^+(\mathfrak{g}) = \emptyset$ and $\mathcal{C}^+(\mathfrak{g}) = \{0\}$. See also Corollary 8.21, Problem 8.37, and the comments following the proof of Corollary 8.27.

REMARK 7.6 Let V be a real Hilbert space with the scalar product denoted by $(\cdot \mid \cdot)$, $q\colon V \times V \to \mathbb{R}$ a skew-symmetric non-degenerate bounded bilinear form, and $\mathfrak{g} = \mathfrak{h}(V, q, \mathbb{R})$ the corresponding *Heisenberg algebra*. We have

$$\mathfrak{g} = V \dotplus \mathbb{R}$$

as an orthogonal sum of real Hilbert spaces, and the bracket of $\mathfrak{g}$ is defined by

$$\big[(v_1, t_1), (v_2, t_2)\big] = (0, q(v_1, v_2))$$

for all $(v_1, t_1), (v_2, t_2) \in \mathfrak{g}$. We check that for the canonically involutive Banach-Lie algebra $\mathfrak{g}$ we have $\mathcal{C}_0^+(\mathfrak{g}) = \emptyset$.

To this end, let $\iota \in \mathcal{C}^+(\mathfrak{g})$. We perform the usual identifications $\mathfrak{g}^\# = \mathfrak{g}$ (with the duality pairing given by the scalar product of $\mathfrak{g}$) and

$$\mathcal{B}(\mathfrak{g}) = \left\{ \begin{pmatrix} A & u \\ v & s \end{pmatrix} \mid A \in \mathcal{B}(V), u, v \in V, s \in \mathbb{R} \right\}$$

acting on $\mathfrak{g} = V \dotplus \mathbb{R}$ by

$$\begin{pmatrix} A & u \\ v & s \end{pmatrix} \begin{pmatrix} x \\ r \end{pmatrix} = \begin{pmatrix} Ax + ru \\ (v \mid x) + sr \end{pmatrix}$$

for all $x \in V$ and $r \in \mathbb{R}$, and with the multiplication

$$\begin{pmatrix} A_1 & u_1 \\ v_1 & s_1 \end{pmatrix} \begin{pmatrix} A_2 & u_2 \\ v_2 & s_2 \end{pmatrix} = \begin{pmatrix} A_1 A_2 + (v_2 \mid \cdot) u_1 & A_1 u_2 + s_2 u_1 \\ A_2^\# v_1 + s_1 v_2 & (v_1 \mid u_2) + s_1 s_2 \end{pmatrix}.$$

Since $\iota = \iota^\# \in \mathcal{B}(\mathfrak{g})$ (according to the identification $\mathfrak{g}^\# = \mathfrak{g}$), it then follows that

$$\iota = \begin{pmatrix} T & v \\ v & t \end{pmatrix}$$

for some $T \in \mathcal{B}(V)$, $v \in V$ and $t \in \mathbb{R}$.

On the other hand, the properties of the bilinear form q imply that there exists $Q \in \mathcal{B}(V)$ such that $Q^\# = -Q$, $\operatorname{Ker} Q = \{0\}$ and

$$(\forall v_1, v_2 \in V) \qquad q(v_1, v_2) = (Q v_1 \mid v_2).$$

Then it is straightforward to check that

$$(\forall w \in V \subseteq \mathfrak{g}) \qquad \mathrm{ad}_{\mathfrak{g}} w = \begin{pmatrix} 0 & 0 \\ Qw & 0 \end{pmatrix} \text{ and } (\mathrm{ad}_{\mathfrak{g}} w)^\# = \begin{pmatrix} 0 & Qw \\ 0 & 0 \end{pmatrix}.$$

Thus the equivariance property of ι is equivalent to the fact that for all $w \in V$ we have

$$\begin{pmatrix} T & v \\ v & t \end{pmatrix} \cdot \begin{pmatrix} 0 & Qw \\ 0 & 0 \end{pmatrix} = -\begin{pmatrix} 0 & 0 \\ Qw & 0 \end{pmatrix} \cdot \begin{pmatrix} T & v \\ v & t \end{pmatrix},$$

that is,

$$\begin{pmatrix} 0 & TQw \\ 0 & (v \mid Qw) \end{pmatrix} = -\begin{pmatrix} 0 & 0 \\ T^{\#}Qw & (Qw \mid v) \end{pmatrix}.$$

it follows that

$$(\forall w \in V) \qquad TQw = 0.$$

But $\operatorname{Ran} Q$ is dense in V (since $Q^{\#} = -Q$ and $\operatorname{Ker} Q = \{0\}$), hence $T = 0$. We thus get

$$\iota = \begin{pmatrix} 0 & v \\ v & t \end{pmatrix}$$

and then it is straightforward to check that the fact that ι is monotone is equivalent to the conditions $v = 0$ and $t \geq 0$.

Consequently, the elements of $\mathcal{C}^{+}(\mathfrak{g})$ are just the nonnegative scalar multiples of the natural projection map

$$\iota_0 \colon \mathfrak{g} = V \dotplus \mathbb{R} \to \mathbb{R} \hookrightarrow \mathfrak{g}.$$

Since none of these multiples is injective, we have $\mathcal{C}_0^{+}(\mathfrak{g}) = \emptyset$.

7.2 H*-algebras and L*-algebras

We are going to have a look at a special class of involutive Banach algebras, namely the H^*-algebras. The Lie H^*-algebras will be called L^*-algebras. We are interested in these algebras since they turn out to play an essential role in order to understand the structure of much more general involutive Banach algebras. See the next chapters for more details in this connection.

DEFINITION 7.7 If $\mathfrak{X}$ is an involutive Banach algebra over $\mathbb{K} \in \{\mathbb{R}, \mathbb{C}\}$, then we say that $\mathfrak{X}$ is an *H^*-algebra* if its underlying Banach space is in fact a Hilbert space over $\mathbb{K}$ defined by a scalar product $(\cdot \mid \cdot)$ such that

$$(\forall a, b, c \in \mathfrak{X}) \qquad (ab \mid c) = (b \mid a^*c) = (a \mid cb^*).$$

An *L^*-algebra* is a Lie H^*-algebra. A subalgebra of an H^*-algebra that is both topologically closed and closed under the involution is called an *H^*-subalgebra*. An H^*-subalgebra of an L^*-algebra is called an *L^*-subalgebra*.

EXAMPLE 7.8 Let $\mathcal{H}$ be a complex Hilbert space and $\mathcal{C}_2(\mathcal{H})$ the ideal of Hilbert-Schmidt operators on $\mathcal{H}$. Then $\mathfrak{X} = \mathcal{C}_2(\mathcal{H})$ with the usual bracket and scalar product is a complex L^*-algebra. This complex L^*-algebra is topologically simple if and only if $\dim \mathcal{H} = \infty$, and in this case it is known as an L^*-algebra of type (A) (see Theorem 7.18 below).

REMARK 7.9

(a) Unlike the usual definition of an H^*-algebra, we include in the definition the condition that the involution is continuous.

(b) For later use we note that *if* $\mathfrak{X}$ *is a real* H^**-algebra such that* $\operatorname{Ann}(\mathfrak{X}) = \{0\}$*, then for every* $a \in \mathfrak{X}$ *we have* $\|a\| = \|a^*\|$ (see 1.1.8 in [CMMR94]), *which is equivalent to* $(a^* \mid b) = (a \mid b^*)$ *for all* $a, b \in \mathfrak{X}$.

REMARK 7.10 If $\mathfrak{X}$ is a real H^*-algebra, then the following assertions hold.

(a) For every closed ideal $\mathfrak{J}$ of $\mathfrak{X}$ the orthogonal complement $\mathfrak{J}^\perp$ is in turn a closed ideal of $\mathfrak{X}$ and $\mathfrak{X} = \mathfrak{J} \oplus \mathfrak{J}^\perp$.

(b) The annihilator $\operatorname{Ann}(\mathfrak{X})$ is a $*$-invariant closed ideal of $\mathfrak{X}$.

(c) Let $P\colon \mathfrak{X} \to (\operatorname{Ann}(\mathfrak{X}))^\perp$ be the orthogonal projection. Then $(\operatorname{Ann}(\mathfrak{X}))^\perp$ is an H^*-algebra with the scalar product inherited from $\mathfrak{X}$ and the involution defined by $x \mapsto P(x^*)$. Moreover $\operatorname{Ann}((\operatorname{Ann}(\mathfrak{X}))^\perp) = \{0\}$.

We collect in the next statement a number of basic properties of the real H^*-algebras without annihilator.

THEOREM 7.11 *Let* $\mathfrak{X}$ *be a real* H^**-algebra with* $\operatorname{Ann}(\mathfrak{X}) = \{0\}$*, and denote by* $\{\mathfrak{X}_i\}_{i\in I}$ *the set of all nonzero minimal closed ideals of* $\mathfrak{X}$*. Then the following assertions hold.*

(a) *We have the Hilbert space orthogonal decomposition* $\mathfrak{X} = \bigoplus_{i\in I} \mathfrak{X}_i$.

(b) *All closed ideals of* $\mathfrak{X}$ *are* $*$*-invariant.*

(c) *For every* $i \in I$ *the ideal* $\mathcal{X}_i$ *is a topologically simple real* H^**-algebra with the scalar product and involution inherited from* $\mathfrak{X}$.

(d) *The linear span of the set* $\{xy \mid x, y \in \mathfrak{X}\}$ *is dense in* $\mathfrak{X}$.

(e) *If* $\mathfrak{J}$ *is a closed ideal of* $\mathfrak{X}$*, then for every* $D \in \operatorname{Der}(\mathfrak{X})$ *we have* $D(\mathfrak{J}) \subseteq \mathfrak{J}$.

PROOF See Theorem 1 in [CS94], Proposition 2 in [CR87], and Subsection 1.1.8 in [CMMR94]. ∎

DEFINITION 7.12 Let $\mathfrak{g}$ be an involutive Banach algebra over $\mathbb{K} \in \{\mathbb{R}, \mathbb{C}\}$ and X^* one of the symbols H^* and L^*. An X^**-ideal* of $\mathfrak{g}$ is an ideal $\mathfrak{J}$ of $\mathfrak{g}$ such that there exists an X^*-algebra $\mathfrak{X}$ over $\mathbb{K}$ and $\varphi \in \operatorname{Hom}^*(\mathfrak{X}, \mathfrak{g})$ with $\operatorname{Ran}\varphi = \mathfrak{J}$. We denote the set of the X^*-ideals of $\mathfrak{g}$ by $X^*\operatorname{Id}(\mathfrak{g})$.

The following simple remark will play an important role in studying the H^*-ideals.

REMARK 7.13 Note that in Definition 7.12, we may always assume $\operatorname{Ker}\varphi = \{0\}$ by restricting φ to $(\operatorname{Ker}\varphi)^\perp$. This remark easily implies that if $\mathfrak{g}$ is an involutive Banach-Lie algebra over $\mathbb{K}$, then $L^*\mathrm{Id}(\mathfrak{g}) = H^*\mathrm{Id}(\mathfrak{g})$. The same remark implies that if $\mathbb{K} = \mathbb{R}$ and $\mathfrak{g}$ is a canonically involutive Banach-Lie algebra, then the L^*-algebra $\mathfrak{X}$ in Definition 7.12 can be assumed canonically involutive.

The following lemma exhibits a connection between the dual $D^\#$ of a derivation D of an H^*-algebra on one hand, and the map D^* defined as in Remark 7.2(b) on the other hand.

LEMMA 7.14 *If $\mathfrak{X}$ is a real H^*-algebra with* $\mathrm{Ann}(\mathfrak{X}) = \{0\}$, *then every* $D \in \mathrm{Der}(\mathfrak{X})$ *has the property* $D^\# = -D^*$.

PROOF Use Theorem 2.1 in [CR85] for the complex-linear extension of D to the complexification of $\mathfrak{X}$. □

LEMMA 7.15 *Let $\mathfrak{A}$ and $\mathfrak{B}$ be real H^*-algebras and $\psi\colon \mathfrak{A} \to \mathfrak{B}$ a continuous $*$-homomorphism with* $\operatorname{Ran}\psi$ *dense in $\mathfrak{B}$. Then for every $b \in B$ we have*

$$(\psi\psi^\#)R_b = R_b(\psi\psi^\#) \text{ and } (\psi\psi^\#)L_b = L_b(\psi\psi^\#),$$

where $R_b, L_b\colon \mathfrak{B} \to \mathfrak{B}$ are defined by $R_b z = zb$ and $L_b z = bz$ for all $z \in \mathfrak{B}$.

PROOF For all $x, y \in \mathfrak{A}$ and $z \in \mathfrak{B}$ we have $(\psi(xy) \mid z) = (xy \mid \psi^\#(z)) = (x \mid \psi^\#(z)y^*)$. On the other hand, $(\psi(xy) \mid z) = (\psi(x)\psi(y) \mid z) = (\psi(x) \mid z\psi(y)^*) = (x \mid \psi^\#(z\psi(y^*)))$, whence $(x \mid \psi^\#(z)y^*) = (x \mid \psi^\#(z\psi(y^*)))$.

Since x is arbitrary in $\mathfrak{A}$, it then follows that $\psi^\#(z)y^* = \psi^\#(z\psi(y^*))$. By replacing y by y^*, we get $\psi^\#(z)y = \psi^\#(z\psi(y))$, whence $\psi\psi^\#(z)\psi(y) = \psi\psi^\#(z\psi(y))$ for all $y \in \mathfrak{A}$ and $z \in \mathfrak{B}$. Since $\operatorname{Ran}\psi$ is dense in $\mathfrak{B}$, it then follows that $(\psi\psi^\#)R_b = R_b(\psi\psi^\#)$ for all $b \in \mathfrak{B}$.

The second of the desired equalities can be proved in a similar manner. □

The next auxiliary fact will play a key role in what follows.

LEMMA 7.16 *Let $\mathfrak{X}_1$ and $\mathfrak{X}_2$ be real H^*-algebras. If both $\mathfrak{X}_1$ and $\mathfrak{X}_2$ are topologically simple, $\psi \in \mathrm{Hom}^*(\mathfrak{X}_1, \mathfrak{X}_2)$ and $\operatorname{Ran}\psi \trianglelefteq \mathfrak{X}_2$, then either $\psi = 0$ or ψ is a positive multiple of an isometry of $\mathfrak{X}_1$ onto $\mathfrak{X}_2$. If moreover $\mathfrak{X}_1 = \mathfrak{X}_2$, then ψ is actually an isometry.*

PROOF First recall that for every $a \in \mathfrak{X}_2$ we denote by $L_a, R_a \in \mathcal{B}(\mathfrak{X}_2)$ the maps $b \mapsto ab$ and $b \mapsto ba$, respectively.

If we assume that $\psi \not\equiv 0$, then it follows that the closure of $\operatorname{Ran}\psi$ is a non-zero closed ideal of $\mathfrak{X}_2$, which is invariant under the involution. Hence $\operatorname{Ran}\psi$ must be dense in $\mathfrak{X}_2$, according to the hypothesis that $\mathfrak{X}_2$ is topologically simple. Since $\psi \in \operatorname{Hom}^*(\mathfrak{X}, \mathfrak{g})$, it then follows by Lemma 7.15 that $\psi\psi^{\#}$ ($\in \mathcal{B}(\mathfrak{X}_2)$) commutes with each element of the set $\{L_a \mid a \in \mathfrak{X}_2\} \cup \{R_a \mid a \in \mathfrak{X}_2\}$ ($\subseteq \mathcal{B}(\mathfrak{X}_2)$).

Consequently, $\psi\psi^{\#}$ commutes with each element of the set

$$\mathcal{S} := \{L_a \mid a \in \mathfrak{X}_2\} \cup \{R_a \mid a \in \mathfrak{X}_2\} \quad (\subseteq \mathcal{B}(\mathfrak{X}_2)).$$

But $\{0\}$ and $\mathfrak{X}_2$ are the only closed linear subspaces of $\mathfrak{X}_2$ which are invariant to each operator in $\mathcal{S}$, for $\mathfrak{X}_2$ is topologically simple. Hence Theorem 7.30 shows that there exists $\lambda \in \mathbb{R}$ such that $\psi\psi^{\#} = \lambda \mathrm{id}_{\mathfrak{X}_2}$. Since $\psi \not\equiv 0$, it follows that $\lambda \neq 0$ (actually $\lambda > 0$), and then $\operatorname{Ran}\psi = \mathfrak{X}_2$.

On the other hand, $\operatorname{Ker}\psi \trianglelefteq \mathfrak{X}_1$, hence $\operatorname{Ker}\psi = \{0\}$ in view of the hypothesis that $\mathfrak{X}_1$ is topologically simple. Consequently, ψ is invertible, and then the relation $\psi\psi^{\#} = \lambda \mathrm{id}_{\mathfrak{X}_2}$ implies that $\psi^{\#}\psi = \lambda \mathrm{id}_{\mathfrak{X}_1}$, hence $(1/\sqrt{\lambda})\psi$ is an isometry of $\mathfrak{X}_1$ onto $\mathfrak{X}_2$.

Now assume that $\mathfrak{X}_1 = \mathfrak{X}_2$ and denote

$$M = \sup\{\|xy\| \mid x, y \in \mathfrak{X}_1,\ \|x\| \le 1,\ \|y\| \le 1\} \quad (\in \mathbb{R}_+^*).$$

For all $x, y \in \mathfrak{X}_1$ we have $\psi(xy) = \psi(x)\psi(y)$, hence

$$\sqrt{\lambda}\|xy\| = \|\psi(xy)\| \le M\|\psi(x)\| \cdot \|\psi(y)\| = M \cdot \sqrt{\lambda}\|x\| \cdot \sqrt{\lambda}\|y\|,$$

whence $\|xy\| \le M \cdot \sqrt{\lambda}\|x\| \cdot \|y\|$. Consequently $\lambda \ge 1$. The same reasoning applied to ψ^{-1} instead of ψ shows that $\lambda \le 1$, hence $\lambda = 1$, and then ψ is an isometry. ∎

PROPOSITION 7.17 *Let $\mathfrak{X}$ be a topologically simple real L^*-algebra. Then either the complexification $\mathfrak{X}_{\mathbb{C}}$ is topologically simple when viewed as a real L^*-algebra or there exist a topologically simple complex L^*-algebra $\mathfrak{Z}$ and a continuous isomorphism $\psi\colon \mathfrak{Z} \to \mathfrak{X}$ of real Lie algebras.*

PROOF Let $\kappa\colon \mathfrak{X}_{\mathbb{C}} \to \mathfrak{X}_{\mathbb{C}}$, $\kappa(x+iy) = x-iy$ for all $x, y \in \mathfrak{X}$ and assume that $\mathfrak{X}_{\mathbb{C}}$ is not topologically simple as a real L^*-algebra. Then, by Theorem 7.11(a), it follows that the real L^*-algebra $\mathfrak{X}_{\mathbb{C}}$ has a closed minimal ideal $\mathfrak{Z}$ with $\{0\} \neq \mathfrak{Z} \neq \mathfrak{X}_{\mathbb{C}}$. Define $p\colon \mathfrak{X}_{\mathbb{C}} \to \mathfrak{X}$ by $p(z) = (z + \kappa(z))/2$ and denote by $\mathfrak{X}_0$ the closure of $p(\mathfrak{Z})$.

Since $\mathfrak{Z}$ is a closed ideal of $\mathfrak{X}_{\mathbb{C}}$, it follows at once that $\mathfrak{X}_0$ is a closed ideal of $\mathfrak{X}$. Then $\mathfrak{X}_0 = \mathfrak{X}$ since $\mathfrak{X}$ is topologically simple. Now, since $\kappa\colon \mathfrak{X}_{\mathbb{C}} \to \mathfrak{X}_{\mathbb{C}}$ is an isomorphism of real Banach-Lie algebras and $\mathfrak{Z}$ is a closed minimal ideal of $\mathfrak{X}_{\mathbb{C}}$, it follows that so is $\kappa(\mathfrak{Z})$. Hence we have either $\mathfrak{Z} = \kappa(\mathfrak{Z})$ or $\mathfrak{Z} \cap \kappa(\mathfrak{Z}) = \{0\}$, and in the latter case $\mathfrak{Z} \perp \kappa(\mathfrak{Z})$ by Theorem 7.11(a) again.

If $\mathfrak{Z} = \kappa(\mathfrak{Z})$, then $p(\mathfrak{Z}) = \mathfrak{Z} \cap \mathfrak{X}$ is closed in $\mathfrak{Z}$, hence $p(\mathfrak{Z}) = \mathfrak{X}$ in view of the hypothesis that $\mathfrak{X}$ is topologically simple. Then we get $\mathfrak{X}_{\mathbb{C}} = \mathfrak{Z}$, a contradiction with the way $\mathfrak{Z}$ was chosen.

If $\mathfrak{Z} \perp \kappa(\mathfrak{Z})$, then, by $p|_{\mathfrak{Z}} = (\mathrm{id}_{\mathfrak{Z}} + \kappa|_{\mathfrak{Z}})/2$, it follows at once that $p(\mathfrak{Z})$ is closed in $\mathfrak{X}$, hence $p(\mathfrak{Z}) = \mathfrak{X}$ just as above. Then $\mathfrak{X}_{\mathbb{C}} = \mathfrak{Z} + \kappa(\mathfrak{Z})$, whence $\mathfrak{X}_{\mathbb{C}} = \mathfrak{Z} \oplus \kappa(\mathfrak{Z})$. Now define

$$\psi \colon \mathfrak{Z} \to \mathfrak{X}, \quad \psi(z) = z + \kappa(z).$$

It follows at once that ψ is an isomorphism of real Banach-Lie algebras, and the proof ends. ∎

We now state the classification theorem on topologically simple complex L^*-algebras of infinite dimension.

THEOREM 7.18 *For every infinite-dimensional topologically simple complex L^*-algebra $\mathfrak{X}_0$ there exists an infinite-dimensional complex Hilbert space $\mathcal{H}$ such that $\mathfrak{X}_0$ is isomorphic to a complex L^*-algebra of one of the following types.*

(A) $\mathfrak{X} = \mathcal{C}_2(\mathcal{H})$.

(B) $\mathfrak{X} = \{A \in \mathcal{C}_2(\mathcal{H}) \mid A = -JA^*J^{-1}\}$ *for some conjugate-linear isometry* $J \colon \mathcal{H} \to \mathcal{H}$ *satisfying* $J^2 = \mathrm{id}_{\mathcal{H}}$.

(C) $\mathfrak{X} = \{A \in \mathcal{C}_2(\mathcal{H}) \mid A = -\widetilde{J}A^*\widetilde{J}^{-1}\}$ *for some conjugate-linear isometry* $\widetilde{J} \colon \mathcal{H} \to \mathcal{H}$ *satisfying* $\widetilde{J}^2 = -\mathrm{id}_{\mathcal{H}}$.

PROOF See [CGM90]. ∎

We now describe the classification of infinite-dimensional topologically simple *real* L^*-algebras.

THEOREM 7.19 *For every infinite-dimensional topologically simple real L^*-algebra $\mathfrak{X}_0$ there exists an infinite-dimensional complex Hilbert space $\mathcal{H}$ such that $\mathfrak{X}_0$ is isomorphic to a real L^*-algebra of one of the following types.*

(AI) $\mathfrak{X} = \{A \in \mathcal{C}_2(\mathcal{H}) \mid JA = AJ\}$ *for some conjugate-linear isometry* $J \colon \mathcal{H} \to \mathcal{H}$ *satisfying* $J^2 = \mathrm{id}_{\mathcal{H}}$.

(AII) $\mathfrak{X} = \{A \in \mathcal{C}_2(\mathcal{H}) \mid JA = AJ\}$ *for some conjugate-linear isometry* $J \colon \mathcal{H} \to \mathcal{H}$ *satisfying* $J^2 = -\mathrm{id}_{\mathcal{H}}$.

(AIII) $\mathfrak{X}_U = \{A \in \mathcal{C}_2(\mathcal{H}) \mid A^*U = -UA\}$ *for some operator* $U \in \mathcal{B}(\mathcal{H})$ *satisfying* $U = U^* = U^{-1}$.

(BI) $\mathfrak{X} = \{A \in \mathcal{C}_2(\mathcal{H}) \mid A = -J_1A^*J_1^{-1}, J_2A = AJ_2\}$ *for some conjugate-linear isometries* $J_1, J_2 \colon \mathcal{H} \to \mathcal{H}$ *satisfying* $J_1^2 = J_2^2 = \mathrm{id}_{\mathcal{H}}$ *and* $J_1J_2 = J_2J_1$.

(BII) $\mathfrak{X} = \{A \in \mathcal{C}_2(\mathcal{H}) \mid A = -J_1A^*J_1^{-1}, J_2A = AJ_2\}$ *for some conjugate-linear isometries* $J_1, J_2 \colon \mathcal{H} \to \mathcal{H}$ *satisfying* $J_1^2 = \mathrm{id}_{\mathcal{H}}$, $J_2^2 = -\mathrm{id}_{\mathcal{H}}$ *and* $J_1J_2 = J_2J_1$.

(CI) $\mathfrak{X} = \{A \in \mathcal{C}_2(\mathcal{H}) \mid A = -J_1A^*J_1^{-1}, J_2A = AJ_2\}$ *for some conjugate-linear isometries* $J_1, J_2 \colon \mathcal{H} \to \mathcal{H}$ *satisfying* $J_1^2 = -\mathrm{id}_{\mathcal{H}}$, $J_2^2 = \mathrm{id}_{\mathcal{H}}$ *and* $J_1J_2 = J_2J_1$.

(CII) $\mathfrak{X} = \{A \in \mathcal{C}_2(\mathcal{H}) \mid A = -J_1A^*J_1^{-1}, J_2A = AJ_2\}$ *for some conjugate-linear isometries* $J_1, J_2 \colon \mathcal{H} \to \mathcal{H}$ *satisfying* $J_1^2 = J_2^2 = -\mathrm{id}_{\mathcal{H}}$ *and* $J_1J_2 = J_2J_1$.

PROOF See [CGM90] and [Ba72]. □

We conclude this section with a couple of results on automorphisms of topologically simple complex L^*-algebras. These results will be needed in the proof of Proposition 8.33.

THEOREM 7.20 *Let* $\mathcal{H}$ *be a complex infinite-dimensional Hilbert space and consider the complex simple* L^**-algebra* $\mathfrak{L} = \mathcal{C}_2(\mathcal{H})$. *For every unitary operator* $U \in \mathcal{B}(\mathcal{H})$ *define*

$$\theta_U \colon \mathfrak{L} \to \mathfrak{L}, \quad \theta_U(A) = UAU^{-1}.$$

Moreover, for every conjugate-linear bijective isometry $J \colon \mathcal{H} \to \mathcal{H}$ *define*

$$\omega_J \colon \mathfrak{L} \to \mathfrak{L}, \quad \omega_J(T) = -JT^*J^{-1}.$$

Then every $*$*-automorphism of* $\mathfrak{L}$ *either is equal to* θ_U *for some unitary operator* $U \in \mathcal{B}(\mathcal{H})$ *or is equal to* ω_J *for a suitable conjugate-linear bijective isometry* $J \colon \mathcal{H} \to \mathcal{H}$.

PROOF See Theorem 5 at page 1235 in [Ba72]. □

THEOREM 7.21 *Assume that* $\mathfrak{X}_0$ *is a topologically simple complex* L^**-algebra and* $\mathcal{H}$ *is an infinite-dimensional complex Hilbert space such that* $\mathfrak{X}_0$ *is represented as an* L^**-subalgebra of* $\mathcal{C}_2(\mathcal{H})$ *as in* Theorem 7.18 *above.*

Then every $*$-automorphism of the complex L^*-algebra $\mathfrak{X}_0$ extends to a $*$-automorphism of the complex L^*-algebra $\mathcal{C}_2(\mathcal{H})$.

PROOF See Corollary 2 to Theorem 6 at page 1240 in [Ba72] □

COROLLARY 7.22 *In the setting of* Theorem 7.21, *let $\psi\colon \mathfrak{X}_0 \to \mathfrak{X}_0$ be a conjugate-linear $*$-automorphism of $\mathfrak{X}_0$, and denote by $\mathfrak{F}$ the set of all finite-rank operators on $\mathcal{H}$. Then $\psi(\mathfrak{F} \cap \mathfrak{X}_0) \subseteq \mathfrak{F} \cap \mathfrak{X}_0$.*

PROOF Let $\varphi\colon \mathfrak{X}_0 \to \mathfrak{X}_0$, $\varphi(A) = -\psi(A^*)$. Then φ is a $*$-automorphism of $\mathfrak{X}_0$, hence it follows by Theorems 7.20 and 7.21 that we have either $\varphi = \theta_U$ or $\varphi = \omega_J$ for suitable U or J. In both cases it is clear that $\varphi(\mathfrak{F}\cap\mathfrak{X}_0) \subseteq \mathfrak{F}\cap\mathfrak{X}_0$, whence the desired conclusion follows in view of the construction of φ. □

7.3 Equivariant monotone operators as reproducing kernels

Throughout this section, $\mathfrak{Y}$ stands for a real Banach space. According to Remark A.14(b), we can think of $\mathfrak{Y}$ as a canonically involutive abelian Banach-Lie algebra and we then have

$$\mathcal{C}^+(\mathfrak{Y}) = \{\iota \in \mathcal{B}(\mathfrak{Y}^\#, \mathfrak{Y}) \mid \iota^\# = \iota \text{ and } \langle f, \iota(f)\rangle \geq 0 \text{ for all } f \in \mathfrak{Y}^\#\}.$$

In particular, when $\mathfrak{Y}$ is a real Hilbert space, $\mathcal{C}^+(\mathfrak{Y})$ is just the cone of the self-adjoint non-negative operators on $\mathfrak{Y}$.

The aim of the present section is to study the L^*-ideals of the abelian canonically involutive Banach-Lie algebra $\mathfrak{Y}$. Each of these ideals is abelian, hence Remark 7.13 shows that

$$L^*\mathrm{Id}(\mathfrak{Y}) = \{\mathcal{I} \subseteq \mathfrak{Y} \mid \text{there exists a real Hilbert space } \mathfrak{X} \text{ and } \varphi \in \mathcal{B}(\mathfrak{X}, \mathfrak{Y}) \text{ with } \operatorname{Ran}\varphi = \mathfrak{I}\}.$$

We will describe a connection between the elements of $L^*\mathrm{Id}(\mathfrak{Y})$ and the ones of $\mathcal{C}^+(\mathfrak{Y})$ (see Proposition 7.23 below).

The following result shows that each monotone operator has a sort of "square root," and the Hilbert space constructed here is just the *reproducing kernel Hilbert space* associated with a monotone operator.

PROPOSITION 7.23 *If $\iota \in \mathcal{B}(\mathfrak{Y}^\#, \mathfrak{Y})$, then we have $\iota \in \mathcal{C}^+(\mathfrak{Y})$ if and only if there exists a real Hilbert space $\mathfrak{X}$ and $\varphi \in \mathcal{B}(\mathfrak{X}, \mathfrak{Y})$ such that*

$$\iota = \varphi \circ \varphi^\#.$$

(The operator φ *can be chosen with* $\operatorname{Ker}\varphi = \{0\}$*.) If this is the case, then the linear subspace* $\operatorname{Ker}\varphi^{\#}$ *of* $\mathfrak{Y}^{\#}$ *depends only on* ι*, actually*

$$\operatorname{Ker}\varphi^{\#} = \operatorname{Ker}\iota;$$

also the linear subspace $\operatorname{Ran}\varphi$ *of* $\mathfrak{Y}$ *depends only on* ι *and we denote*

$$\operatorname{Ran}\varphi =: \Phi(\iota).$$

PROOF 1° If there exist $\mathfrak{X}$ and φ as indicated such that $\iota = \varphi \circ \varphi^{\#}$, then for all $f, g \in \mathfrak{Y}^{\#}$ we have

$$\langle f, \iota(g)\rangle = \langle f, \varphi(\varphi^{\#}(g))\rangle = \left(\varphi^{\#}(f) \mid \varphi^{\#}(g)\right)_{\mathfrak{X}}$$

hence, since the scalar product $(\cdot \mid \cdot)_{\mathfrak{X}}$ of $\mathfrak{X}$ is symmetric, we get $\langle f, \iota(g)\rangle = \langle g, \iota(f)\rangle$. Consequently $\iota = \iota^{\#}$. The same computation shows that for all $f \in \mathfrak{Y}^{\#}$ we have

$$\langle f, \iota(f)\rangle = \|\varphi^{\#}(f)\|_{\mathfrak{X}} \geq 0.$$

This shows that $\iota \in \mathcal{C}^{+}(\mathfrak{Y})$, and also that $\operatorname{Ker}\iota = \operatorname{Ker}\varphi^{\#}$ (see Remark 7.3, or relation (7.3) below).

2° Conversely, assume that $\iota \in \mathcal{C}^{+}(\mathfrak{Y})$. Recall from Remark 7.3 the bilinear form

$$\mathfrak{Y}^{\#} \times \mathfrak{Y}^{\#} \to \mathbb{R}, \quad (f, g) \mapsto \langle f, \iota(g)\rangle, \tag{7.1}$$

for which we used Schwartz' inequality

$$(\forall f, g \in \mathfrak{Y}^{\#}) \qquad \langle f, \iota(g)\rangle^2 \leq \langle f, \iota(f)\rangle \cdot \langle g, \iota(g)\rangle \tag{7.2}$$

in order to deduce that

$$\operatorname{Ker}\iota = \{g \in \mathfrak{Y}^{\#} \mid \langle g, \iota(g)\rangle = 0\}. \tag{7.3}$$

It follows by (7.3) that the bilinear form (7.1) induces a real scalar product on $\mathfrak{Y}^{\#}/\operatorname{Ker}\iota$. Let $\mathfrak{X}$ be the Hilbert space obtained as the completion of $\mathfrak{Y}^{\#}/\operatorname{Ker}\iota$ with respect to that scalar product, and denote by

$$\eta \colon \mathfrak{Y}^{\#}/\operatorname{Ker}\iota \to \mathfrak{X}$$

the corresponding inclusion map. Also denote by

$$\pi \colon \mathfrak{Y}^{\#} \to \mathfrak{Y}^{\#}/\operatorname{Ker}\iota$$

the projection map, and by

$$\varphi_0 \colon \mathfrak{Y}^{\#}/\operatorname{Ker}\iota \to \mathfrak{Y}$$

the map induced by $\iota \colon \mathfrak{Y}^{\#} \to \mathfrak{Y}$. In particular, $\iota = \varphi_0 \circ \pi$.

For $g \in \mathfrak{Y}^\#$, Schwartz' inequality (7.2) implies that

$$(\forall f \in \mathfrak{Y}^\#) \qquad |\langle f, \iota(g)\rangle| \le \|\iota\|^{1/2} \cdot \|f\| \cdot \|\eta(\pi(g))\|_{\mathfrak{X}},$$

whence, by the Hahn-Banach theorem, $\|\iota(g)\| \le \|\iota\|^{1/2} \cdot \|\eta(\pi(g))\|_{\mathfrak{X}}$. It then follows that for each $g \in \mathfrak{Y}^\#$ we have $\|\varphi_0(\pi(g))\| \le \|\iota\|^{1/2} \cdot \|\eta(\pi(g))\|_{\mathfrak{X}}$, that is,

$$(\forall v \in \mathfrak{Y}^\#/\mathrm{Ker}\,\iota) \qquad \|\varphi_0(v)\| \le \|\iota\|^{1/2} \cdot \|\eta(v)\|_{\mathfrak{X}}.$$

Consequently, φ_0 extends to a bounded linear map from $\mathfrak{X}$ into $\mathfrak{Y}$. More precisely, there exists a unique $\varphi \in \mathcal{B}(\mathfrak{X}, \mathfrak{Y})$ such that $\varphi \circ \eta = \varphi_0$ (and $\|\varphi\| \le \|\iota\|^{1/2}$). Since $\iota = \varphi_0 \circ \pi$, we also deduce that

$$\varphi \circ \eta \circ \pi = \iota. \tag{7.4}$$

Hence, to prove the equality $\varphi \circ \varphi^\# = \iota$, we have to check that

$$\varphi^\# = \eta \circ \pi.$$

To this end, take $f, g \in \mathfrak{Y}^\#$ arbitrary. We then have

$$\begin{aligned}\big((\eta \circ \pi)(f) \mid (\eta \circ \pi)(g)\big)_{\mathfrak{X}} &= \big(\eta(\pi(f)) \mid \eta(\pi(g))\big)_{\mathfrak{X}} \\ &= \langle f, \iota(g)\rangle \\ &= \langle f, \varphi((\eta \circ \pi)(g))\rangle \\ &= \big(\varphi^\#(f) \mid (\eta \circ \pi)(g)\big)_{\mathfrak{X}},\end{aligned}$$

where the third equality is a consequence of (7.4)). Since $\{(\eta \circ \pi)(g) \mid g \in \mathfrak{Y}^\#\}$ ($= \mathrm{Ran}\,\eta$) is a dense subset of $\mathfrak{X}$, it then follows that $\varphi^\#(f) = (\eta \circ \pi)(f)$, as desired. Also note that

$$\mathrm{Ker}\,\varphi = (\mathrm{Ran}\,\varphi^\#)^\perp = (\mathrm{Ran}\,(\eta \circ \pi))^\perp = \{0\}.$$

3° Finally, let $\mathfrak{X}_j$ be a real Hilbert space and $\varphi_j \in \mathcal{B}(\mathfrak{X}_j, \mathfrak{Y})$ such that $\iota = \varphi_j \circ \varphi_j^\#$ for $j = 1, 2$. We are going to show that $\mathrm{Ran}\,\varphi_1 = \mathrm{Ran}\,\varphi_2$ ($\subseteq \mathfrak{Y}$).
To this end, first note that, as at the stage 1° of the proof, we have

$$(\forall f \in \mathfrak{Y}^\#) \quad \langle f, \iota(f)\rangle = \|\varphi_1^\#(f)\|^2_{\mathfrak{X}_1} = \|\varphi_2^\#(f)\|^2_{\mathfrak{X}_2} \tag{7.5}$$

and

$$\mathrm{Ker}\,\iota = \mathrm{Ker}\,\varphi_1^\# = \mathrm{Ker}\,\varphi_2^\#. \tag{7.6}$$

It follows by (7.5) that the map $\mathrm{Ran}\,\varphi_1^\# \to \mathrm{Ran}\,\varphi_2^\#$, $\varphi_1^\#(f) \mapsto \varphi_2^\#(f)$, is well defined and extends to a unitary operator

$$U \colon \overline{\mathrm{Ran}\,\varphi_1^\#} \to \overline{\mathrm{Ran}\,\varphi_2^\#}.$$

Next note that $\overline{\mathrm{Ran}\,\varphi_j^\#}^\perp = \mathrm{Ker}\,\varphi_j$, hence

$$\mathrm{Ran}\,\varphi_j = \varphi_j\big(\overline{\mathrm{Ran}\,\varphi_j^\#}\big) \qquad \text{for } j = 1, 2. \tag{7.7}$$

For every $f \in \mathfrak{Y}^{\#}$, we have $\varphi_1(\varphi_1^{\#}(f)) = \varphi_2(\varphi_2^{\#}(f))$ $(= \iota(f))$, hence

$$\varphi_2 \circ U = \varphi_1 \mid_{\overline{\operatorname{Ran} \varphi_1^{\#}}} \quad \text{and} \quad \varphi_1 \circ U^{-1} = \varphi_2 \mid_{\overline{\operatorname{Ran} \varphi_2^{\#}}} .$$

Now (7.7) implies at once that $\operatorname{Ran} \varphi_1 = \operatorname{Ran} \varphi_2$, as desired. □

In the remainder of the present section, we are concerned with some basic properties of the surjective map

$$\Phi \colon \mathcal{C}^{+}(\mathfrak{Y}) \to L^{*}\mathrm{Id}(\mathfrak{Y})$$

constructed in Proposition 7.23 above.

REMARK 7.24 For every $\iota \in \mathcal{C}^{+}(\mathfrak{Y})$, we have $\operatorname{Ran} \iota \subseteq \Phi(\iota)$, and this inclusion can be strict (as the case when $\mathfrak{Y}$ is a Hilbert space shows).

REMARK 7.25 For later use, we now state an inequality obtained at the stage 2° in the proof of Proposition 7.23: if $\mathfrak{Y}$ is a real Banach space, then for each $\iota \in \mathcal{C}^{+}(\mathfrak{Y})$ we have

$$(\forall g \in \mathfrak{Y}^{\#}) \qquad \|\iota(g)\|^2 \leq \|\iota\| \cdot \langle g, \iota(g) \rangle .$$

More generally, *if* $\mathfrak{A}$ *denotes the complexification of* $\mathfrak{Y}$ *and* $a \mapsto \bar{a}$ *stands for the conjugation of* $\mathfrak{A}$ *whose fixed points are the elements of* $\mathfrak{Y}$*, then*

$$(\forall w \in \mathfrak{A}^{\#}) \qquad \|\iota(w)\|^2 \leq \|\iota\| \cdot \langle w, \overline{\iota(w)} \rangle . \tag{7.8}$$

This inequality can be obtained just as the above one, by making use of Schwartz' inequality for the non-negative hermitian form

$$\mathfrak{A}^{\#} \times \mathfrak{A}^{\#} \to \mathbb{C}, \quad (v, w) \mapsto \langle v, \overline{\iota(w)} \rangle ,$$

to deduce that

$$(\forall v, w \in \mathfrak{A}^{\#}) \quad |\langle v, \overline{\iota(w)} \rangle|^2 \leq \langle v, \overline{\iota(v)} \rangle \cdot \langle w, \overline{\iota(w)} \rangle \leq \|\iota\| \cdot \|v\|^2 \cdot \langle w, \overline{\iota(w)} \rangle .$$

(Recall that $\|\bar{a}\| = \|a\|$ for all $a \in \mathfrak{A}$; cf. Notation A.11.) It then follows by the Hahn-Banach theorem that inequality (7.8) holds.

It is clear that for all $\iota \in \mathcal{C}^{+}(\mathfrak{Y})$ and $\lambda > 0$ we have $\lambda\iota \in \mathcal{C}^{+}(\mathfrak{Y})$ and $\Phi(\iota) = \Phi(\lambda\iota)$, hence Φ is by no means injective. In order to study "how non-injective" Φ is, we first show that it is in some sense non-increasing.

PROPOSITION 7.26 *If* $\iota_1, \iota_2 \in \mathcal{C}^{+}(\mathfrak{Y})$*, then the following assertions are equivalent:*

(i) *We have* $\Phi(\iota_1) \subseteq \Phi(\iota_2)$.

(ii) *There exists* $\lambda > 0$ *such that* $\lambda\iota_2 \in \iota_1 + \mathcal{C}^+(\mathfrak{Y})$.

PROOF For $j = 1, 2$, let $\mathfrak{X}_j$ be a real Hilbert space and $\varphi_j \in \mathcal{B}(\mathfrak{X}_j, \mathfrak{Y})$ with

$$\varphi_j \circ \varphi_j^{\#} = \iota_j \quad \text{and} \quad \operatorname{Ran} \varphi_j = \Phi(\iota_j)$$

(cf. Proposition 7.23). As in Remark 7.13, we may assume that $\operatorname{Ker} \varphi_j = \{0\}$ for $j = 1, 2$. With these preparations, we now begin to prove the desired equivalence.

"(i) $\Rightarrow$ (ii)" If $\Phi(\iota_1) \subseteq \Phi(\iota_2)$, that is, $\operatorname{Ran} \varphi_1 \subseteq \operatorname{Ran} \varphi_2$, then

$$\psi := \varphi^{-1} \circ \varphi_2 \colon \mathfrak{X}_2 \to \mathfrak{X}_1$$

is a closed operator, hence it is continuous by the closed graph theorem. Next note that

$$\varphi_1 \circ \psi = \varphi_2,$$

whence $\psi^{\#} \circ \varphi_1^{\#} = \varphi_2^{\#}$. Consequently, $\varphi_1 \circ \psi \circ \psi^{\#} \circ \varphi_1^{\#} = \varphi_2 \circ \varphi_2^{\#} = \iota_2$. Then for every $f \in \mathfrak{Y}^{\#}$ we have

$$\begin{aligned}
\langle f, \iota_2(f)\rangle &= \langle f, (\varphi_1 \circ \psi \circ \psi^{\#} \circ \varphi_1^{\#})(f)\rangle \\
&= \big((\psi^{\#} \circ \varphi_1^{\#})(f) \mid (\psi^{\#} \circ \varphi_1^{\#})(f)\big)_{\mathfrak{X}_2} \\
&= \|\psi^{\#}(\varphi_1^{\#}(f))\|^2_{\mathfrak{X}_2} \\
&\le \|\psi^{\#}\|^2 \cdot \|\varphi_1^{\#}(f)\|^2_{\mathfrak{X}_1} \\
&= \|\psi^{\#}\|^2 \big(\varphi_1^{\#}(f) \mid \varphi_1^{\#}(f)\big)_{\mathfrak{X}_1} \\
&= \|\psi^{\#}\|^2 \cdot \langle f, \varphi_1(\varphi_1^{\#}(f))\rangle \\
&= \|\psi^{\#}\|^2 \cdot \langle f, \iota_1(f)\rangle,
\end{aligned}$$

hence the assertion (ii) holds with $\lambda = 1/\|\psi^{\#}\|^2$.

"(i) $\Leftarrow$ (ii)" If the assertion (ii) holds, then we have $\lambda\langle f, \iota_2(f)\rangle \ge \langle f, \iota_1(f)\rangle$ for all $f \in \mathfrak{Y}^{\#}$. Since $\iota_j = \varphi_j \circ \varphi_j^{\#}$ for $j = 1, 2$, it follows that

$$(\forall f \in \mathfrak{Y}^{\#}) \qquad \|\varphi_1^{\#}(f)\|^2_{\mathfrak{X}_1} \le \lambda \cdot \|\varphi_2^{\#}(f)\|^2_{\mathfrak{X}_2}.$$

Hence the linear map

$$\operatorname{Ran} \varphi_2^{\#} \to \operatorname{Ran} \varphi_1^{\#}, \quad \varphi_2^{\#}(f) \mapsto \varphi_1^{\#}(f),$$

is well defined and continuous. We extend it by continuity to the closure of $\operatorname{Ran} \varphi_2^{\#}$, and then to a linear operator $\theta \in \mathcal{B}(\mathfrak{X}_1, \mathfrak{X}_2)$ vanishing on $(\operatorname{Ran} \varphi_2^{\#})^{\perp}$. Then $\theta \circ \varphi_2^{\#} = \varphi_2^{\#}$, whence

$$\varphi_1 = \varphi_2 \circ \theta^{\#}.$$

This relation shows that $\operatorname{Ran}\varphi_1 \subseteq \operatorname{Ran}\varphi_2$, that is, $\Phi(\iota_1) \subseteq \Phi(\iota_2)$. ∎

COROLLARY 7.27 *If $\iota_1, \iota_2 \in \mathcal{C}^+(\mathfrak{Y})$, then the following assertions hold.*

(a) *We have $\Phi(\iota_1) \subseteq \Phi(\iota_2)$ if and only if*

$$\mathbb{R}_+^* \cdot \iota_2 \subseteq \mathbb{R}_+^* \cdot \iota_1 + \mathcal{C}^+(\mathfrak{Y}).$$

(b) *The equality $\Phi(\iota_1) = \Phi(\iota_2)$ holds if and only if there exist $\lambda, \mu > 0$ such that*

$$(\forall f \in \mathfrak{Y}^\#) \qquad \lambda\langle f, \iota_1(f)\rangle \leq \langle f, \iota_2(f)\rangle \leq \mu\langle f, \iota_1(f)\rangle.$$

PROOF Just use Proposition 7.26. ∎

We now prove an additivity property of our map $\Phi\colon \mathcal{C}^+(\mathfrak{Y}) \to L^*\mathrm{Id}(\mathfrak{Y})$.

PROPOSITION 7.28 *We have*

$$\Phi(\iota_1 + \iota_2) = \Phi(\iota_1) + \Phi(\iota_2) \quad (\subseteq \mathfrak{Y})$$

for all $\iota_1, \iota_2 \in \mathcal{C}^+(\mathfrak{Y})$.

PROOF For $j = 1, 2$, let $\mathfrak{X}$ be a real Hilbert space and $\varphi_j \in \mathcal{B}(\mathfrak{X}_j, \mathfrak{Y})$ with $\varphi_j \circ \varphi_j^\# = \iota_j$ and $\operatorname{Ran}\varphi_j = \Phi(\iota_j)$, according to Proposition 7.23. Consider the Hilbert space direct sum

$$\mathfrak{X} = \mathfrak{X}_1 \oplus \mathfrak{X}_2$$

and define $\varphi \in \mathcal{B}(\mathfrak{X}, \mathfrak{Y})$ by

$$\varphi(x_1, x_2) = \varphi_1(x_1) + \varphi_2(x_2)$$

for all $x_1 \in \mathfrak{X}_1$ and $x_2 \in \mathfrak{X}_2$. Then it is easy to check that the dual operator $\varphi^\# \in \mathcal{B}(\mathfrak{Y}^\#, \mathfrak{X})$ is defined by

$$\varphi^\#(f) = (\varphi_1^\#(f), \varphi_2^\#(f))$$

for all $f \in \mathfrak{Y}^\#$. Consequently

$$\varphi \circ \varphi^\# = \varphi_1 \circ \varphi_1^\# + \varphi_2 \circ \varphi_2^\# = \iota_1 + \iota_2,$$

whence

$$\Phi(\iota_1 + \iota_2) = \operatorname{Ran}\varphi = \operatorname{Ran}\varphi_1 + \operatorname{Ran}\varphi_2 = \Phi(\iota_1) + \Phi(\iota_2),$$

and the proof ends. ∎

The following result is meant as a characterization of the cone $\mathcal{C}_0^+(\mathfrak{Y})$ $(= \{\iota \in \mathcal{C}^+(\mathfrak{Y}) \mid \operatorname{Ker}\iota = \{0\}\}$; see Definition 7.1) in terms of the map Φ.

PROPOSITION 7.29 *If $\iota \in \mathcal{C}^+(\mathfrak{Y})$, then the following assertions are equivalent.*

(i) *We have $\iota \in \mathcal{C}_0^+(\mathfrak{Y})$.*

(ii) *The linear subspace $\Phi(\iota)$ is dense in $\mathfrak{Y}$.*

PROOF Let $\mathfrak{X}$ be a real Hilbert space and $\varphi \in \mathcal{B}(\mathfrak{X}, \mathfrak{Y})$ with $\operatorname{Ker}\varphi = \{0\}$, $\iota = \varphi \circ \varphi^\#$ and $\operatorname{Ran}\iota = \Phi(\iota)$. We then have $\operatorname{Ker}\iota = \operatorname{Ker}\varphi^\#$ (see Proposition 7.23) and moreover

$$\operatorname{Ker}\varphi^\# = (\operatorname{Ran}\varphi)^\perp \quad (\subseteq \mathfrak{Y}^\#).$$

Consequently

$$\big(\Phi(\iota)\big)^\perp = \operatorname{Ker}\iota \quad (\subseteq \mathfrak{Y}^\#),$$

which at once implies the desired conclusion. ∎

7.4 H*-ideals of H*-algebras

Our aim in this section is to determine the set $H^*\mathrm{Id}(\mathfrak{g})$ of all H^*-ideals of an arbitrary real H^*-algebra $\mathfrak{g}$. To this end, we need the following result concerning *real* Hilbert spaces.

THEOREM 7.30 *Let $\mathfrak{X}$ be a real Hilbert space and $\mathcal{S} \subseteq \mathcal{B}(\mathfrak{X})$ such that $\{0\}$ and $\mathfrak{X}$ are the only closed linear subspaces of $\mathfrak{X}$ which are invariant to all operators in $\mathcal{S}$. If $A \in \mathcal{B}(\mathfrak{X})$, $A = A^\#$, and A commutes with each element of $\mathcal{S}$, then A is a real scalar multiple of the identity operator on $\mathfrak{X}$.*

PROOF Use e.g., the method of proof of Theorem 3.11 in the Appendix of [La01]. ∎

Using Lemma 7.16, we can now determine the H^*-ideals of a topologically simple H^*-algebra.

PROPOSITION 7.31 *If $\mathfrak{g}$ is a topologically simple real H^*-algebra, then*

$$H^*\mathrm{Id}(\mathfrak{g}) = \{\{0\}, \mathfrak{g}\}.$$

PROOF Let $\mathfrak{X}$ be a real H^*-algebra and $0 \not\equiv \varphi \in \mathrm{Hom}^*(\mathfrak{X}, \mathfrak{g})$ with $\mathrm{Ran}\,\varphi \trianglelefteq \mathfrak{g}$. What we have to prove is that $\mathrm{Ran}\,\varphi = \mathfrak{g}$.

Replacing φ by $\varphi|_{(\mathrm{Ker}\,\varphi)^\perp}$, we may assume that $\mathrm{Ker}\,\varphi = \{0\}$. Furthermore the closure of $\mathrm{Ran}\,\varphi$ is a closed ideal of $\mathfrak{g}$, which is moreover closed under the involution. Since $\mathfrak{g}$ is topologically simple, it then follows that $\mathrm{Ran}\,\varphi$ is dense in $\mathfrak{g}$.

Now let us assume that the real H^*-algebra $\mathfrak{X}$ is not topologically simple. Then there exist some closed self-adjoint ideals $\mathfrak{X}_1$, $\mathfrak{X}_2$ of $\mathfrak{X}$ with $\mathfrak{X}_1 \neq \{0\} \neq \mathfrak{X}_2$ and $\mathfrak{X}_1 \cdot \mathfrak{X}_2 = \{0\}$ (see Remark 7.10(a) above). Then $\overline{\varphi(\mathfrak{X}_1)} \cdot \overline{\varphi(\mathfrak{X}_2)} = \{0\}$. In particular,

$$\overline{\varphi(\mathfrak{X}_1)} \neq \mathfrak{g}.$$

(Otherwise we get $\varphi(\mathfrak{X}_2) \subseteq \mathrm{Ann}(\mathfrak{g})$. Since $\mathfrak{g}$ is topologically simple we have $\mathrm{Ann}(\mathfrak{g}) = \{0\}$, whence $\{0\} \neq \mathfrak{X}_2 \subseteq \mathrm{Ker}\,\varphi$, thus contradicting the assumption $\mathrm{Ker}\,\varphi = \{0\}$.)

Next note that since $\mathfrak{X}_1$ is an ideal of $\mathfrak{X}$, we have $\mathfrak{X}_1 \cdot \mathfrak{X} \subseteq \mathfrak{X}_1$, hence $\varphi(\mathfrak{X}_1) \cdot \varphi(\mathfrak{X}) \subseteq \varphi(\mathfrak{X}_1)$. Then $\overline{\varphi(\mathfrak{X}_1)} \cdot \overline{\varphi(\mathfrak{X})} \subseteq \overline{\varphi(\mathfrak{X}_1)}$. Since we have seen above that $\overline{\varphi(\mathfrak{X})} = \mathfrak{g}$, we get $\overline{\varphi(\mathfrak{X}_1)} \cdot \mathfrak{g} \subseteq \overline{\varphi(\mathfrak{X}_1)}$. Similarly, $\mathfrak{g} \cdot \overline{\varphi(\mathfrak{X}_1)} \subseteq \overline{\varphi(\mathfrak{X}_1)}$, hence $\overline{\varphi(\mathfrak{X}_1)}$ is a closed proper self-adjoint ideal of $\mathfrak{g}$, thus contradicting the hypothesis.

Consequently, $\mathfrak{X}$ is topologically simple, and then $\mathrm{Ran}\,\varphi = \mathfrak{g}$ according to Lemma 7.16. ∎

REMARK 7.32 For later use, we note the following fact which can be proved by means of an argument in the proof of Proposition 7.31 above: *If* $\mathfrak{g}$ *is a real involutive Banach-Lie algebra,* $\mathfrak{X}$ *is a real* H^**-algebra,* $\varphi \in \mathrm{Hom}^*(\mathfrak{X}, \mathfrak{g})$, $\mathrm{Ker}\,\varphi = \{0\}$, $\mathrm{Ran}\,\varphi$ *is a dense ideal of* $\mathfrak{g}$*, and* $\mathfrak{X}_1$ *stands for a proper closed self-adjoint ideal of* $\mathfrak{X}$*, then* $\overline{\varphi(\mathfrak{X}_1)}$ *is a proper closed self-adjoint ideal of* $\mathfrak{g}$.

Before proceeding with the general result on H^*-ideals of an H^*-algebra (see Theorem 7.34 below), let us recall the following simple fact (see the beginning of Section 7.3).

REMARK 7.33 If $\mathfrak{g}$ is a real involutive Banach algebra with $\mathfrak{g} \cdot \mathfrak{g} = \{0\}$ then $H^*\mathrm{Id}(\mathfrak{g}) = L^*\mathrm{Id}(\mathfrak{g})$, and the latter set consists of all linear subspaces $\mathfrak{I} \subseteq \mathfrak{g}$ such that $\mathfrak{I}$ is closed under the involution and there exist a real Hilbert space $\mathfrak{X}$ and $\varphi \in \mathcal{B}(\mathfrak{X}, \mathfrak{g})$ with $\mathrm{Ran}\,\varphi = \mathfrak{I}$.

Now we are ready to prove the main result of the present section.

THEOREM 7.34 *If* $\mathfrak{g}$ *is a real* H^**-algebra and* $\{\mathfrak{g}_i\}_{i \in I}$ *is the family of all minimal closed self-adjoint ideals of* $\mathfrak{g}$ *with* $\mathfrak{g}_i \cap \mathrm{Ann}(\mathfrak{g}) = \{0\}$ *for each* $i \in I$*, then* $\mathfrak{I} \in H^*\mathrm{Id}(\mathfrak{g})$ *if and only if there exist* $\mathfrak{I}_0 \in H^*\mathrm{Id}(\mathrm{Ann}(\mathfrak{g}))$*,* $J \subseteq I$*, and a* *bounded family* $\{M_i\}_{i\in I}$ *of positive numbers such that*

$$\mathfrak{I}_0 = \Big\{x_0 + \sum_{i\in J} x_i \mid x_0 \in \mathfrak{I}_0; (\forall i \in J)\ x_i \in \mathfrak{g}_i \text{ and } \sum_{i\in J}(1/M_i)^2\|x_i\|^2 < \infty\Big\}.$$

PROOF The proof has several stages.

1° To begin with, let us recall that, denoting $\mathrm{Ann}(\mathfrak{g}) = \mathfrak{g}_0$ and $I_0 = I \sqcup \{0\}$, we have

$$\mathfrak{g} = \bigoplus_{i\in I_0} \mathfrak{g}_i \tag{7.9}$$

as a Hilbert space direct sum, with $\mathfrak{g}_i \cdot \mathfrak{g}_j = \{0\}$ whenever $i, j \in I_0$ and $i \neq j$ (see Remark 7.10 and Theorem 7.11(a) above).

Now take $\mathfrak{I} \in H^*\mathrm{Id}(\mathfrak{g})$ arbitrary. Then there exists a real H^*-algebra $\mathfrak{X}$ and $\varphi \in \mathrm{Hom}^*(\mathfrak{X}, \mathfrak{g})$ with $\mathrm{Ker}\,\varphi = \{0\}$ and $\mathrm{Ran}\,\varphi = \mathfrak{I}$.

2° We show at this stage that if $i \in I$ and $\mathfrak{g}_i \cap \mathfrak{I} \neq \{0\}$, then $\mathfrak{g}_i \subseteq \mathfrak{I}$. Since $\mathrm{Ker}\,\varphi = \{0\}$, the fact that $\mathfrak{g}_i \cap \mathfrak{I} \neq \{0\}$ is equivalent to

$$\mathfrak{X}_i := \varphi^{-1}(\mathfrak{g}_i) \neq \{0\}.$$

Note that $\mathfrak{X}_i$ is a closed self-adjoint ideal of $\mathfrak{X}$, because $\varphi \in \mathrm{Hom}^*(\mathfrak{X}, \mathfrak{g})$ and $\mathfrak{g}_i$ is a closed self-adjoint ideal of $\mathfrak{g}$. We then have

$$\varphi_i := \varphi|_{\mathfrak{X}_i} \in \mathrm{Hom}^*(\mathfrak{X}_i, \mathfrak{g}_i)$$

and thus $\mathrm{Ran}\,\varphi_i$ is a non-zero H^*-ideal of $\mathfrak{g}_i$, that is, $\{0\} \neq \mathfrak{g}_i \cap \mathfrak{I} \in H^*\mathrm{Id}(\mathfrak{g}_i)$. Since $\mathfrak{g}_i$ is a topologically simple real H^*-algebra, Proposition 7.31 implies that we have $\mathfrak{g}_i \cap \mathfrak{I} = \mathfrak{g}_i$, hence $\mathfrak{g}_i \subseteq \mathfrak{I}$, as claimed.

3° Let $x \in \mathfrak{I}$. According to (7.9), for each $i \in I_0$ there exists $x_i \in \mathfrak{g}_i$ such that

$$x = \sum_{i\in I_0} x_i \tag{7.10}$$

(the sum being convergent in $\mathfrak{g}$). Let $i_0 \in I$ be arbitrary such that $\mathfrak{I} \cap \mathfrak{g}_{i_0} = \{0\}$. Since both $\mathfrak{I}$ and $\mathfrak{g}_{i_0}$ are ideals, we get $\mathfrak{I}\cdot\mathfrak{g}_{i_0} \subseteq \mathfrak{I}\cap\mathfrak{g}_{i_0} = \{0\}$. In particular, $x \cdot \mathfrak{g}_{i_0} = \{0\}$. Since for every $i \in I_0 \setminus \{i_0\}$ we have $x_i \cdot \mathfrak{g}_{i_0} \in \mathfrak{g}_i \cdot \mathfrak{g}_{i_0} = \{0\}$, it then follows by (7.10) that $x_{i_0} \cdot \mathfrak{g}_{i_0} = \{0\}$. Similarly $\mathfrak{g}_{i_0} \cdot x_{i_0} = \{0\}$ and thus $x_{i_0} \in \mathrm{Ann}(\mathfrak{g}_{i_0}) = \{0\}$.

Consequently, for each $i \in I$, if we have $x_i \neq 0$ in (7.10), then $\mathfrak{g}_i \cap \mathfrak{I} \neq \{0\}$. In such a case, we have $\mathfrak{g}_i \subseteq \mathfrak{I}$ according to the stage 2° of the proof, hence $x_i \in \mathfrak{I}$.

4° Let

$$J := \{i \in I \mid \mathfrak{g}_i \subseteq \mathfrak{I}\}.$$

For each $i \in J$ we have by stage 2° the isomorphism of topologically simple H^*-algebras $\varphi_i\colon \mathfrak{X}_i \to \mathfrak{g}_i$, hence Lemma 7.16 shows that there exists $M_i > 0$ such that $(1/M_i)\varphi_i$ is an isometry. Then for every $i \in J$ we have

$$M_i = \|\varphi_i\| \leq \|\varphi\|,$$

hence the family of positive numbers $\{M_i\}_{i\in I}$ is bounded.

5° Now observe that, denoting by $\mathfrak{X}_0$ the orthogonal complement in $\mathfrak{X}$ of the closed subspace generated by $\bigcup\limits_{i\in J} \mathfrak{X}_i$, we have

$$\mathfrak{X} = \mathfrak{X}_0 \oplus \bigoplus_{i\in J} \mathfrak{X}_i.$$

Take $x \in \mathfrak{X}$, denote $y = \varphi^{-1}(x) \in \mathfrak{X}$ and for every $i \in J_0 := J \sqcup \{0\}$ let $y_i \in \mathfrak{X}_i$ with $y = \sum\limits_{i\in J_0} y_i$. Then

$$x = \varphi(y) = \sum_{i\in J_0} \varphi(y_i)$$

and for each $i \in J_0$ we have $\varphi(y_i) \in \mathfrak{g}_i$. But x can be uniquely written under the form (7.10), hence $x_i = \varphi(y_i)$ for each $i \in J_0$. In particular $x_0 \in \operatorname{Ran}\varphi = \mathfrak{I}$, hence $x_0 \in \mathfrak{I} \cap \operatorname{Ann}(\mathfrak{g}) =: \mathfrak{I}_0 \in H^*\operatorname{Id}(\operatorname{Ann}(\mathfrak{g}))$.

On the other hand,

$$\sum_{i\in J} \Bigl(\frac{1}{M_i}\Bigr)^2 \|x_i\|^2 = \sum_{i\in J} \Bigl(\frac{1}{M_i}\Bigr)^2 \|\varphi(y_i)\|^2 = \sum_{i\in J} \|y_i\|^2 \le \|y\|^2 < \infty.$$

Conversely, if $\{x_i\}_{i\in J}$ is a family of vectors in $\mathfrak{g}$ with $x_i \in \mathfrak{g}_i$ for each $i \in J$ and $\sum\limits_{i\in J} (1/M_i)^2\|x_i\|^2 < \infty$, we take $y_i := \varphi^{-1}(x_i) \in \mathfrak{X}_i$, and then $y_i \perp y_j$ whenever $i \neq j$. By the above computation we have $\sum\limits_{i\in J} \|y_i\|^2 < \infty$, hence $\sum\limits_{i\in J} y_i =: y'$ is convergent in $\mathfrak{X}$. Then $\sum\limits_{i\in J} x_i = \varphi(y) \in \operatorname{Ran}\varphi = \mathfrak{I}$.

6° Now assume that $\mathfrak{I}$ has the desired form for some $\mathfrak{I}_0 \in H^*\operatorname{Id}(\operatorname{Ann}(\mathfrak{g}))$, $J \subseteq I$ and $\{M_i\}_{i\in J}$ as indicated in the statement of the theorem. Let $\mathfrak{X}_0$ be a real H^*-algebra and $\varphi_0 \in \operatorname{Hom}^*(\mathfrak{X}_0, \mathfrak{g})$ with $\operatorname{Ker}\varphi_0 = \{0\}$ and $\operatorname{Ran}\varphi_0 = \mathfrak{I}_0$. We let

$$J_0 = J \sqcup \{0\}$$

and define

$$\mathfrak{X} = \Bigl\{x = (x_i)_{i\in J_0} \in \prod_{i\in J_0} \mathfrak{X}_i \mid \|x\|^2 := \|x_0\|^2 + \sum_{i\in J} (1/M_i)^2 \|x_i\|^2 < \infty\Bigr\}$$

and

$$\varphi\colon \mathfrak{X} \to \mathfrak{g}, \quad \varphi\bigl((x_i)_{i\in J_0}\bigr) = x_0 + \sum_{i\in J} x_i.$$

Note that the last sum is convergent in $\mathfrak{g}$ because $M := \sup\limits_{i\in J} M_i < \infty$. Also,

$$(\forall x \in \mathfrak{X}) \qquad \|\varphi(x)\|^2 = \|x_0\|^2 + \sum_{i\in J} \|x_i\|^2 \le \max(M^2, 1)\|x\|^2,$$

hence $\varphi \in \mathcal{B}(\mathfrak{X}, \mathfrak{g})$. It is easy to check that $\mathfrak{X}$ is in fact an H^*-algebra (with the multiplication and the involution defined pointwise), $\varphi \in \operatorname{Hom}^*(\mathfrak{X}, \mathfrak{g})$, and $\operatorname{Ran}\varphi = \mathfrak{I}$, hence $\mathfrak{I} \in H^*\operatorname{Id}(\mathfrak{g})$. ∎

7.5 Elementary properties of H*-ideals

We will develop in Chapter 8 a method to study the L^*-ideals of real involutive Banach-Lie algebras by means of equivariant monotone operators. Before constructing that machinery, it is worth knowing what properties of the L^*-ideals can be obtained by elementary means, that is to say, by arguments involving the L^*-ideals in a direct way. It turns out that the corresponding reasonings make no special use of the Lie algebra structure, and that is the reason why we formulate them in natural generality, for arbitrary algebras, thus preserving one moment more the "arbitrarily non-associative" flavor of the previous section.

PROPOSITION 7.35 *If $\mathfrak{g}$ is a real involutive Banach algebra and X^* is either of the symbols L^* and H^*, then for every $\mathfrak{I}_1, \mathfrak{I}_2 \in X^*\mathrm{Id}(\mathfrak{g})$ we have $\mathfrak{I}_1 \cap \mathfrak{I}_2 \in X^*\mathrm{Id}(\mathfrak{g})$.*

PROOF For $j = 1, 2$, let $\mathfrak{X}_j$ be a real X^*-algebra and $\varphi_j \in \mathrm{Hom}^*(\mathfrak{X}_j, \mathfrak{g})$ such that $\mathrm{Ker}\,\varphi_j = \{0\}$ and $\mathrm{Ran}\,\varphi_j = \mathfrak{I}_j$. Denote

$$\mathfrak{X} = \varphi_1^{-1}(\mathfrak{I}_1 \cap \mathfrak{I}_2) \quad (\subseteq \mathfrak{X}_1)$$

and define

$$\psi\colon \mathfrak{X} \to \mathfrak{X}_2, \quad \psi = \varphi_2^{-1} \circ (\varphi_1|_{\mathfrak{X}}).$$

Moreover, endow $\mathfrak{X}$ with the scalar product defined by

$$(\forall x, y \in \mathfrak{X}) \qquad (x \mid y)_{\mathfrak{X}} = (x \mid y)_{\mathfrak{X}_1} + (\psi(x) \mid \psi(y))_{\mathfrak{X}_2}.$$

Since both φ_1 and φ_2 are continuous, it follows that ψ is a closed operator. This implies that $\mathfrak{X}$ is a real Hilbert space with the above defined scalar product. Furthermore, since $\mathfrak{I}_1 \cap \mathfrak{I}_2$ is a self-adjoint subalgebra of $\mathfrak{g}$, it follows that $\mathfrak{X}$ $(= \varphi_1^{-1}(\mathfrak{I}_1 \cap \mathfrak{I}_2))$ is in turn a self-adjoint subalgebra of $\mathfrak{X}_2$.

Now note that for every $x, y, z \in \mathfrak{X}$ we have

$$\begin{aligned}
(xy \mid z)_{\mathfrak{X}} &= (xy \mid z)_{\mathfrak{X}_1} + (\psi(xy) \mid \psi(z))_{\mathfrak{X}_2} \\
&= (xy \mid z)_{\mathfrak{X}_1} + (\psi(x)\psi(y) \mid \psi(z))_{\mathfrak{X}_2} \\
&= (y \mid x^*z)_{\mathfrak{X}_1} + (\psi(y) \mid \psi(x)^*\psi(z))_{\mathfrak{X}_2} \\
&= (y \mid x^*z)_{\mathfrak{X}_1} + (\psi(y) \mid \psi(x^*z))_{\mathfrak{X}_2} \\
&= (y \mid x^*z)_{\mathfrak{X}}
\end{aligned}$$

and similarly $(xy \mid z)_{\mathfrak{X}} = (x \mid zy^*)_{\mathfrak{X}}$. Also,

$$\|x^*\|^2_{\mathfrak{X}} = \|x^*\|^2_{\mathfrak{X}_1} + \|\psi(x^*)\|^2_{\mathfrak{X}_2}$$

hence the involution of $\mathfrak{X}$ is continuous. Consequently, $\mathfrak{X}$ is a real H^*-algebra. (Recall Remark 7.9(a).) Moreover, the inclusion map $\eta\colon \mathfrak{X} \to \mathfrak{X}_1$ belongs to $\mathrm{Hom}^*(\mathfrak{X}, \mathfrak{X}_1)$. (Note that since η is injective, if $\mathfrak{X}_1$ is a Lie algebra, then $\mathfrak{X}$ is in turn a Lie algebra.)

Now $\varphi_1 \circ \eta \in \mathrm{Hom}^*(\mathfrak{X}, \mathfrak{g})$ and $\mathrm{Ran}\,(\varphi_1 \circ \eta) = \mathfrak{J}_1 \cap \mathfrak{J}_2 \trianglelefteq \mathfrak{g}$, hence the proof ends. ∎

REMARK 7.36 For later use, let us state the following by-product of the proof of Proposition 7.35: *Let $\mathfrak{g}$ be a real involutive Banach algebra, X^* be either of the symbols L^* and H^*, $\mathfrak{J} \in X^*\mathrm{Id}(\mathfrak{g})$, $\mathfrak{X}$ a real X^*-algebra and $\varphi \in \mathrm{Hom}^*(\mathfrak{X}, \mathfrak{g})$ with $\mathrm{Ker}\,\varphi = \{0\}$ and $\mathrm{Ran}\,\varphi = \mathfrak{J}$. Then for every $\mathfrak{J} \in X^*\mathrm{Id}(\mathfrak{g})$ we have $\varphi^{-1}(\mathfrak{J}) \in X^*\mathrm{Id}(\mathfrak{X})$.*

REMARK 7.37 The preceding Proposition 7.35 is the version in the present framework of the fact that the intersection of two operator ranges in a complex Hilbert space is again an operator range (see Corollary 2 to Theorem 2.2 in [FW71]).

THEOREM 7.38 *If $\mathfrak{g}$ is a topologically simple involutive real Banach algebra, then it has at most one non-zero H^*-ideal. If such an ideal exists, then it is dense in $\mathfrak{g}$.*

PROOF Assume that we have $\mathfrak{J}_1, \mathfrak{J}_2 \in H^*\mathrm{Id}(\mathfrak{g})$ with $\mathfrak{J}_1 \neq \{0\} \neq \mathfrak{J}_2$. For $j = 1, 2$, let $\mathfrak{X}_j$ be a real H^*-algebra and

$$\varphi_j \in \mathrm{Hom}^*(\mathfrak{X}_j, \mathfrak{g})$$

with $\mathrm{Ker}\,\varphi_j = \{0\}$ and $\mathrm{Ran}\,\varphi_j = \mathfrak{J}_j$. Since the closure of $\mathfrak{J}_j$ is a closed self-adjoint non-zero ideal of $\mathfrak{g}$, while $\mathfrak{g}$ is topologically simple, we must have $\mathfrak{J}_j$ dense in $\mathfrak{g}$ for $j = 1, 2$. Then, using once again the fact that $\mathfrak{g}$ is topologically simple, it follows by Remark 7.32 that each of the H^*-algebras $\mathfrak{X}_1$ and $\mathfrak{X}_2$ is topologically simple.

Now note that since both $\mathfrak{J}_1$ and $\mathfrak{J}_2$ are dense in $\mathfrak{g}$, we cannot have $\mathfrak{J}_1 \cap \mathfrak{J}_2 = \{0\}$. Indeed, if the latter relation holds, then $\mathfrak{J}_1 \cdot \mathfrak{J}_2 \subseteq \mathfrak{J}_1 \cap \mathfrak{J}_2 = \{0\}$, hence $\mathfrak{g} \cdot \mathfrak{g} = \{0\}$, thus contradicting the fact that $\mathfrak{g}$ is topologically simple. Consequently $\mathfrak{J}_1 \cap \mathfrak{J}_2 \neq \{0\}$, that is, $\varphi^{-1}(\mathfrak{J}_2) \neq \{0\}$. But $\varphi^{-1}(\mathfrak{J}_2) \in H^*\mathrm{Id}(\mathfrak{X}_1)$ by the above Remark 7.36, and we have seen that $\mathfrak{X}_1$ is topologically simple, hence Proposition 7.31 shows that we must have $\varphi^{-1}(\mathfrak{J}_2) = \mathfrak{X}_1$, that is, $\mathfrak{J}_1 \subseteq \mathfrak{J}_2$.

We can similarly show that $\mathfrak{J}_2 \subseteq \mathfrak{J}_1$, hence $\mathfrak{J}_1 = \mathfrak{J}_2$, and the proof ends. ∎

EXAMPLE 7.39 For an infinite-dimensional complex Hilbert space $\mathfrak{H}$ and $1 \leq p \leq \infty$, let $\mathfrak{u}_p$ be the canonically involutive Banach-Lie algebra of all skew-adjoint operators belonging to the Schatten ideal $\mathcal{C}_p(\mathfrak{H})$, where $\mathcal{C}_\infty(\mathfrak{H})$ stands for the set of all compact operators on $\mathfrak{H}$.

For $p \geq 2$ we have

$$\mathfrak{u}_2 \subseteq \mathfrak{u}_p,$$

and it follows by the above Theorem 7.38 and Remark 7.13 that $\mathfrak{u}_2$ is the only non-zero L^*-ideal of $\mathfrak{u}_p$. That is,

$$L^*\mathrm{Id}(\mathfrak{u}_p) = \{\{0\}, \mathfrak{u}_2\} \text{ whenever } 2 \leq p \leq \infty.$$

Recall that the real Banach-Lie algebra $\mathfrak{u}_p$ is topologically simple for $1 < p \leq \infty$. (In this connection, see e.g., Proposition II.8 at page 92 in [dlH72].)

See also Example 8.12 and Example 8.23 for further discussion on the above Example 7.39.

Notes

The equivariant monotone operators were introduced in [Be03]; compare also hypotheses 2° and 5° in Section 2 of [Be05a].

A good survey of the theory of H^*-algebras can be found in Section 1 of [CMMR94]. The L^*-algebras were introduced in the papers [Sch60] and [Sch61]. For the classification of L^*-algebras see e.g., [Ba72] and [CGM90]. See also the book [dlH72] for more information on the role of L^*-algebras in operator theory.

A version of the preceding Proposition 7.23 where $\mathfrak{Y}$ is a complex Hilbert space can be found in Theorem 3.1 in [GG04]. (See also Proposition 10 on page 154 in [Scw64].) We note that in the case when $\mathfrak{Y}$ has a structure of complex Banach space, the cone $\mathcal{C}^+(\mathfrak{Y})$ clearly contains the cone $\mathcal{L}^+(\mathfrak{Y})$ of non-negative reproducing kernels on $\mathfrak{Y}$ (see page 141 in [Scw64]).

It is also worth mentioning that a similar structure occurs in the theory of abstract Wiener spaces, with the additional assumption that the operator φ is absolutely 2-summable. (See e.g., either §1 in Chapter 1 in [BlDa90], or [Kuo75], or page 3 in [dlH72].)

Proposition 7.26 is suggested by Theorem 2.1 in [FW71] (see also Proposition 13 on page 160 in [Scw64]), while Proposition 7.28 is a version of Theorem 2.2 in [FW71] (see also Proposition 12 on page 160 in [Scw64]).

The characterization of the H^*-ideals of an H^*-algebra, as given in the above Theorem 7.34, is similar to the one of the operator ranges in a complex Hilbert space given by condition (5) in Theorem 1.1 in [FW71]. The reasoning used in stage 3° of the proof of Theorem 7.34 is classical; see e.g., the proof of Proposition 2.9 in [Va72]. The preceding Lemma 7.16 is a version of

Corollary 3.5 in [CR85] for real H^*-algebras. A similar comment can be made on the preceding Proposition 7.31 versus Theorem 3.1 in [CR85].

Chapter 8

L^*-ideals and Equivariant Monotone Operators

Abstract. In this chapter we investigate the close connection that exists between the equivariant monotone operators and the L^*-ideals of a real involutive Banach-Lie algebra. Actually this is a special instance of the correspondence between abstract reproducing kernels and reproducing kernel Hilbert spaces. In the present setting, it is fairly easy to show that if the reproducing kernel Hilbert space is actually an L^*-ideal of a Banach-Lie algebra, then the corresponding abstract reproducing kernel is an equivariant monotone operator. Conversely, it is much harder to build the structure of L^*-algebra out of an equivariant monotone operator. After accomplishing this task, we proceed to a systematic investigation of L^*-ideals. We thus single out the class of adequate algebras, which are involutive Banach-Lie algebras whose L^*-ideals all come from equivariant monotone operators. We then show that L^*-ideals allow us to construct Hilbert space representations of automorphism groups of topologically simple Banach-Lie algebras. The chapter concludes by an application of L^*-ideals to enlargibility questions.

8.1 From ideals to operators

In the following we use the notation $\mathcal{C}^+(\mathfrak{g})$ from Definition 7.1. The following result shows that, to an L^*-ideal of a real involutive Banach-Lie algebra, one can "often" associate in a natural way an equivariant monotone operator.

PROPOSITION 8.1 *Let $\mathfrak{g}$ be a real involutive Banach-Lie algebra. Consider a real L^*-algebra $\mathfrak{X}$ and $\varphi \in \mathrm{Hom}^*(\mathfrak{X}, \mathfrak{g})$ with* $\mathrm{Ker}\,\varphi = \{0\}$ *and*

$$\mathfrak{J} := \mathrm{Ran}\,\varphi \trianglelefteq \mathfrak{g},$$

and assume that one of the following conditions holds:

(i) *the center of* $\mathfrak{J}$ *reduces to* $\{0\}$*;*

(ii) $\mathfrak{J}$ *is dense in* $\mathfrak{g}$ *and the involution of* $\mathfrak{X}$ *is isometric; or*

(iii) $\mathfrak{J}$ *is contained in the center of* $\mathfrak{g}$ *and the involution of* $\mathfrak{X}$ *is isometric.*

Then for the operator

$$\iota := \varphi \circ \varphi^{\#} : \mathfrak{g}^{\#} \to \mathfrak{g}$$

we have $\iota \in \mathcal{C}^{+}(\mathfrak{g})$.

PROOF According to Proposition 7.23, we have $\iota^{\#} = \iota$ and $\langle f, \iota(f) \rangle \geq 0$ for all $f \in \mathfrak{g}^{\#}$. In view of Definition 7.1, what remains to be proved is that if one of the conditions (i)–(iii) holds, then

$$\iota \in \mathrm{Hom}^{*}_{\mathfrak{g}}(\mathfrak{g}^{\#}, \mathfrak{g}). \tag{8.1}$$

First note that since $\varphi \in \mathrm{Hom}^{*}(\mathfrak{X}, \mathfrak{g})$, we have for every $f \in \mathfrak{g}^{\#}$ and $x \in \mathfrak{X}$,

$$(\varphi^{\#}(f^{*}) \mid x)_{\mathfrak{X}} = \langle f^{*}, \varphi(x) \rangle = \langle f, \varphi(x)^{*} \rangle = \langle f, \varphi(x^{*}) \rangle = (\varphi^{\#}(f) \mid x^{*})_{\mathfrak{X}} \tag{8.2}$$

Then for all $g \in \mathfrak{g}^{\#}$,

$$\langle g, \iota(f^{*}) \rangle = \langle g, \varphi(\varphi^{\#}(f^{*})) \rangle = (\varphi^{\#}(g) \mid \varphi^{\#}(f^{*}))_{\mathfrak{X}} = ((\varphi^{\#}(g))^{*} \mid \varphi^{\#}(f))_{\mathfrak{X}}$$

where the latter equality follows by (8.2). On the other hand,

$$\langle g, \iota(f)^{*} \rangle = \langle g, \varphi(\varphi^{\#}(f))^{*} \rangle = (\varphi^{\#}(g) \mid (\varphi^{\#}(f))^{*})_{\mathcal{X}}.$$

Since the involution of $\mathfrak{X}$ is in all cases isometric (in the case (i), this follows by Remark 7.9(b)), we get $\langle g, \iota(f^{*}) \rangle = \langle g, \iota(f)^{*} \rangle$ for all $f, g \in \mathfrak{g}^{\#}$. The Hahn-Banach theorem then implies that

$$(\forall f \in \mathfrak{g}^{\#}) \quad \iota(f^{*}) = \iota(f)^{*}.$$

Consequently, to complete the proof of (8.1), it remains to check that for each $a \in \mathfrak{g}$ we have

$$(\mathrm{ad}_{\mathfrak{g}} a) \circ \iota = \iota \circ (\mathrm{ad}_{\mathfrak{g}} a^{*})^{\#}. \tag{8.3}$$

To this end, first use Lemma A.16 to deduce that there exists a homomorphism of Banach-Lie algebras $\rho \colon \mathfrak{g} \to \mathrm{Der}(\mathfrak{X})$ such that

$$(\forall a \in \mathfrak{g}) \quad (\mathrm{ad}_{\mathfrak{g}} a) \circ \varphi = \varphi \circ \rho(a). \tag{8.4}$$

Consequently, for all $a \in \mathfrak{g}$ we have $\rho(a^{*})^{\#} \circ \varphi^{\#} = \varphi^{\#} \circ (\mathrm{ad}_{\mathfrak{g}} a^{*})^{\#}$, hence

$$\iota \circ (\mathrm{ad}_{\mathfrak{g}} a^{*})^{\#} = \varphi \circ \varphi^{\#} \circ (\mathrm{ad}_{\mathfrak{g}} a^{*})^{\#} = \varphi \circ \rho(a^{*})^{\#} \circ \varphi^{\#}. \tag{8.5}$$

On the other hand, we have by (8.4)

$$(\mathrm{ad}_{\mathfrak{g}} a) \circ \iota = (\mathrm{ad}_{\mathfrak{g}} a) \circ \varphi \circ \varphi^{\#} = \varphi \circ \rho(a) \circ \varphi^{\#}. \tag{8.6}$$

Thus, relation (8.3) will follow by (8.5) and (8.6) as soon as we prove that

$$(\forall a \in \mathfrak{g}) \qquad \rho(a)^{\#} = \rho(a^*). \tag{8.7}$$

To prove this fact, first note that for every $a \in \mathfrak{g}$ and $x \in \mathfrak{X}$ we have

$$\begin{aligned}\varphi(\rho(a)^* x) &= \varphi((\rho(a)x^*)^*) = (\varphi(\rho(a)x^*))^* = ((\mathrm{ad}_{\mathfrak{g}} a)(\varphi(x^*)))^* \\ &= ((\mathrm{ad}_{\mathfrak{g}} a)(\varphi(x)^*))^* = [a, \varphi(x)^*]^* = -[a^*, \varphi(x)] \\ &= -((\mathrm{ad}_{\mathfrak{g}} a^*) \circ \varphi)(x) = -\varphi(\rho(a^*)x)\end{aligned}$$

where both the third and the last equalities follow by (8.4). Since $\operatorname{Ker}\varphi = \{0\}$, we get

$$(\forall a \in \mathfrak{g}) \quad \rho(a)^* = -\rho(a^*) \tag{8.8}$$

(see Remark 7.2(b) for the way $\rho(a)^*$ is defined).

Now let us consider the conditions (i), (ii), and (iii) separately.

(i) If the center of $\mathfrak{J}$ is $\{0\}$ then, since $\operatorname{Ker}\varphi = \{0\}$, the center of $\mathfrak{X}$ must be $\{0\}$. Lemma 7.14 then implies that for each $a \in \mathfrak{g}$ we have $\rho(a)^{\#} = -\rho(a)^*$ which, by (8.8), implies (8.7).

(ii) Now assume that $\mathfrak{J}$ is dense in $\mathfrak{g}$ and the involution of $\mathfrak{X}$ is isometric. Note that for all $x \in \mathfrak{X}$ we have $\varphi \circ (\mathrm{ad}_{\mathfrak{X}} x) = (\mathrm{ad}_{\mathfrak{g}}(\varphi(x))) \circ \varphi$, because $\varphi\colon \mathfrak{X} \to \mathfrak{g}$ is a Lie algebra homomorphism. Then (8.4) implies that $\varphi \circ (\mathrm{ad}_{\mathfrak{X}} x) = \varphi \circ \rho(\varphi(x))$ for all $x \in \mathfrak{X}$. Since $\operatorname{Ker}\varphi = \{0\}$, we deduce that

$$(\forall x \in \mathfrak{X}) \qquad \rho(\varphi(x)) = \mathrm{ad}_{\mathfrak{X}} x. \tag{8.9}$$

On the other hand, since $\mathfrak{X}$ is an L^*-algebra, we have

$$(\mathrm{ad}_{\mathfrak{X}} x)^{\#} = \mathrm{ad}_{\mathfrak{X}} x^*$$

for all $x \in \mathfrak{X}$. But $\varphi \in \operatorname{Hom}^*(\mathfrak{X}, \mathfrak{g})$, hence it follows by (8.9) that for $a = \varphi(x) \in \mathfrak{J}$ we have $\rho(a)^{\#} = \rho(a^*)$. Since $\mathfrak{J}$ is dense in $\mathfrak{g}$, we deduce that $\rho(a)^{\#} = \rho(a^*)$ for all $a \in \mathfrak{g}$, which is just (8.7).

(iii) If $\mathfrak{J}$ is contained in the center of $\mathfrak{g}$, then $\rho(a) = 0$ for each $a \in \mathfrak{g}$, hence (8.7) holds trivially. ∎

REMARK 8.2 For later use, let us state the following by-product of the proof of Proposition 8.1 (see (8.7)). *Let $\mathfrak{g}$ be a real involutive Banach-Lie algebra, $\mathfrak{X}$ a real L^*-algebra, and $\varphi \in \operatorname{Hom}^*(\mathfrak{X}, \mathfrak{g})$ such that $\operatorname{Ker}\varphi = \{0\}$, $\operatorname{Ran}\varphi \trianglelefteq \mathfrak{g}$, and either the center of $\mathfrak{X}$ reduces to $\{0\}$ or $\operatorname{Ran}\varphi$ is dense in $\mathfrak{g}$. Consider the representation $\rho\colon \mathfrak{g} \to \operatorname{Der}(\mathfrak{X})$ such that $(\mathrm{ad}_{\mathfrak{g}} a) \circ \varphi = \varphi \circ \rho(a)$ for all $a \in \mathfrak{g}$. Then we have*

$$\rho(a)^{\#} = \rho(a^*)$$

for all $a \in \mathfrak{g}$. (Note that the involution of $\mathfrak{X}$ is isometric according to Remark 7.9(b).)

8.2 From operators to ideals

We now reverse what we have previously done. More specifically, we show how one can use the equivariant monotone operators in order to construct L^*-ideals. The main result in this connection is contained in Theorem 8.8.

NOTATION 8.3 Until the end of the present section, $\mathfrak{g}$ stands for a real involutive Banach-Lie algebra, and $\mathfrak{A}$ denotes the complex Banach-Lie algebra which is the complexification of $\mathfrak{g}$. Then $\mathfrak{A}$ comes with two involutions: one of them, denoted by

$$a \mapsto \bar{a},$$

is the unique involution of $\mathfrak{A}$ whose set of fixed points is just $\mathfrak{g}$; this involution is isometric according to the convention of Notation A.11 concerning the norm of the complexification. The other involution of $\mathfrak{A}$, denoted by

$$a \mapsto a^*,$$

is the unique extension of the involution of $\mathfrak{g}$ to an involution of the complex Banach-Lie algebra $\mathfrak{A}$. When we say that $\mathfrak{A}$ is a complex involutive Banach-Lie algebra, we think of $\mathfrak{A}$ endowed with the involution $a \mapsto a^*$.

We always denote by the same symbol both a linear map between real vector spaces and its complex-linear extension between the complexified spaces.

We pick

$$\iota \in \mathcal{C}^+(\mathfrak{g})$$

and use the notation in the stage 2° of the proof of Proposition 7.23 (with $\mathfrak{Y}$ replaced by $\mathfrak{g}$). More precisely, there exists on $\mathfrak{g}^\#/\mathrm{Ker}\,\iota$ a real scalar product defined by

$$(\forall f, g \in \mathfrak{g}^\#) \qquad \big(\pi(f) \mid \pi(g)\big) = \langle f, \iota(g)\rangle,$$

where $\pi\colon \mathfrak{g}^\# \to \mathfrak{g}^\#/\mathrm{Ker}\,\iota$ is the natural projection. Then $\mathfrak{X}$ stands for the real Hilbert space obtained as the completion of $\mathfrak{g}^\#/\mathrm{Ker}\,\iota$ with respect to that scalar product, and $\eta\colon \mathfrak{g}^\#/\mathrm{Ker}\,\iota \to \mathfrak{X}$ denotes the natural inclusion map. Moreover, ι induces an operator $\varphi \in \mathcal{B}(\mathfrak{X}, \mathfrak{g})$ such that $\mathrm{Ker}\,\varphi = \{0\}$, $\varphi^\# = \eta \circ \pi$ and $\iota = \varphi \circ \varphi^\#$.

A symbol which did *not* appear in the proof of Proposition 7.23, and which we introduce just now, is

$$\mathfrak{L},$$

by which we denote the complex Hilbert space obtained by complexifying $\mathfrak{X}$.

We now prove a key fact.

LEMMA 8.4 *Let $\alpha \in \mathcal{B}(\mathfrak{g})$ such that for some $\varepsilon \in \{-1, 1\}$ we have*

$$\alpha \circ \iota = \varepsilon \cdot \iota \circ \alpha^{\#}. \tag{8.10}$$

Then there exists a unique operator $\rho_\alpha \in \mathcal{B}(\mathfrak{X})$ such that

$$\rho_\alpha^{\#} = \varepsilon \cdot \rho_\alpha, \quad \|\rho_\alpha\| \leq \|\alpha\|$$

and the diagram

$$\begin{array}{ccc} \mathfrak{g}^{\#} & \xrightarrow{\eta \circ \pi} & \mathfrak{X} \\ {\scriptstyle \alpha^{\#}} \downarrow & & \downarrow {\scriptstyle \rho_\alpha} \\ \mathfrak{g}^{\#} & \xrightarrow{\eta \circ \pi} & \mathfrak{X} \end{array}$$

is commutative.

PROOF First note that the uniqueness of a continuous operator ρ_α which makes the above diagram commutative follows by the fact that $\operatorname{Ran} \eta$ is dense in $\mathfrak{X}$. Next, by repeatedly making use of Schwartz' inequality (see e.g., Remark 7.3) and of (8.10), we get for each $f \in \mathfrak{g}^{\#}$

$$\begin{aligned} &|\langle \alpha^{\#}(f), \iota(\alpha^{\#}(f))\rangle| \\ &\quad = |\langle f, \iota((\alpha^2)^{\#}(f))\rangle| \\ &\quad \leq |\langle f, \iota(f)\rangle|^{1/2} \cdot |\langle (\alpha^2)^{\#}(f), \iota((\alpha^2)^{\#}(f))\rangle|^{1/2} \\ &\quad = |\langle f, \iota(f)\rangle|^{1/2} \cdot |\langle f, \iota((\alpha^4)^{\#}(f))\rangle|^{1/2} \\ &\quad \leq |\langle f, \iota(f)\rangle|^{(1/2)+(1/4)} \cdot |\langle (\alpha^4)^{\#}(f), \iota((\alpha^4)^{\#}(f))\rangle|^{1/4} \\ &\quad \cdots\cdots\cdots\cdots\cdots\cdots\cdots\cdots\cdots\cdots\cdots\cdots \\ &\quad \leq |\langle f, \iota(f)\rangle|^{(1/2)+(1/4)+\cdots+(1/2^{n-1})} \cdot |\langle f, \iota((\alpha^{2^n})^{\#}(f))\rangle|^{1/2^{n-1}}, \end{aligned}$$

whence

$$|\langle \alpha^{\#}(f), \iota(\alpha^{\#}(f))\rangle| \leq |\langle f, \iota(f)\rangle|^{\sum_{j=1}^{n-1} 1/2^j} \cdot \|f\|^{1/2^{n-2}} \cdot \|\iota\|^{1/2^{n-1}} \cdot \|\alpha\|^2$$

for every $n \geq 2$. Letting $n \to \infty$, we get

$$(\forall f \in \mathfrak{g}^{\#}) \qquad |\langle \alpha^{\#}(f), \iota(\alpha^{\#}(f))\rangle| \leq \|\alpha\|^2 \cdot |\langle f, \iota(f)\rangle|. \tag{8.11}$$

On the other hand, it follows by (8.10) that $\operatorname{Ker} \iota$ is invariant to $\alpha^{\#}$, and thus there exists $\rho_\alpha^0 \in \mathcal{B}(\mathfrak{g}^{\#}/\operatorname{Ker} \iota)$ with $\rho_\alpha^0 \circ \pi = \pi \circ \alpha^{\#}$. Then (8.11) reads

$$(\forall f \in \mathfrak{g}^{\#}) \qquad \|\rho_\alpha^0(\pi(f))\|_{\mathfrak{X}}^2 \leq \|\alpha\|^2 \cdot \|\pi(f)\|_{\mathfrak{X}}^2$$

(see the above Notation 8.3), hence ρ_α^0 extends to a bounded linear operator on $\mathfrak{X}$. More precisely, there exists $\rho_\alpha \in \mathcal{B}(\mathfrak{X})$ with $\|\rho_\alpha\| \leq \|\alpha\|$ and $\rho_\alpha \circ \eta = \eta \circ \rho_\alpha^0$, whence

$$\rho_\alpha \circ (\eta \circ \pi) = \eta \circ \rho_\alpha^0 \circ \pi = (\eta \circ \pi) \circ \alpha^{\#}.$$

It only remained to check that

$$\rho_\alpha^{\#} = \varepsilon \rho_\alpha.$$

Since ρ_α is bounded on $\mathfrak{X}$, it suffices to show that for all $x_1, x_2 \in \operatorname{Ran}(\eta \circ \pi)$ we have

$$(\rho_\alpha(x_1) \mid x_2) = \varepsilon(x_1 \mid \rho_\alpha(x_2)).$$

Let $f_1, f_2 \in \mathfrak{g}^{\#}$ such that $x_j = \eta(\pi(f_j))$ for $j = 1, 2$. Then

$$\begin{aligned}(\rho_\alpha(x_1) \mid x_2) &= \langle \alpha^{\#}(f_1), \iota(f_2)\rangle = \langle f_1, \alpha(\iota(f_2))\rangle = \varepsilon\langle f_1, \iota(\alpha^{\#}(f_2))\rangle \\ &= \varepsilon(x_1 \mid \rho_\alpha(x_2))\end{aligned}$$

where the third equality follows by (8.10). The proof is finished. ∎

REMARK 8.5 In the framework of the above Lemma 8.4, we also have a commutative diagram

$$\begin{array}{ccc} \mathfrak{A}^{\#} & \xrightarrow{\eta\circ\pi} & \mathfrak{L} \\ {\scriptstyle \alpha^{\#}}\downarrow & & \downarrow{\scriptstyle \rho_\alpha} \\ \mathfrak{A}^{\#} & \xrightarrow{\eta\circ\pi} & \mathfrak{L} \end{array}$$

of the complexified spaces.

PROPOSITION 8.6 *There exists a uniquely determined, continuous $*$-representation $\rho\colon \mathfrak{g} \to \mathcal{B}(\mathfrak{X})$ such that, for every $a \in \mathfrak{g}$, the diagram*

$$\begin{array}{ccc} \mathfrak{X} & \xrightarrow{\varphi} & \mathfrak{g} \\ {\scriptstyle \rho(a)}\downarrow & & \downarrow{\scriptstyle \operatorname{ad}_{\mathfrak{g}} a} \\ \mathfrak{X} & \xrightarrow{\varphi} & \mathfrak{g} \end{array}$$

is commutative.

PROOF First note that each $a \in \mathfrak{g}$ can be uniquely written as

$$a = a_+ + a_- \quad \text{with} \quad a_\pm^* = \pm a_\pm$$

(namely $a_\pm = (a \pm a^*)/2$) and

$$\|a_\pm\| \le M\|a\|$$

with some $M > 0$ independent on a, because the involution of $\mathfrak{g}$ is continuous. Since $\iota \in \operatorname{Hom}_{\mathfrak{g}}^*(\mathfrak{g}^{\#}, \mathfrak{g})$, we then have

$$(\operatorname{ad}_{\mathfrak{g}} a_\pm) \circ \iota = \iota \circ (\operatorname{ad}_{\mathfrak{g}} a_\pm^*)^{\#} = \pm \iota \circ (\operatorname{ad}_{\mathfrak{g}} a_\pm)^{\#}$$

and then Lemma 8.4 shows that there exist $\rho_{\mathrm{ad}_{\mathfrak{g}} a_\pm} \in \mathcal{B}(\mathfrak{X})$ with

$$\rho_{\mathrm{ad}_{\mathfrak{g}} a_\pm} \circ (\eta \circ \pi) = (\eta \circ \pi) \circ (\mathrm{ad}_{\mathfrak{g}} a_\pm)^{\#}$$

and

$$(\rho_{\mathrm{ad}_{\mathfrak{g}} a_\pm})^{\#} = \pm \rho_{\mathrm{ad}_{\mathfrak{g}} a_\pm}.$$

We define

$$\rho(a) = \rho_{\mathrm{ad}_{\mathfrak{g}} a_+} + \rho_{\mathrm{ad}_{\mathfrak{g}} a_-} \qquad (\in \mathcal{B}(\mathfrak{X})).$$

Since $a^* = a_+ - a_-$, it follows that

$$\rho(a^*) = \rho(a)^{\#}. \tag{8.12}$$

Also, according to Lemma 8.4, we have

$$\|\rho(a)\| \leq \|\rho_{\mathrm{ad}_{\mathfrak{g}} a_+}\| + \|\rho_{\mathrm{ad}_{\mathfrak{g}} a_-}\| \leq \|\mathrm{ad}_{\mathfrak{g}} a_+\| + \|\mathrm{ad}_{\mathfrak{g}} a_-\| \leq \|\mathrm{ad}_{\mathfrak{g}}\|(\|a_+\| + \|a_-\|),$$

whence

$$\|\rho(a)\| \leq 2M\|a\|. \tag{8.13}$$

Next note that

$$\begin{aligned}\varphi \circ \rho(a) \circ (\eta \circ \pi) &= \varphi \circ (\eta \circ \pi) \circ (\mathrm{ad}_{\mathfrak{g}} a^*)^{\#} = \varphi \circ \varphi^{\#} \circ (\mathrm{ad}_{\mathfrak{g}} a^*)^{\#} \\ &= \iota \circ (\mathrm{ad}_{\mathfrak{g}} a^*)^{\#} = (\mathrm{ad}_{\mathfrak{g}} a) \circ \iota = (\mathrm{ad}_{\mathfrak{g}} a) \circ \varphi \circ (\eta \circ \pi).\end{aligned}$$

Since $\mathrm{Ran}\,(\eta \circ \pi)$ is dense in $\mathfrak{X}$, it follows that

$$\varphi \circ \rho(a) = (\mathrm{ad}_{\mathfrak{g}} a) \circ \varphi. \tag{8.14}$$

Now, since the map $\rho\colon \mathfrak{g} \to \mathcal{B}(\mathfrak{X})$ is clearly linear, it follows by (8.12)–(8.14) that in order to prove that ρ is a bounded $*$-representation, it only remains to check that for all $a, b \in \mathfrak{g}$ we have

$$\rho([a,b]) = [\rho(a), \rho(b)]. \tag{8.15}$$

To this end, note that (8.14) (or Lemma 8.4) implies that

$$(\forall a \in \mathfrak{g}) \qquad \rho(a)^{\#} \circ (\eta \circ \pi) = (\eta \circ \pi) \circ (\mathrm{ad}_{\mathfrak{g}} a)^{\#}.$$

It then follows by (8.12) that

$$(\forall a \in \mathfrak{g}) \qquad \rho(a) \circ (\eta \circ \pi) = (\eta \circ \pi) \circ (\mathrm{ad}_{\mathfrak{g}} a^*)^{\#}.$$

Since

$$\mathfrak{g} \to \mathcal{B}(\mathfrak{g}^{\#}), \quad a \mapsto (\mathrm{ad}_{\mathfrak{g}} a^*)^{\#},$$

is a representation, it then follows that equality (8.15) holds on $\mathrm{Ran}\,(\eta \circ \pi)$. But the latter space is dense in $\mathfrak{X}$, hence (8.15) is completely proved.

Finally, the uniqueness of ρ follows by the fact that $\mathrm{Ker}\,\rho = \{0\}$. ∎

REMARK 8.7 It follows by the above Proposition 8.6 that there exists a bounded $*$-representation

$$\rho\colon \mathfrak{A} \to \mathcal{B}(\mathfrak{L})$$

such that for every $a \in \mathfrak{A}$ the diagram of the complexified spaces

$$\begin{array}{ccc} \mathfrak{L} & \xrightarrow{\varphi} & \mathfrak{A} \\ {\scriptstyle \rho(a)}\downarrow & & \downarrow{\scriptstyle \mathrm{ad}_{\mathfrak{A}} a} \\ \mathfrak{L} & \xrightarrow{\varphi} & \mathfrak{A} \end{array}$$

is commutative.

We now come to one of the main results of the present chapter.

THEOREM 8.8 *Assume that $\rho\colon \mathfrak{g} \to \mathcal{B}(\mathfrak{X})$ is the bounded $*$-representation from* Proposition 8.6. *Then the real Hilbert space $\mathfrak{X}$ has a unique structure of real L^*-algebra with isometric involution such that $\varphi \in \mathrm{Hom}^*(\mathfrak{X}, \mathfrak{g})$. The corresponding bracket on $\mathfrak{X}$ has the property*

$$(\forall x_1, x_2 \in \mathfrak{X}) \qquad [x_1, x_2] = \big(\rho(\varphi(x_1))\big)(x_2). \tag{8.16}$$

Moreover, $\mathrm{Ran}\,\varphi \trianglelefteq \mathfrak{g}$ *and* $\mathrm{Ran}\,\rho \subseteq \mathrm{Der}(\mathfrak{X})$.

PROOF The proof has several stages.

1° By Proposition 8.6, we have $(\mathrm{ad}_{\mathfrak{g}} a) \circ \varphi = \varphi \circ \rho(a)$ for every $a \in \mathfrak{g}$, which easily implies that $\mathrm{Ran}\,\varphi \trianglelefteq \mathfrak{g}$, and then

$$[\mathrm{Ran}\,\varphi, \mathrm{Ran}\,\varphi] \subseteq \mathrm{Ran}\,\varphi,$$

hence we can define a unique Lie algebra bracket on $\mathfrak{X}$ such that $\varphi\colon \mathfrak{X} \to \mathfrak{g}$ is a Lie algebra homomorphism, namely

$$(\forall x_1, x_2 \in \mathfrak{X}) \qquad [x_1, x_2] = \varphi^{-1}([\varphi(x_1), \varphi(x_2)]).$$

Then

$$\varphi([x_1, x_2]) = [\varphi(x_1), \varphi(x_2)] = \big((\mathrm{ad}_{\mathfrak{g}}(\varphi(x_1))) \circ \varphi\big)(x_2) = \big(\varphi \circ \rho(\varphi(x_1))\big)(x_2),$$

where the latter equality follows by the commutative diagram of Proposition 8.6. Since $\mathrm{Ker}\,\varphi = \{0\}$, relation (8.16) follows.

Now (8.16) implies that for all $x_1, x_2 \in \mathfrak{X}$ we have by the continuity of $\rho\colon \mathfrak{g} \to \mathcal{B}(\mathfrak{X})$ that

$$\|[x_1, x_2]\| \le \|\rho\| \cdot \|\varphi\| \cdot \|x_1\| \cdot \|x_2\|,$$

hence the above defined bracket on $\mathfrak{X}$ is continuous with respect to the norm of $\mathfrak{X}$.

2° We now define the involution of $\mathfrak{X}$. To this end, first note that $\operatorname{Ker}\iota$ is invariant to the map

$$\mathfrak{g}^{\#} \to \mathfrak{g}^{\#}, \quad f \mapsto f^*.$$

Hence this map induces a (similarly denoted) map on $\mathfrak{g}^{\#}/\operatorname{Ker}\iota$ such that $\pi(f^*) = \pi(f)^*$ for all $f \in \mathfrak{g}^{\#}$. (Recall that

$$\pi \colon \mathfrak{g}^{\#} \to \mathfrak{g}^{\#}/\operatorname{Ker}\iota$$

is the natural projection map.) Moreover, for every $f \in \mathfrak{g}^{\#}$ we have

$$\|\eta(\pi(f)^*)\|^2_{\mathfrak{X}} = \|\eta(\pi(f^*))\|^2_{\mathfrak{X}} = \langle f^*, \iota(f^*)\rangle = \langle f, \iota(f)\rangle = \|\eta(\pi(f))\|^2_{\mathfrak{X}}.$$

Hence the involution on $\mathfrak{g}^{\#}/\operatorname{Ker}\iota$ extends by continuity to an isometric involutive map on $\mathfrak{X}$. More precisely, there exists a unique isometric involutive linear map $x \mapsto x^*$ on $\mathfrak{X}$ such that $\eta(v^*) = \eta(v)^*$ for all $v \in \mathfrak{g}^{\#}/\operatorname{Ker}\iota$. Now for every $f \in \mathfrak{g}^{\#}$ we have

$$\varphi(\eta(\pi(f))^*) = \varphi(\eta(\pi(f^*))) = \iota(f^*) = \iota(f)^* = \varphi(\eta(\pi(f)))^*,$$

hence

$$\varphi(x^*) = \varphi(x)^* \quad \text{for all} \quad x \in \mathfrak{X}.$$

According to what we got at stage 1°, we then have

$$\begin{aligned}\varphi([x_1, x_2]^*) &= \varphi([x_1, x_2])^* = [\varphi(x_1), \varphi(x_2)]^* = [\varphi(x_2)^*, \varphi(x_1)^*] \\ &= [\varphi(x_2^*), \varphi(x_1^*)] = \varphi([x_2^*, x_1^*]),\end{aligned}$$

and thus $[x_1, x_2]^* = [x_2^*, x_1^*]$ (since $\operatorname{Ker}\varphi = \{0\}$) for all $x_1, x_2 \in \mathfrak{X}$. Consequently, $\mathfrak{X}$ is an involutive Banach-Lie algebra and $\varphi \in \operatorname{Hom}^*(\mathfrak{X}, \mathfrak{g})$.

3° We prove at this stage that $\mathfrak{X}$ is actually an L^*-algebra. Since we have already seen that the bracket of $\mathfrak{X}$ is continuous and $\operatorname{Ran}(\eta \circ \pi)$ is dense in $\mathfrak{X}$, it suffices to check that for all $f, g, h \in \mathfrak{g}^{\#}$ we have

$$\big([(\eta\circ\pi)(f), (\eta\circ\pi)(g)] \mid (\eta\circ\pi)(h)\big)_{\mathfrak{X}} = \big((\eta\circ\pi)(g) \mid [(\eta\circ\pi)(f)^*, (\eta\circ\pi)(h)]\big)_{\mathfrak{X}}. \tag{8.17}$$

In fact, since $\varphi^{\#} = \eta\circ\pi$ and $\iota = \varphi\circ\varphi^{\#} = \varphi\circ\eta\circ\pi$, we have

$$\begin{aligned}\big([(\eta\circ\pi)(f), (\eta\circ\pi)(g)] \mid (\eta\circ\pi)(h)\big)_{\mathfrak{X}} &= \big([(\eta\circ\pi)(f), (\eta\circ\pi)(g)] \mid \varphi^{\#}(h)\big)_{\mathfrak{X}} \\ &= \langle h, \varphi([(\eta\circ\pi)(f), (\eta\circ\pi)(g)])\rangle \\ &= \langle h, [(\varphi\circ\eta\circ\pi)(f), (\varphi\circ\eta\circ\pi)(g)]\rangle \\ &= \langle h, [\iota(f), \iota(g)]\rangle \\ &= \langle h, ((\operatorname{ad}_{\mathfrak{g}}(\iota(f)))\circ\iota)(g)\rangle \\ &= \langle h, (\iota\circ(\operatorname{ad}_{\mathfrak{g}}(\iota(f))^*)^{\#})(g)\rangle,\end{aligned}$$

where the latter equality follows, since $\iota \in \mathrm{Hom}^*_{\mathfrak{g}}(\mathfrak{g}^\#, \mathfrak{g})$. Thence, using the fact that $\iota = \iota^\#$, we get

$$\begin{aligned}
([(\eta\circ\pi)(f), (\eta\circ\pi)(g)] \mid (\eta\circ\pi)(h))_{\mathfrak{X}} &= \langle (\mathrm{ad}_{\mathfrak{g}}(\iota(f))^*)^\#(g), \iota(h)\rangle \\
&= \langle g, ((\mathrm{ad}_{\mathfrak{g}}(\iota(f))^*)\circ\iota)(h)\rangle \\
&= \langle g, [\iota(f^*), \iota(h)]\rangle \\
&= \langle g, \varphi([(\eta\circ\pi)(f)^*, (\eta\circ\pi)(h)])\rangle \\
&= (\varphi^\#(g) \mid [(\eta\circ\pi)(f)^*, (\eta\circ\pi)(h)])_{\mathfrak{X}}
\end{aligned}$$

where the next-to-last equality follows, since $\varphi\circ\eta\circ\pi = \iota$. Thus (8.17) is proved.

4° Now take $a \in \mathfrak{g}$ arbitrary. To prove that $\rho(a) \in \mathrm{Der}(\mathfrak{X})$, note that for all $x_1, x_2 \in \mathfrak{X}$ we have by Proposition 5.6 that

$$\begin{aligned}
\varphi(\rho(a)([x_1, x_2])) &= (\mathrm{ad}_{\mathfrak{g}} a)(\varphi([x_1, x_2])) \\
&= (\mathrm{ad}_{\mathfrak{g}} a)([\varphi(x_1), \varphi(x_2)]) \\
&= [(\mathrm{ad}_{\mathfrak{g}} a)(\varphi(x_1)), \varphi(x_2)] + [\varphi(x_1), (\mathrm{ad}_{\mathfrak{g}} a)(\varphi(x_2))] \\
&= [\varphi(\rho(a)x_1), \varphi(x_2)] + [\varphi(x_1), \varphi(\rho(a)x_2)] \\
&= \varphi([\rho(a)x_1, x_2] + [x_1, \rho(a)x_2]).
\end{aligned}$$

Since $\mathrm{Ker}\,\varphi = \{0\}$, we get $\rho(a) \in \mathrm{Der}(\mathfrak{X})$, and the proof ends. ∎

REMARK 8.9 It follows by Remark 8.7 and Theorem 8.8 that there exists on the complex Hilbert space $\mathfrak{L}$ a unique structure of complex L^*-algebra with isometric involution such that $\varphi \in \mathrm{Hom}^*(\mathfrak{L}, \mathfrak{A})$ and

$$(\forall x_1, x_2 \in \mathfrak{L}) \qquad [x_1, x_2] = \rho(\varphi(x_1))x_2.$$

Moreover, $\mathrm{Ran}\,\varphi \trianglelefteq \mathfrak{A}$ and $\mathrm{Ran}\,\rho \subseteq \mathrm{Der}(\mathfrak{L})$.

REMARK 8.10 In connection with Proposition 8.6 and Theorem 8.8, note that there occur representations of $\mathfrak{g}$ in $\mathfrak{g}^\#$, $\mathfrak{X}$ and $\mathfrak{g}$, namely $a \mapsto (\mathrm{ad}^*_{\mathfrak{g}})^\#$, $a \mapsto \rho(a)$ and $a \mapsto \mathrm{ad}_{\mathfrak{g}} a$, respectively. These representations are related by the commutative diagrams

$$\begin{array}{ccccc}
\mathfrak{g}^\# & \xrightarrow{\varphi^\#} & \mathfrak{X} & \xrightarrow{\varphi} & \mathfrak{g} \\
{\scriptstyle (\mathrm{ad}_{\mathfrak{g}} a^*)^\#}\downarrow & & \downarrow{\scriptstyle \rho(a)} & & \downarrow{\scriptstyle \mathrm{ad}_{\mathfrak{g}} a} \\
\mathfrak{g}^\# & \xrightarrow{\varphi^\#} & \mathfrak{X} & \xrightarrow{\varphi} & \mathfrak{g}
\end{array}$$

for each $a \in \mathfrak{g}$. Similarly, we have commutative diagrams of the complexified spaces

$$
\begin{array}{ccccc}
\mathfrak{A}^{\#} & \xrightarrow{\varphi^{\#}} & \mathfrak{L} & \xrightarrow{\varphi} & \mathfrak{A} \\
{\scriptstyle (\mathrm{ad}_{\mathfrak{A}} a^*)^{\#}}\downarrow & & \downarrow{\scriptstyle \rho(a)} & & \downarrow{\scriptstyle \mathrm{ad}_{\mathfrak{A}} a} \\
\mathfrak{A}^{\#} & \xrightarrow{\varphi^{\#}} & \mathfrak{L} & \xrightarrow{\varphi} & \mathfrak{A}
\end{array}
$$

for all $a \in \mathfrak{A}$ (see Remark 8.5, Remark 8.7, and Remark 8.9).

REMARK 8.11 We now note that the real Banach space $\mathfrak{g}^{\#}/\mathrm{Ker}\,\iota$ possesses a natural structure of involutive Banach-Lie algebra such that the injective operator

$$\varphi \circ \eta \colon \mathfrak{g}^{\#}/\mathrm{Ker}\,\iota \to \mathfrak{g}$$

(induced by $\iota \colon \mathfrak{g}^{\#} \to \mathfrak{g}$) is a continuous homomorphism.

Indeed, for each $a \in \mathfrak{g}$, the relation $(\mathrm{ad}_{\mathfrak{g}} a) \circ \iota = \iota \circ (\mathrm{ad}_{\mathfrak{g}} a^*)^{\#}$ shows that $\mathrm{Ker}\,\iota$ is invariant to $(\mathrm{ad}_{\mathfrak{g}} a^*)^{\#}$, and thus $(\mathrm{ad}_{\mathfrak{g}} a^*)^{\#}$ induces an operator $\tilde{\rho}(a)$ on $\mathfrak{g}^{\#}/\mathrm{Ker}\,\iota$. Hence we have a representation

$$\tilde{\rho} \colon \mathfrak{g} \to \mathcal{B}(\mathfrak{g}^{\#}/\mathrm{Ker}\,\iota)$$

such that for each $a \in \mathfrak{g}$ the diagram

$$
\begin{array}{ccc}
\mathfrak{g}^{\#}/\mathrm{Ker}\,\iota & \xrightarrow{\varphi\circ\eta} & \mathfrak{g} \\
{\scriptstyle \tilde{\rho}(a)}\downarrow & & \downarrow{\scriptstyle \mathrm{ad}_{\mathfrak{g}} a} \\
\mathfrak{g}^{\#}/\mathrm{Ker}\,\iota & \xrightarrow{\varphi\circ\eta} & \mathfrak{g}
\end{array}
$$

is commutative. Then, as at the beginning of the proof of Theorem 8.8 above, we can define the desired bracket on $\mathfrak{g}^{\#}/\mathrm{Ker}\,\iota$. Moreover, the following version of (8.16) holds:

$$(\forall v_1, v_2 \in \mathfrak{g}^{\#}/\mathrm{Ker}\,\iota) \qquad [v_1, v_2] = \tilde{\rho}((\varphi \circ \eta)(v_1))v_2$$

or, according to the way $\tilde{\rho}$ is constructed,

$$(\forall f_1, f_2 \in \mathfrak{g}^{\#}) \qquad [f_1 + \mathrm{Ker}\,\iota, f_2 + \mathrm{Ker}\,\iota] = (\mathrm{ad}_{\mathfrak{g}}(\iota(f_1^*)))^{\#} f_2 + \mathrm{Ker}\,\iota.$$

We also have an involution on $\mathfrak{g}^{\#}/\mathrm{Ker}\,\iota$, defined as in the stage 2° of the proof of Theorem 8.8 by $(f + \mathrm{Ker}\,\iota)^* = f^* + \mathrm{Ker}\,\iota$ for each element $f + \mathrm{Ker}\,\iota$ of $\mathfrak{g}^{\#}/\mathrm{Ker}\,\iota$.

Furthermore, $\mathrm{Ran}\,(\varphi \circ \eta) = \mathrm{Ran}\,\iota \trianglelefteq \mathfrak{g}$ and

$$\mathrm{Ran}\,\tilde{\rho} \subseteq \mathrm{Der}(\mathfrak{g}^{\#}/\mathrm{Ker}\,\iota).$$

(The latter inclusion can be proved by the method of stage 4° of the proof of Theorem 8.8.)

Finally, we note that similar facts can be said in connection with the complexification $\mathfrak{A}^{\#}/\mathrm{Ker}\,\iota$ of $\mathfrak{g}^{\#}/\mathrm{Ker}\,\iota$.

We conclude this section by an example illustrating the constructions performed above.

EXAMPLE 8.12 Let $2 \leq p < \infty$ and $q = p/(p-1)$, so that

$$1 < q \leq 2 \leq p < \infty \text{ and } \frac{1}{p} + \frac{1}{q} = 1.$$

Then consider $\mathfrak{g} = \mathfrak{u}_p$, where $\mathfrak{u}_p$ is the canonically involutive Banach-Lie algebra of all skew-adjoint operators belonging to the Schatten ideal $\mathcal{C}_p(\mathfrak{H})$, as in Example 7.39. It is well known that $(\mathfrak{u}_p)^{\#} = \mathfrak{u}_q$, the natural pairing being

$$\langle \cdot, \cdot \rangle \colon \mathfrak{u}_q \times \mathfrak{u}_p \to \mathbb{R}, \quad \langle f, a \rangle = -\mathrm{Tr}\,(fa),$$

where $\mathrm{Tr} : \mathcal{C}_1(\mathcal{H}) \to \mathbb{C}$ is the usual trace on the ideal of trace-class operators. Further denote by

$$\varphi \colon \mathfrak{u}_2 \to \mathfrak{u}_p$$

the inclusion map $\mathfrak{u}_2 \hookrightarrow \mathfrak{u}_p$. It is obvious that $\varphi \in \mathrm{Hom}^*(\mathfrak{u}_2, \mathfrak{u}_p)$ and for every $x \in \mathfrak{u}_2$ and $f \in \mathfrak{u}_q$ we have

$$\langle f, \varphi(x) \rangle = -\mathrm{Tr}\,(fx) = \mathrm{Tr}\,(fx^*) = (f \mid x)_{\mathfrak{u}_2}$$

hence $\varphi^{\#} : \mathfrak{u}_q \to \mathfrak{u}_2$ is just the inclusion map. Thus

$$\iota := \varphi \circ \varphi^{\#} : \mathfrak{u}_q \to \mathfrak{u}_p$$

is the inclusion map $\mathfrak{u}_q \hookrightarrow \mathfrak{u}_p$. Then

$$\iota \in \mathcal{C}^+(\mathfrak{u}_p)$$

according to the above Proposition 8.1.

This fact can be checked directly as well: for all $a, b \in \mathfrak{g} = \mathfrak{u}_p$ and $f \in \mathfrak{g}^{\#} = \mathfrak{u}_q$, we have

$$\begin{aligned}\langle f, (\mathrm{ad}_{\mathfrak{g}} a^*)b \rangle &= -\mathrm{Tr}\,(f[-a, b]) = \mathrm{Tr}\,(fab) - \mathrm{Tr}\,(fba) = \mathrm{Tr}\,(fab) - \mathrm{Tr}\,(afb) \\ &= -\mathrm{Tr}\,([a, f]b) = \langle [a, f], b \rangle.\end{aligned}$$

Hence

$$(\mathrm{ad}_{\mathfrak{g}} a^*)^{\#} f = [a, f]$$

for all $a \in \mathfrak{u}_p$ and $f \in \mathfrak{u}_q$. This formula shows at once that $\iota \colon \mathfrak{u}_q \to \mathfrak{u}_p$ is equivariant (see Definition 7.1). Moreover, ι is also monotone: for every $f \in \mathfrak{u}_q$ we have

$$\langle f, \iota(f) \rangle = -\mathrm{Tr}\,(f^2) \geq 0$$

for f is skew-adjoint.

8.3 Parameterizing L*-ideals

We now use the results of the previous sections in order to parameterize the L^*-ideals of a real involutive Banach-Lie algebra by means of the equivariant monotone operators. The main idea of this approach is that, in general, the equivariant monotone operators are easier to determine than the L^*-ideals. This point is particularly illustrated in Example 8.23 below, where we prove that the canonically involutive real form $\mathfrak{u}_p$ of the Schatten ideal $\mathcal{C}_p(\mathfrak{H})$ has no non-zero L^*-ideals when $1 < p < 2$. (Compare with Example 7.39.)

The following theorem concerns the map Φ constructed in Proposition 7.23.

THEOREM 8.13 *Let $\mathfrak{g}$ be a real involutive Banach-Lie algebra. Then for every $\iota \in \mathcal{C}^+(\mathfrak{g})$ we have $\Phi(\iota) \in L^*\mathrm{Id}(\mathfrak{g})$, and the map*

$$\Phi \colon \mathcal{C}^+(\mathfrak{g}) \to L^*\mathrm{Id}(\mathfrak{g})$$

has the following properties:

(i) *For all $\iota_1, \iota_2 \in \mathcal{C}^+(\mathfrak{g})$,*

$$\Phi(\iota_1 + \iota_2) = \Phi(\iota_1) + \Phi(\iota_2).$$

(ii) *If $\iota_1, \iota_2 \in \mathcal{C}^+(\mathfrak{g})$, then*

$$\Phi(\iota_1) \subseteq \Phi(\iota_2) \iff \mathbb{R}_+^* \cdot \iota_2 \subseteq \mathbb{R}_+^* \cdot \iota_1 + \mathcal{C}^+(\mathfrak{g}).$$

(iii) *We have*

$$\mathcal{C}_0^+(\mathfrak{g}) = \{\iota \in \mathcal{C}^+(\mathfrak{g}) \mid \Phi(\iota) \text{ is dense in } \mathfrak{g}\}.$$

(iv) *Each of the sets*

$$\{\mathfrak{I} \in L^*\mathrm{Id}(\mathfrak{g}) \mid \mathfrak{I} \text{ is dense in } \mathfrak{g}\},$$
$$\{\mathfrak{I} \in L^*\mathrm{Id}(\mathfrak{g}) \mid \text{the center of } \mathfrak{I} \text{ reduces to } \{0\}\},$$
$$\{\mathfrak{I} \in L^*\mathrm{Id}(\mathfrak{g}) \mid \mathfrak{I} \text{ is contained in the center of } \mathfrak{g}\}$$

is contained in $\mathrm{Ran}\,\Phi$.

PROOF It follows by Theorem 8.8 that $\Phi(\iota) \in L^*\mathrm{Id}(\mathfrak{g})$ whenever $\iota \in \mathcal{C}^+(\mathfrak{g})$. The other assertions of the theorem can be proved as follows.

(i) Use Proposition 7.28.

(ii) Use Corollary 7.27(a).

(iii) See Proposition 7.29.

(iv) Use Proposition 8.1 (see also Proposition 7.23). ∎

COROLLARY 8.14 *Let $\mathfrak{g}$ be a real involutive Banach-Lie algebra. If $\mathfrak{g}$ is topologically simple then the following assertions hold.*

(j) *We have*

$$\mathcal{C}^+(\mathfrak{g}) = \{0\} \cup \mathcal{C}_0^+(\mathfrak{g}).$$

(jj) *We have $\mathcal{C}_0^+(\mathfrak{g}) \neq \emptyset$ if and only if $\mathfrak{g}$ has a non-zero L^*-ideal.*

PROOF First note that, by Theorem 7.38, every non-zero L^*-ideal of $\mathfrak{g}$ is dense in $\mathfrak{g}$. In view of the obvious fact that $\Phi(\iota) = 0$ if and only if $\iota = 0$, the desired equality in (j) then follows by the above Theorem 8.13(iii).

Then assertion (jj) follows by Theorem 7.38 along with the above Theorem 8.13(iii). ∎

Since, in the above Theorem 8.13, it is not clear that the map Φ is always onto, it is convenient to introduce the following concept.

DEFINITION 8.15 We say that a real involutive Banach-Lie algebra $\mathfrak{g}$ is *adequate* if the corresponding map $\Phi\colon \mathcal{C}^+(\mathfrak{g}) \to L^*\mathrm{Id}(\mathfrak{g})$ is onto.

We now consider several examples of adequate Banach-Lie algebras.

EXAMPLE 8.16 If a real involutive Banach-Lie algebra is abelian, then it is adequate. (See the last of the sets in Theorem 8.13(iv).)

EXAMPLE 8.17 If a real involutive Banach-Lie algebra is topologically simple, then it is adequate.

Indeed, according to Theorem 7.38, every L^*-ideal of such an algebra is dense; then use the first of the sets in Theorem 8.13(iv).

EXAMPLE 8.18 Every real L^*-algebra $\mathfrak{g}$ is adequate.

To prove this fact, denote by $\mathfrak{z}$ the center of $\mathfrak{g}$ and consider an arbitrary L^*-ideal $\mathfrak{J}$ of $\mathfrak{g}$. It easily follows by Theorem 7.34 that

$$\mathfrak{J} = \mathfrak{J}_0 \oplus \mathfrak{J}_1,$$

where $\mathfrak{J}_0 := \mathfrak{J} \cap \mathfrak{z}$ and $\mathfrak{J}_1 := \mathfrak{J} \cap \mathfrak{z}^\perp$, and $\mathfrak{J}_0, \mathfrak{J}_1 \in L^*\mathrm{Id}(\mathfrak{g})$.

Since $\mathfrak{J}_0 \subseteq \mathfrak{z}$ and the center of $\mathfrak{J}_1$ is $\{0\}$, the above Theorem 8.13(iv) shows that there exist $\iota_0, \iota_1 \in \mathcal{C}^+(\mathfrak{g})$ with $\Phi(\iota_0) = \mathfrak{J}_0$ and $\Phi(\iota_1) = \mathfrak{J}_1$. Then

$$\mathfrak{J} = \mathfrak{J}_0 + \mathfrak{J}_1 = \Phi(\iota_0) + \Phi(\iota_1) = \Phi(\iota_0 + \iota_1)$$

according to Theorem 8.13(i), and we are done.

Other examples of adequate Banach-Lie algebras will be encountered in what follows (see Corollary 8.27 below).

PROPOSITION 8.19 *If $\mathfrak{g}$ is an adequate Banach-Lie algebra, then the set $L^*\mathrm{Id}(\mathfrak{g})$ is a lattice with respect to vector addition and set intersection.*

PROOF It follows by the above Theorem 8.13(i) that $L^*\mathrm{Id}(\mathfrak{g})$ ($= \mathrm{Ran}\,\Phi$) is closed under vector addition. On the other hand, Proposition 7.35 shows that $L^*\mathrm{Id}(\mathfrak{g})$ is closed under set intersection as well. ∎

Here is another interesting property of the adequate algebras.

PROPOSITION 8.20 *If $\mathfrak{g}$ is an adequate Banach-Lie algebra and $\iota \in \mathcal{C}^+(\mathfrak{g})$, then the L^*-ideal $\Phi(\iota)$ is maximal in $L^*\mathrm{Id}(\mathfrak{g})$ if and only if $\mathbb{R}_+ \cdot \iota$ is an extreme ray in the convex cone $\mathcal{C}^+(\mathfrak{g})$.*

PROOF Use Theorem 8.13(ii). ∎

COROLLARY 8.21 *Let $\mathfrak{g}$ be a real involutive Banach-Lie algebra. If $\mathfrak{g}$ is topologically simple, then either*

$$\mathcal{C}^+(\mathfrak{g}) = \{0\}$$

or

$$\mathcal{C}^+(\mathfrak{g}) = \mathbb{R}_+ \cdot \iota$$

for some $\iota \in \mathcal{C}_0^+(\mathfrak{g})$.

PROOF Assume $\mathcal{C}^+(\mathfrak{g}) \neq \{0\}$. Then $\mathfrak{g}$ has precisely one non-zero L^*-ideal, according to Theorem 7.38 and Corollary 8.14 above. Since $\mathfrak{g}$ is adequate (see the above Example 8.17), Proposition 8.20 implies that every ray of the closed convex cone $\mathcal{C}^+(\mathfrak{g})$ (see Lemma 7.4) is extreme, which easily implies the desired conclusion (see also Corollary 8.14(j)). ∎

We continue with a characterization of the real L^*-algebras in terms of equivariant monotone operators.

PROPOSITION 8.22 *Let $\mathfrak{g}$ be a real involutive Banach-Lie algebra. Then $\mathfrak{g}$ is $*$-isomorphic to a real L^*-algebra if and only if $\mathcal{C}^+(\mathfrak{g})$ contains surjective operators.*

PROOF If $\mathfrak{g}$ is a real L^*-algebra, then

$$\mathfrak{g}^{\#} = \mathfrak{g}$$

and we have

$$(\mathrm{ad}_{\mathfrak{g}} a^*)^{\#} = \mathrm{ad}_{\mathfrak{g}} a$$

for all $a \in \mathfrak{g}$, hence the identity map $\mathrm{id}_{\mathfrak{g}}$ belongs to $\mathcal{C}^+(\mathfrak{g})$.

Conversely, assume that there exists $\iota \in \mathcal{C}^+(\mathfrak{g})$ with $\mathrm{Ran}\,\iota = \mathfrak{g}$. Since $\iota^{\#} = \iota$, we get

$$\mathrm{Ker}\,\iota = (\mathrm{Ran}\,\iota)^{\perp} = \{0\}.$$

Next let $\mathfrak{X}$ be a real L^*-algebra and $\varphi \in \mathrm{Hom}^*(\mathfrak{X}, \mathfrak{g})$ with $\mathrm{Ker}\,\varphi = \{0\}$ and $\iota = \varphi \circ \varphi^{\#}$ (see Notation 8.3 and Theorem 8.8). Then

$$\mathfrak{g} = \mathrm{Ran}\,\iota \subseteq \mathrm{Ran}\,\varphi \subseteq \mathfrak{g},$$

hence $\mathrm{Ran}\,\varphi = \mathfrak{g}$. Consequently $\varphi\colon \mathfrak{X} \to \mathfrak{g}$ is a $*$-isomorphism of a real L^*-algebra onto $\mathfrak{g}$. ∎

We now use the above Corollary 8.14(jj) to complete the discussion of Example 7.39.

EXAMPLE 8.23 If $1 < p < 2$, then

$$L^*\mathrm{Id}(\mathfrak{u}_p) = \{\{0\}\}.$$

Indeed, recall that $\mathfrak{u}_p$ is topologically simple, as noted in Example 7.39. Then $\mathfrak{u}_p$ is adequate by the above Example 8.17.

Now assume that $L^*\mathrm{Id}(\mathfrak{u}_p)$ contains elements different from $\{0\}$. Since every non-zero L^*-ideal of $\mathfrak{u}_p$ must be dense (see Theorem 7.38), it follows by the above Theorem 8.13(iii) that there exists $\iota \in \mathcal{C}_0^+(\mathfrak{u}_p)$. Since $(\mathfrak{u}_p)^{\#} = \mathfrak{u}_q$, we have

$$\iota\colon \mathfrak{u}_q \to \mathfrak{u}_p,$$

where $q = p/(p-1) > 2 > p$. Denote by

$$j\colon \mathfrak{u}_p \to \mathfrak{u}_q$$

the natural inclusion map. The equivariance property of ι (see Definition 7.1) reads

$$(\forall a \in \mathfrak{u}_p)(\forall f \in \mathfrak{u}_q) \qquad \iota([a, f]) = [a, \iota(f)]. \tag{8.18}$$

(See Example 8.12.) For $f \in \mathfrak{u}_p (\subseteq \mathfrak{u}_q)$ fixed, this equality extends by continuity to all $a \in \mathfrak{u}_q$ hence we have also

$$(\forall a \in \mathfrak{u}_q)(\forall f \in \mathfrak{u}_p) \qquad \iota([a, f]) = [a, \iota(f)].$$

This implies that

$$j \circ \iota|_{\mathfrak{u}_p} \in \mathcal{C}^+(\mathfrak{u}_p).$$

Since $\mathfrak{u}_p$ is dense in $\mathfrak{u}_q$ and $\iota \not\equiv 0$, we must have $0 \not\equiv j \circ \iota|_{\mathfrak{u}_p}$ and then

$$\{0\} \neq \Phi(j \circ \iota|_{\mathfrak{u}_p}) \in L^*\mathrm{Id}(\mathfrak{u}_q)$$

by Theorem 8.13 (where $\Phi\colon \mathcal{C}^+(\mathfrak{u}_q) \to L^*\mathrm{Id}(\mathfrak{u}_q)$ as in Theorem 8.13). Since $2 < q$, it then follows by Example 7.39 that

$$\Phi(j \circ \iota|_{\mathfrak{u}_p}) = \mathfrak{u}_2 = \Phi(j).$$

Then Corollary 7.27(b) implies that there exist $\lambda, \mu > 0$ such that

$$(\forall a \in \mathfrak{u}_p) \qquad -\lambda \mathrm{Tr}\,(a^2) \leq -\mathrm{Tr}\,(a \cdot \iota(a)) \leq -\mu \mathrm{Tr}\,(a^2).$$

Since $\mathfrak{u}_p$ is dense in $\mathfrak{u}_2$, we deduce that

$$(\forall a \in \mathfrak{u}_2) \qquad \lambda \|a\|_2^2 \leq (a \mid \iota(a))_{\mathfrak{u}_2} \leq \mu \|a\|_2^2.$$

Since $\iota^{\#} = \iota$, it then follows that $\iota|_{\mathfrak{u}_2}\colon \mathfrak{u}_2 \to \mathfrak{u}_2$ is an invertible operator. In particular $\mathfrak{u}_2 = \iota(\mathfrak{u}_2) \subseteq \iota(\mathfrak{u}_q) \subseteq \mathfrak{u}_p$, thus contradicting the fact that $p < 2$.

Consequently, $\mathfrak{u}_p$ has no non-zero L^*-ideals when $1 < p < 2$.

Elliptic Banach-Lie algebras

We concude the present section by investigating the relationship between adequate algebras and the elliptic Banach-Lie algebras introduced in the following definition.

DEFINITION 8.24 A real Banach-Lie algebra $\mathfrak{g}$ is *elliptic* if the automorphism

$$\exp(\mathrm{ad}_{\mathfrak{g}} a)\colon \mathfrak{g} \to \mathfrak{g}$$

is an isometry for each $a \in \mathfrak{g}$.

The following simple property of elliptic algebras will prove to be very useful in what follows.

PROPOSITION 8.25 *If $\mathfrak{g}$ is an elliptic Banach-Lie algebra and $\mathfrak{J}$ is an abelian ideal of $\mathfrak{g}$, then $\mathfrak{J}$ is contained in the center of $\mathfrak{g}$.*

PROOF Let $a \in \mathfrak{J}$. We have $[a,[a,\mathfrak{J}]] \subseteq [a,\mathfrak{J}] \subseteq [\mathfrak{J},\mathfrak{J}] = \{0\}$. Hence, denoting by $\mathfrak{A}$ the complexification of $\mathfrak{g}$ we have $(\mathrm{ad}_{\mathfrak{A}} a)^2 = 0$.

On the other hand, since $\mathfrak{g}$ is elliptic, it follows that $\mathrm{ad}_{\mathfrak{A}} a$ is a hermitian-equivalent operator on $\mathfrak{A}$, hence $\mathrm{ad}_{\mathfrak{A}} a = 0$ (see e.g., Corollary 3 in §14 in [BS01]). In particular, a belongs to the center of $\mathfrak{g}$. ∎

COROLLARY 8.26 *If $\mathfrak{g}$ is an elliptic Banach-Lie algebra, then for every ideal $\mathfrak{J}$ of $\mathfrak{g}$ we have $\mathcal{Z}(\mathfrak{J}) \subseteq \mathcal{Z}(\mathfrak{g})$.*

PROOF In view of the above Proposition 8.25, it suffices to prove that the center $\mathcal{Z}(\mathfrak{J})$ of $\mathfrak{J}$ is an ideal of $\mathfrak{g}$.

We have $[\mathfrak{J}, \mathcal{Z}(\mathfrak{J})] = \{0\}$, hence for all $a \in \mathfrak{g}$, $b \in \mathfrak{J}$ and $z \in \mathcal{Z}(\mathfrak{J})$,

$$0 = [a, [b, z]] = [[a, b], z] + [b, [a, z]]. \tag{8.19}$$

But $[a, b] \in [\mathfrak{g}, \mathfrak{J}] \subseteq \mathfrak{J}$ hence $[[a, b], z] = 0$. Then (8.19) implies that $[b, [a, z]] = 0$ for all $b \in \mathfrak{J}$, that is, $[a, z] \in \mathcal{Z}(\mathfrak{J})$.

Consequently $[\mathfrak{g}, \mathcal{Z}(\mathfrak{J})] \subseteq \mathcal{Z}(\mathfrak{J})$ and the proof ends, in view of the beginning remark. ∎

We now use Corollary 8.26 to show that many elliptic Banach-Lie algebras are adequate.

COROLLARY 8.27 *If $\mathfrak{g}$ is an involutive elliptic Banach-Lie algebra with $\mathcal{Z}(\mathfrak{g}) = \{0\}$, then the following assertions hold:*

(i) *The algebra $\mathfrak{g}$ is adequate.*

(ii) *For every L^*-ideal $\mathfrak{J}$ of $\mathfrak{g}$, the closures of the subsets $\mathfrak{J}$ and $[\mathfrak{J}, \mathfrak{J}]$ of $\mathfrak{g}$ are equal.*

PROOF (i) It follows by the above Corollary 8.26 that the center of every ideal of $\mathfrak{g}$ reduces to $\{0\}$. Then use the second of the sets in Theorem 8.13(iv).

(ii) If $\mathfrak{J} \in L^*\mathrm{Id}(\mathfrak{g})$, then the already proved assertion (i) shows that there exist a real L^*-algebra $\mathfrak{X}$ and $\varphi \in \mathrm{Hom}^*(\mathfrak{X}, \mathfrak{g})$ with $\mathrm{Ran}\,\varphi = \mathfrak{J}$. As usual, we may assume that $\mathrm{Ker}\,\varphi = \{0\}$ (see Remark 7.13).

Since $\mathcal{Z}(\mathfrak{J}) = \{0\}$ (cf. the proof of (i)), it then follows that $\mathcal{Z}(\mathfrak{X}) = \{0\}$, which in turn implies that $[\mathfrak{X}, \mathfrak{X}]$ is dense in $\mathfrak{X}$, according to Theorem 7.11(d). Since $\varphi \colon \mathfrak{X} \to \mathfrak{g}$ is continuous, we then deduce that $\mathfrak{J} = \varphi(\mathfrak{X})$ and its subset $[\mathfrak{J}, \mathfrak{J}] = \varphi([\mathfrak{X}, \mathfrak{X}])$ have the same closure in $\mathfrak{g}$. ∎

REMARK 8.28 It is worth noting the following by-product of the proof of Corollary 8.27(ii) (obtained by making use also of Theorem 8.13(iii)): *If $\mathfrak{g}$ is a real involutive Banach-Lie algebra with $\mathcal{Z}(\mathfrak{g}) = \{0\}$ and $C_0^+(\mathfrak{g}) \neq \emptyset$, then $[\mathfrak{g}, \mathfrak{g}]$ is dense in $\mathfrak{g}$. Equivalently, if $\mathcal{Z}(\mathfrak{g}) \neq \{0\}$ and $[\mathfrak{g}, \mathfrak{g}]$ is not dense in $\mathfrak{g}$, then $C_0^+(\mathfrak{g}) = \emptyset$.*

Let us also note that the latter fact sometimes holds with one of the two hypotheses removed; see the example of Heisenberg algebras in Remark 7.6.

8.4 Representations of automorphism groups

THEOREM 8.29 *Let $\mathfrak{g}$ be a canonically involutive Banach-Lie algebra such that $\mathfrak{g}$ is topologically simple and $\mathcal{C}_0^+(\mathfrak{g}) \neq \emptyset$. Pick $\iota \in \mathcal{C}_0^+(\mathfrak{g})$ and let $\mathfrak{X}$ be a real L^*-algebra and $\varphi \in \mathrm{Hom}^*(\mathfrak{X}, \mathfrak{g})$ with $\mathrm{Ker}\,\varphi = \{0\}$ and $\iota = \varphi \circ \varphi^{\#}$. Then there exists a unique irreducible continuous group representation*

$$\rho\colon \mathrm{Aut}(\mathfrak{g}) \to \mathcal{B}(\mathfrak{X})$$

such that

$$(\forall \alpha \in \mathrm{Aut}(\mathfrak{g})) \qquad \varphi \circ (\rho(\alpha)) = \alpha \circ \varphi.$$

Moreover, for every $\alpha \in \mathrm{Aut}(\mathfrak{g})$, we have

$$\alpha \circ \iota = \iota \circ (\alpha^{\#})^{-1},$$

$\rho(\alpha) \in \mathrm{Aut}(\mathfrak{X})$ and $\rho(\alpha)$ is an isometry.

PROOF First note that $\mathfrak{X}$ and φ exist by Theorem 8.13 (see also Notation 8.3). Denote

$$\mathfrak{I} := \mathrm{Ran}\,\varphi = \Phi(\iota) \in L^*\mathrm{Id}(\mathfrak{g}).$$

Now, if $\alpha \in \mathrm{Aut}(\mathfrak{g})$, we have obviously $\{0\} \neq \alpha(\mathfrak{I}) \in L^*\mathrm{Id}(\mathfrak{g})$. But $\mathfrak{I}$ is the only non-zero L^*-ideal of $\mathfrak{g}$ (by Theorem 7.38), hence $\alpha(\mathfrak{I}) = \mathfrak{I}$. It then follows by the closed graph theorem that there exists $\rho(\alpha) \in \mathcal{B}(\mathfrak{X})$ with

$$\varphi \circ \rho(\alpha) = \alpha \circ \varphi.$$

Since $\mathrm{Ker}\,\varphi = \{0\}$ and $\varphi \in \mathrm{Hom}^*(\mathfrak{X}, \mathfrak{g})$, it easily follows that $\alpha \mapsto \rho(\alpha)$ is a group representation and $\rho(\alpha) \in \mathrm{Aut}(\mathfrak{X})$ (see e.g., the argument at stage 4° in the proof of Theorem 8.8).

Now assume that $\mathfrak{X}_0$ is a closed subspace of $\mathfrak{X}$ such that $\mathfrak{X}_0$ is invariant to $\rho(\alpha)$ for all $\alpha \in \mathrm{Aut}(\mathfrak{g})$. Note that since $\varphi \in \mathrm{Hom}^*(\mathfrak{X}, \mathfrak{g})$, we have $\varphi \circ (\mathrm{ad}_{\mathfrak{X}} x) = (\mathrm{ad}_{\mathfrak{g}}(\varphi(x))) \circ \varphi$, hence $\varphi \circ (\exp(\mathrm{ad}_{\mathfrak{X}} x)) = (\exp(\mathrm{ad}_{\mathfrak{g}}(\varphi(x)))) \circ \varphi$ for each $x \in \mathfrak{X}$. In other words,

$$\rho(\alpha) = \exp(\mathrm{ad}_{\mathfrak{X}} x) \text{ whenever } \alpha = \exp(\mathrm{ad}_{\mathfrak{g}}(\varphi(x))) \text{ with } x \in \mathfrak{X}.$$

Then $\mathfrak{X}_0$ is invariant to each automorphism $\exp(\mathrm{ad}_{\mathfrak{X}} x)$ of $\mathfrak{X}$ with $x \in \mathfrak{X}$, which implies at once that

$$\mathfrak{X}_0 \trianglelefteq \mathfrak{X}.$$

On the other hand, since $\mathfrak{g}$ is topologically simple, it follows by Remark 7.32 that $\mathfrak{X}$ is in turn topologically simple. Hence we have either $\mathfrak{X}_0 = \{0\}$ or $\mathfrak{X}_0 = \mathfrak{X}$. Thus the representation ρ is irreducible.

Next, if $\alpha \in \mathrm{Aut}(\mathfrak{g})$, then $\varphi \circ (\rho(\alpha)) = \alpha \circ \varphi$, hence

$$\varphi \circ (\rho(\alpha)) \circ (\rho(\alpha))^{\#} \circ \varphi^{\#} = \alpha \circ \varphi \circ \varphi^{\#} \circ \alpha^{\#} = \alpha \circ \iota \circ \alpha^{\#}.$$

But we actually know by Lemma 7.16 that $\rho(\alpha)$ is an isometry, because we have seen above that $\mathfrak{X}$ is topologically simple. Hence the above equality implies that

$$\alpha \circ \iota \circ \alpha^{\#} = \varphi \circ \varphi^{\#} = \iota, \tag{8.20}$$

as desired.

It only remained to show that ρ is continuous. To this end, note that since (8.20) holds for all $\alpha \in \mathrm{Aut}(\mathfrak{g})$, we have

$$(\forall \delta \in \mathrm{Der}(\mathfrak{g})) \qquad \delta \circ \iota + \iota \circ \delta^{\#} = 0.$$

Then Lemma 8.4 shows that for each $\delta \in \mathrm{Der}(\mathfrak{g})$ there exists $\rho_\delta \in \mathcal{B}(\mathfrak{X})$ with

$$\delta \circ \varphi = \varphi \circ \rho_\delta, \; \rho_\delta^{\#} = -\rho_\delta \text{ and } \|\rho_\delta\| \le \|\delta\|.$$

Furthermore, as at the stage 4° in the proof of Theorem 8.8, we deduce that $\rho_\delta \in \mathrm{Der}(\mathfrak{X})$ for all $\delta \in \mathrm{Der}(\mathfrak{g})$. Moreover,

$$\mathrm{Der}(\mathfrak{g}) \to \mathrm{Der}(\mathfrak{X}), \quad \delta \mapsto \rho_\delta$$

is a Lie algebra homomorphism and

$$(\forall \delta \in \mathrm{Der}(\mathfrak{g})) \quad \exp(\rho_\delta) = \rho(\exp \delta) \qquad (\in \mathrm{Aut}(\mathfrak{X})). \tag{8.21}$$

Now recall that the exponential mapping

$$\exp\colon \mathrm{Der}(\mathfrak{g}) \to \mathrm{Aut}(\mathfrak{g})$$

is a local homeomorphism at $0 \in \mathrm{Der}(\mathfrak{g})$. In fact, see Example 4.10; the idea is that $\mathrm{Aut}(\mathfrak{g})$ is an algebraic subgroup of degree ≤ 2 of $\mathcal{B}(\mathfrak{g})^{\times}$ with Lie algebra $\mathrm{Der}(\mathfrak{g})$, and then Proposition 4.4 and the version of Theorem 4.13 for real Banach algebras apply.

It then follows by (8.21) that the representation ρ of the real Banach-Lie group $\mathrm{Aut}(\mathfrak{g})$ is continuous on a neighborhood of $1 \in \mathrm{Aut}(\mathfrak{g})$. Consequently, this representation is continuous throughout on $\mathrm{Aut}(\mathfrak{g})$. ∎

EXAMPLE 8.30 Let $2 \le p < \infty$, $\mathfrak{g} = \mathfrak{u}_p$, $\mathfrak{X} = \mathfrak{u}_2$ and $\varphi\colon \mathfrak{u}_2 \hookrightarrow \mathfrak{u}_p$ as in Example 8.12. In this special case, the above Theorem 8.29 explains why every automorphism of $\mathfrak{u}_p$ induces an isometric automorphism of $\mathfrak{u}_2$, and this "induction process" is continuous (see Proposition II.12 at page 93 in [dlH72]).

8.5 Applications to enlargibility

In this section we describe an application of equivariant monotone operators to the enlargibility theory presented in Chapter 3. In particular we show that a real involutive Banach-Lie algebra $\mathfrak{g}$ is enlargible provided $\mathcal{C}_0^+(\mathfrak{g}) \neq \emptyset$. Actually we prove that the latter condition on $\mathcal{C}_0^+(\mathfrak{g})$ has much stronger consequences on enlargibility of all split central extensions of $\mathfrak{g}$; see Corollary 8.36 below.

DEFINITION 8.31 A *locally finite* Lie algebra is a Lie algebra such that each of its finite subsets belongs to some finite-dimensional subalgebra.

LEMMA 8.32 *If a real Banach-Lie algebra has a dense locally finite subalgebra, then it is enlargible.*

PROOF See Corollary 3.5 in [Pe92] (or Corollaire 2 in [Pe88]). See also [Be04] for a "standard" proof of the local enlargibility theorem of [Pe88] and [Pe92]. ∎

PROPOSITION 8.33 *Every L^*-algebra over $\mathbb{K} \in \{\mathbb{R}, \mathbb{C}\}$ has a dense locally finite ideal.*

To prove this fact, we need the following auxiliary result.

LEMMA 8.34 *Let $\mathfrak{L}$ be a real Banach space and $\mathfrak{F}$ a dense subspace of $\mathfrak{L}$. If $\tau\colon \mathfrak{L} \to \mathfrak{L}$ is a bounded linear operator with the properties*

(a) $\tau^2 = \mathrm{id}_{\mathfrak{L}}$, *and*

(b) $\tau(\mathfrak{F}) \subseteq \mathfrak{F}$,

then $\mathfrak{F} \cap \mathrm{Ker}\,(\tau - \mathrm{id}_{\mathfrak{L}})$ is dense in $\mathrm{Ker}\,(\tau - \mathrm{id}_{\mathfrak{L}})$.

PROOF First note that

$$\mathrm{Ker}\,(\tau - \mathrm{id}_{\mathfrak{L}}) = \mathrm{Ran}\,(\tau + \mathrm{id}_{\mathfrak{L}}), \tag{8.22}$$

as an easy consequence of the property (a) of τ. It then follows by (b) that $(\tau + \mathrm{id}_{\mathfrak{L}})\mathfrak{F} \subseteq \mathfrak{F} \cap \mathrm{Ker}\,(\tau - \mathrm{id}_{\mathfrak{L}})$, hence it suffices to show that $(\tau + \mathrm{id}_{\mathfrak{L}})\mathfrak{F}$ is dense in $\mathrm{Ker}\,(\tau - \mathrm{id}_{\mathfrak{L}})$.

To this end, we remark that since $\mathfrak{F}$ is dense in $\mathfrak{L}$ and τ is continuous, it follows that $(\tau + \mathrm{id}_{\mathfrak{L}})\mathfrak{F}$ is dense in $\mathrm{Ran}\,(\tau + \mathrm{id}_{\mathfrak{L}})$, whence the desired fact follows by (8.22). ∎

PROOF (of Proposition 8.33) Since every L^*-algebra is the Hilbert space orthogonal sum of its center and its closed simple ideals (see Remark 7.10 and Theorem 7.11(a)) it clearly suffices to prove the assertion in the case of topologically simple L^*-algebras. The proof in this case has two stages.

1° First assume $\mathbb{K} = \mathbb{C}$ and let $\mathfrak{L}$ be an infinite-dimensional topologically simple L^*-algebra over $\mathbb{C}$. According to Theorem 7.18, there exists an infinite-dimensional complex Hilbert space $\mathfrak{H}$ such that either $\mathfrak{L} = \mathcal{C}_2(\mathfrak{H})$ or

$$\mathfrak{L} = \{T \in \mathcal{C}_2(\mathfrak{H}) \mid \tau(T) = T\}$$

for some continuous bijective real-linear operator $J\colon \mathfrak{H} \to \mathfrak{H}$, where

$$\tau\colon \mathcal{C}_2(\mathfrak{H}) \to \mathcal{C}_2(\mathfrak{H}), \qquad T \mapsto -JT^*J^{-1}$$

and $\tau^2 = \mathrm{id}_{\mathcal{C}_2(\mathfrak{H})}$. (We denote by T^* the Hilbert-space adjoint of an operator T on $\mathfrak{H}$.) Since it is easy to see that the ideal $\mathfrak{F}$ of finite-rank operators on $\mathfrak{H}$ is locally finite and is dense in $\mathcal{C}_2(\mathfrak{H})$, it then follows by Lemma 8.34 that the locally finite ideal $\mathfrak{F} \cap \mathfrak{L}$ is always dense in $\mathfrak{L}$.

2° Now assume $\mathbb{K} = \mathbb{R}$ and let $\mathfrak{X}$ be an infinite-dimensional topologically simple real L^*-algebra. It then follows by Proposition 7.17 that there exists a topologically simple complex L^*-algebra $\mathfrak{L}$ such that either $\mathfrak{X}$ is topologically isomorphic to $\mathfrak{L}$ with the scalars restricted to $\mathbb{R}$, or $\mathfrak{X}_{\mathbb{C}} \simeq \mathfrak{L}$. In the first case, the desired property of $\mathfrak{X}$ clearly follows by what we have already proved at the stage 1°.

We next assume that $\mathfrak{X}$ is a real form of $\mathfrak{L}$, and we use the notation of the stage 1°. Then there exists an involutive conjugate-linear $*$-automorphism $\widetilde{\sigma}$ of the complex L^*-algebra $\mathfrak{L}$ (the complex-conjugation of $\mathfrak{L}$ with respect to $\mathfrak{X}$) such that

$$\mathfrak{X} = \mathrm{Ker}\,(\widetilde{\sigma} - \mathrm{id}_{\mathfrak{L}}).$$

On the other hand, according to Corollary 7.22, every conjugate-linear $*$-automorphism of $\mathfrak{L}$ leaves invariant the ideal $\mathfrak{F} \cap \mathfrak{L}$ of finite-rank operators in $\mathfrak{L}$. Since we proved at stage 1° that the locally finite ideal $\mathfrak{F} \cap \mathfrak{L}$ is dense in $\mathfrak{L}$, it then follows by the above Lemma 8.34 that the locally finite ideal $\mathfrak{F} \cap \mathfrak{X}$ is dense in $\mathfrak{X}$, and the proof ends. ∎

THEOREM 8.35 *If $\mathfrak{g}$ is a real involutive Banach-Lie algebra such that $C_0^+(\mathfrak{g}) \neq \emptyset$, then both $\mathfrak{g}$ and its complexification $\mathfrak{g}_{\mathbb{C}}$ possess dense locally finite subalgebras.*

PROOF It clearly suffices to prove the assertion concerning $\mathfrak{g}$.

Since $C_0^+(\mathfrak{g}) \neq \emptyset$, it follows by Theorem 8.13 that there exist a real L^*-algebra $\mathfrak{X}$ and $\varphi \in \mathrm{Hom}^*(\mathfrak{X}, \mathfrak{g})$ with $\mathrm{Ker}\,\varphi = \{0\}$ and $\mathrm{Ran}\,\varphi$ dense in $\mathfrak{g}$. Now Proposition 8.34 implies that there exists a dense locally finite subalgebra $\mathfrak{X}_0$ of $\mathfrak{X}$. It then easily follows that $\varphi(\mathfrak{X}_0)$ is a dense locally finite subalgebra of the Banach-Lie algebra $\mathfrak{g}$. ∎

For the next statement we recall that a real Banach-Lie algebra $\mathfrak{g}_1$ is said to be a *linearly split central extension* of another real Banach-Lie algebra $\mathfrak{g}$ if there exist a closed central ideal $\mathfrak{z}$ of $\mathfrak{g}_1$, a continuous surjective homomorphism of Banach-Lie algebras $q\colon \mathfrak{g}_1 \to \mathfrak{g}$, and a continuous linear operator $\sigma\colon \mathfrak{g} \to \mathfrak{g}_1$ such that we have the short exact sequence

$$0 \to \mathfrak{z} \hookrightarrow \mathfrak{g}_1 \xrightarrow{q} \mathfrak{g} \to 0$$

and $q \circ \sigma = \mathrm{id}_{\mathfrak{g}}$.

COROLLARY 8.36 *If $\mathfrak{g}$ is a real involutive Banach-Lie algebra with $\mathcal{C}_0^+(\mathfrak{g}) \neq \emptyset$, then every real Banach-Lie algebra which is a linearly split central extension of either $\mathfrak{g}$ or its complexification $\mathfrak{g}_{\mathbb{C}}$ is enlargible.*

PROOF First use Theorem 8.35 to get locally finite subalgebras of $\mathfrak{g}$ and $\mathfrak{g}_{\mathbb{C}}$. Next note that every central extension of a locally finite Lie algebra is in turn locally finite. This allows us to construct a locally finite subalgebra of the linearly split central extension under consideration. Then the desired enlargibility follows by Lemma 8.32. $\square$

PROBLEM 8.37 It follows by Corollary 8.36 that if $\mathfrak{g}$ is a non-enlargible Banach-Lie algebra (see Example 3.35), then $\mathcal{C}_0^+(\mathfrak{g}) = \emptyset$ for every involution on $\mathfrak{g}$. What about $\mathcal{C}^+(\mathfrak{g})$? Is there any connection between $\mathcal{C}^+(\mathfrak{g})$ (respectively $\mathcal{C}_0^+(\mathfrak{g})$) and the period group of $\mathfrak{g}$, at least in the special case when $\mathfrak{g}$ is elliptic?

A positive answer to the latter question seems to be suggested by the result contained in Lemma IV.13 in [GN03]. Namely, there it is proved that if $\mathfrak{g}$ is an elliptic Banach-Lie algebra, then $\mathfrak{g}$ is enlargible if and only if its complexification is enlargible.

Notes

This is the first time that the results of Sections 8.1 and 8.2 on the relationship between L^*-ideals and equivariant monotone operators have been published. Their importance is pointed out by consequences like Corollary 8.21 or Example 8.23. Proposition 8.22 extends Proposition 2.12 in [Be05a] by removing the hypothesis that the Banach space underlying $\mathfrak{g}$ should be reflexive.

For further properties of elliptic Lie algebras, see [Ne02c] and [GN03]. It is interesting to compare our Proposition 8.25 with Proposition VII.23(i) in [Ne00].

The result contained in Theorem 8.29 seems to be new. Corollary 8.36 was announced as Theorem 2.3 in [Be03].

In the special case of complex separable L^*-algebras, the result contained in Proposition 8.33 is noted also in [Wo77].

Chapter 9

Homogeneous Spaces of Pseudo-restricted Groups

Abstract. The point of the present chapter is that the complex structures constructed in Chapter 6 are compatible with certain natural symplectic structures. More precisely, on the Lie algebraic level, the story is thus: consider the class of pseudo-restricted algebras (Definition 9.4). Each of these algebras comes naturally endowed with a continuous vector-valued 2-cocycle (Theorem 9.10), and it turns out that the methods of Chapter 6 actually lead to complex polarizations in those 2-cocycles (Theorem 9.12). Eventually, these turn out to be weakly Kähler polarizations under certain additional assumptions, including in particular the fact that the map ι involved in the definition of the pseudo-restricted algebra is an equivariant monotone operator (Theorem 9.15). The latter condition implies additionally that all the Lie algebras involved in these constructions correspond to certain Lie groups. This is a consequence of Corollary 8.36 and Theorem 3.32. Thus the complex polarizations constructed in this chapter (see Theorems 9.12 and 9.15) point to a general construction principle for Kähler homogeneous spaces. In the last part of the chapter we describe a class of examples of homogeneous spaces falling under that general principle. The background of these examples is the theory of operator ideals.

9.1 Pseudo-restricted algebras and groups

The goal of this section is to introduce a class of Banach-Lie algebras and Banach-Lie groups that are very important for the applications we want to make in this chapter. We mean the pseudo-restricted groups and algebras. Such an algebra will be constructed out of an equivariant monotone operator, and the specific properties of these operators will turn out to have deep consequences for certain homogeneous spaces of groups associated with pseudo-restricted algebras.

In Propositions 9.1 and 9.3 below we prepare the definition of pseudo-restricted algebras (Definition 9.4).

PROPOSITION 9.1 *Let $\mathfrak{z}_0$ be a real Banach space, $\mathfrak{Y}$ and $\mathfrak{V}$ Banach spaces over $\mathbb{K} \in \{\mathbb{R}, \mathbb{C}\}$, and $\iota \in \mathcal{B}(\mathfrak{V}, \mathfrak{Y})$ with $\operatorname{Ker} \iota = \{0\}$. We associate to a bounded linear map*

$$\Psi \colon \mathfrak{z}_0 \to \mathcal{B}(\mathfrak{Y})$$

the linear subspace of $\mathfrak{Y}$

$$\mathfrak{Y}(\Psi, \iota) := \{y \in \mathfrak{Y} \mid (\forall \gamma \in \mathfrak{z}_0) \quad \Psi(\gamma)y \in \operatorname{Ran} \iota\} = \bigcap_{\gamma \in \mathfrak{z}_0} \Psi(\gamma)^{-1}(\operatorname{Ran} \iota)$$

endowed with the norm $\| \cdot \|_{\mathfrak{Y}(\Psi,\iota)}$ defined by

$$(\forall y \in \mathfrak{Y}(\Psi, \iota)) \qquad \|y\|_{\mathfrak{Y}(\Psi,\iota)} = \|y\|_{\mathfrak{Y}} + \|\iota^{-1}(\Psi(\cdot)y)\|_{\mathcal{B}(\mathfrak{z}_0,\mathfrak{V})}.$$

Then $\mathfrak{Y}(\Psi, \iota)$ is a Banach space over $\mathbb{K}$ and the following assertions are equivalent:

(i) *There exists a bounded linear map*

$$\widetilde{\Psi} \colon \mathfrak{z}_0 \to \mathcal{B}(\mathfrak{V})$$

such that for all $\gamma \in \mathfrak{z}_0$ we have $\iota \circ \widetilde{\Psi}(\gamma) = \widetilde{\Psi}(\gamma) \circ \iota$.

(ii) *We have $\operatorname{Ran} \iota \subseteq \mathfrak{Y}(\Psi, \iota)$.*

(iii) *The linear subspace $\operatorname{Ran} \iota$ is invariant to $\Psi(\gamma)$ for all $\gamma \in \mathfrak{z}_0$.*

Moreover, if the assertions (i)–(iii) *hold, then*

(iv) *the map $\widetilde{\Psi}$ is uniquely determined;*

(v) *the linear space $\mathfrak{Y}(\Psi, \iota)$ is invariant to $\Psi(\gamma)$ for all $\gamma \in \mathfrak{z}_0$; and*

(vi) *there exists a linear isometry*

$$\psi \colon \mathfrak{Y}(\Psi, \iota) \to \mathfrak{Y} \times \mathcal{B}(\mathfrak{z}_0, \mathfrak{V})$$

such that, if $[\operatorname{Ran} \Psi, \operatorname{Ran} \Psi] = \{0\}$, then

$$(\forall \gamma \in \mathfrak{z}_0) \qquad \psi \circ (\Psi(\gamma)|_{\mathfrak{Y}(\Psi,\iota)}) = (\Psi(\gamma) \times M_{\widetilde{\Psi}(\gamma)}) \circ \psi,$$

where

$$M_{\widetilde{\Psi}(\gamma)} \colon \mathcal{B}(\mathfrak{z}_0, \mathfrak{V}) \to \mathcal{B}(\mathfrak{z}_0, \mathfrak{V}), \quad \chi \mapsto \widetilde{\Psi}(\gamma) \circ \chi,$$

and the Banach space $\mathfrak{Y} \times \mathcal{B}(\mathfrak{z}_0, \mathfrak{V})$ is endowed with the sum-norm, i.e.,

$$(\forall y \in \mathfrak{Y})(\forall \chi \in \mathcal{B}(\mathfrak{z}_0, \mathfrak{V})) \quad \|(y, \chi)\|_{\mathfrak{Y} \times \mathcal{B}(\mathfrak{z}_0,\mathfrak{V})} = \|y\|_{\mathfrak{Y}} + \|\chi\|_{\mathcal{B}(\mathfrak{z}_0,\mathfrak{V})}; \text{ and}$$

(vii) *in the case when* $\mathbb{K} = \mathbb{C}$ *and* $[\operatorname{Ran}\Psi, \operatorname{Ran}\Psi] = \{0\}$, *for each holomorphic function* $f\colon \mathbb{C} \to \mathbb{C}$ *we have*

$$(\forall \gamma \in \mathfrak{z}_0) \qquad \|f(\Psi(\gamma)|_{\mathfrak{Y}(\Psi,\iota)})\| \le \max\{\|f(\widetilde{\Psi}(\gamma))\|, \|f(\Psi(\gamma)\|\}.$$

PROOF The proof has several stages.

1° In order to check that $\mathfrak{Y}(\Psi, \iota)$ is a Banach space, assume that

$$\lim_{m,n\to\infty} \|y_m - y_n\|_{\mathfrak{Y}(\Psi,\iota)} = 0,$$

that is,

$$\lim_{m,n\to\infty} \|y_m - y_n\|_{\mathfrak{Y}} = 0 \tag{9.1}$$

and

$$\lim_{m,n\to\infty} \|\iota^{-1}(\Psi(\cdot)(y_m - y_n))\|_{\mathcal{B}(\mathfrak{z}_0,\mathfrak{V})} = 0. \tag{9.2}$$

Then (9.1) shows that $\lim\limits_{n\to\infty} \|y_n - y\|_{\mathfrak{Y}} = 0$ for some $y \in \mathfrak{Y}$, hence

$$\lim_{n\to\infty} \|\Psi(\cdot)y_n - \Psi(\cdot)y\|_{\mathcal{B}(\mathfrak{z}_0,\mathfrak{Y})} = 0. \tag{9.3}$$

On the other hand, by (9.2), there exists $\chi \in \mathcal{B}(\mathfrak{z}_0, \mathfrak{V})$ with

$$\lim_{n\to\infty} \|\iota^{-1}(\Psi(\cdot)y_n) - \chi\|_{\mathcal{B}(\mathfrak{z}_0,\mathfrak{V})} = 0.$$

Then $\lim\limits_{n\to\infty} \|\iota(\iota^{-1}(\Psi(\cdot)y_n) - \chi)\|_{\mathcal{B}(\mathfrak{z}_0,\mathfrak{Y})} = 0$, that is,

$$\lim_{n\to\infty} \|\Psi(\cdot)y_n - \iota \circ \chi\|_{\mathcal{B}(\mathfrak{z}_0,\mathfrak{Y})} = 0.$$

Hence we get by (9.3) that $\iota \circ \chi = \Psi(\cdot)y$, whence $y \in \mathfrak{Y}(\Psi, \iota)$. Moreover,

$$\begin{aligned}\|y_n - y\|_{\mathfrak{Y}(\Psi,\iota)} &= \|y_n - y\|_{\mathfrak{Y}} + \|\iota^{-1}(\Psi(\cdot)(y_n - y))\|_{\mathcal{B}(\mathfrak{z}_0,\mathfrak{V})} \\ &= \|y_n - y\|_{\mathfrak{Y}} + \|\iota^{-1}(\Psi(\cdot)y_n) - \chi\|_{\mathcal{B}(\mathfrak{z}_0,\mathfrak{V})} \underset{n\to\infty}{\longrightarrow} 0.\end{aligned}$$

2° Concerning the fact that (i) $\iff$ (ii) $\iff$ (iii), first note that the implication (i) $\Rightarrow$ (ii) is obvious, while (ii) $\iff$ (iii) by the very definition of $\mathfrak{Y}(\Psi, \iota)$. To get (ii) $\Rightarrow$ (i), we need only apply Lemma A.15 to the bounded bilinear map

$$\mathfrak{z}_0 \times \mathfrak{Y} \to \mathfrak{Y}, \quad (\gamma, y) \mapsto \Psi(\gamma)y.$$

3° Now assume that the equivalent assertions (i)–(iii) hold. Then (iv) follows, since $\operatorname{Ker}\iota = \{0\}$.

To prove (v), take $\gamma \in \mathfrak{z}_0$ arbitrary. By the very definition of $\mathfrak{Y}(\Psi, \iota)$, we then have $\Psi(\gamma)\big(\mathfrak{Y}(\Psi, \iota)\big) \subseteq \operatorname{Ran}\iota$, and now (ii) implies the desired conclusion.

For (vi), define

$$\psi\colon \mathfrak{Y}(\Psi, \iota) \to \mathfrak{Y} \times \mathcal{B}(\mathfrak{z}_0, \mathfrak{V}), \quad y \mapsto (y, \iota^{-1}(\Psi(\cdot)y)).$$

Then ψ is obviously an isometry when $\mathfrak{Y} \times \mathcal{B}(\mathfrak{z}_0, \mathfrak{V})$ is normed as indicated in the statement. Next, if $[\operatorname{Ran}\Psi, \operatorname{Ran}\Psi] = \{0\}$, then for all $y \in \mathfrak{Y}(\Psi, \iota)$ and $\gamma \in \mathfrak{z}_0$ we have

$$\begin{aligned}\psi(\Psi(\gamma)y) &= (\Psi(\gamma)y, \iota^{-1}(\Psi(\cdot)\Psi(\gamma)y)) = (\Psi(\gamma)y, \iota^{-1}(\Psi(\gamma)\Psi(\cdot)y)) \\ &= (\Psi(\gamma)y, \widetilde{\Psi}(\gamma)\iota^{-1}(\Psi(\cdot)y)) = (\Psi(\gamma) \times M_{\widetilde{\Psi}(\gamma)})\psi(y),\end{aligned}$$

and the proof of (vi) is complete.

The inequality in (vii) follows by (vi) in view of the fact that since the Banach space $\mathfrak{Y} \times \mathcal{B}(\mathfrak{z}_0, \mathfrak{V})$ is endowed with the sum-norm, we have

$$\begin{aligned}\|f(\Psi(\gamma) \times M_{\widetilde{\Psi}(\gamma)})\| &= \|f(\Psi(\gamma)) \times M_{f(\widetilde{\Psi}(\gamma))}\| = \max\{\|f(\Psi(\gamma))\|, \|M_{f(\widetilde{\Psi}(\gamma))}\|\} \\ &\leq \max\{\|f(\Psi(\gamma))\|, \|f(\widetilde{\Psi}(\gamma))\|\},\end{aligned}$$

and the proof is finished. ∎

REMARK 9.2

(a) In the setting of the above Proposition 9.1, we have

$$\bigcap_{\gamma \in \mathfrak{z}_0} \operatorname{Ker}\Psi(\gamma) \subseteq \mathfrak{Y}(\Psi, \iota).$$

Actually, if $y \in \mathfrak{Y}(\Psi, \iota)$, then

$$\|y\|_{\mathfrak{Y}(\Psi,\iota)} = \|y\|_{\mathfrak{Y}} \iff y \in \bigcap_{\gamma \in \mathfrak{z}_0} \operatorname{Ker}\Psi(\gamma).$$

(b) For later use, we now record still another remark. *In the setting of* Proposition 9.1, *if* $\mathbb{K} = \mathbb{R}$, $n := \dim \mathfrak{z}_0 < \infty$, *both* $\mathfrak{Y}$ *and* $\mathfrak{V}$ *are reflexive Banach spaces,* $[\operatorname{Ran}\Psi, \operatorname{Ran}\Psi] = \{0\}$ *and* $\mathrm{i}\Psi\colon \mathfrak{z}_0 \to \mathcal{B}(\mathfrak{Y}^{\mathbb{C}})$ *is a hermitian map, then* $\mathfrak{Y}(\Psi, \iota)$ *is a reflexive Banach space and the closed subspace*

$$\mathfrak{Y}_0 := \bigcap_{\gamma \in \mathfrak{z}_0} \operatorname{Ker}\Psi(\gamma)$$

is complemented in $\mathfrak{Y}(\Psi, \iota)$. To prove this assertion, first note that $\mathfrak{Y}(\Psi, \iota)$ is isometrically isomorphic to a closed subspace of a direct sum of $\mathfrak{Y}$ and n copies of $\mathfrak{V}$ (see Proposition 9.1(vi)), hence $\mathfrak{Y}(\Psi, \iota)$ is reflexive in view of the fact that both $\mathfrak{Y}$ and $\mathfrak{V}$ are reflexive. For the assertion concerning $\mathfrak{Y}_0$, let $z_1, \dots, z_n$ be a basis of $\mathfrak{z}_0$ and denote $A_j = \mathrm{i}\Psi(z_j)$ for $j = 1, \dots, n$. Then $A_1, \dots, A_n$ are hermitian operators on the reflexive Banach space $\mathfrak{Y}^{\mathbb{C}}$, hence by Corollary 4.5 in [Ma78] we get

$$\mathfrak{Y}^{\mathbb{C}} = \operatorname{Ker} A_1 \dotplus \operatorname{cl}(\operatorname{Ran} A_1),$$

where $\mathrm{cl}\,(\cdot)$ stands for the closure of a subset of $\mathfrak{Y}^{\mathbb{C}}$. But $A_1A_2 = A_2A_1$, hence A_2 leaves invariant both $\mathrm{Ker}\,A_1$ and $\mathrm{cl}\,(\mathrm{Ran}\,A_1)$, and it induces hermitian operators on each of these (reflexive!) Banach spaces. By Corollary 4.5 in [Ma78] again, we get

$$\begin{aligned}\mathfrak{Y}^{\mathbb{C}} =& (\mathrm{Ker}\,A_1 \cap \mathrm{Ker}\,A_2) \dotplus \mathrm{Ran}\,(A_1|_{\mathrm{Ker}\,A_1}) \\ &\dotplus \big(\mathrm{Ker}\,A_2 \cap \mathrm{cl}\,(\mathrm{Ran}\,A_1)\big) \dotplus \mathrm{cl}\,(\mathrm{Ran}\,A_2|_{\mathrm{cl}\,(\mathrm{Ran}\,A_1)}),\end{aligned}$$

and so on. One eventually gets

$$\mathfrak{Y}^{\mathbb{C}} = \Big(\bigcap_{j=1}^{n} \mathrm{Ker}\,A_j\Big) \dotplus \mathfrak{A}$$

for some closed subspace $\mathfrak{A}$ of $\mathfrak{Y}^{\mathbb{C}}$. Since $\bigcap_{j=1}^{n} \mathrm{Ker}\,A_j$ is invariant to that conjugation of $\mathfrak{Y}^{\mathbb{C}}$ whose set of fixed points is $\mathfrak{Y}$, it then easily follows that $\mathfrak{Y} = \mathfrak{Y}_0 \dotplus (\mathfrak{A} \cap \mathfrak{Y})$. But $\mathfrak{Y}_0 \subseteq \mathfrak{Y}(\Psi, \iota)$ (see the above Remark 9.2(a)), hence

$$\mathfrak{Y}(\Psi, \iota) = \mathfrak{Y}_0 \dotplus \big(\mathfrak{A} \cap \mathfrak{Y}(\Psi, \iota)\big).$$

Now it only remains to note that $\mathfrak{A} \cap \mathfrak{Y}(\Psi, \iota)$ is a closed subspace of $\mathfrak{Y}(\Psi, \iota)$, as a consequence of the fact that the inclusion map $\mathfrak{Y}(\Psi, \iota) \hookrightarrow \mathfrak{Y}$ is continuous.

PROPOSITION 9.3 *In the setting of* Proposition 9.1, *assume that* $\mathfrak{Y}$ *has a structure of Banach-Lie algebra, that* $\mathrm{Ran}\,\Psi \subseteq \mathrm{Der}(\mathfrak{Y})$, *and that we have a bounded representation*

$$\rho\colon \mathfrak{Y} \to \mathcal{B}(\mathfrak{V})$$

such that for every $y \in \mathfrak{Y}$ *the diagram*

$$\begin{array}{ccc} \mathfrak{V} & \xrightarrow{\ \iota\ } & \mathfrak{Y} \\ {\scriptstyle \rho(y)}\big\downarrow & & \big\downarrow{\scriptstyle \mathrm{ad}_{\mathfrak{Y}}\, y} \\ \mathfrak{V} & \xrightarrow{\ \iota\ } & \mathfrak{Y} \end{array}$$

is commutative. Then

(j) $\mathfrak{Y}(\Psi, \iota)$ *is a subalgebra of* $\mathfrak{Y}$ *which is a Banach-Lie algebra under the norm* $\|\cdot\|_{\mathfrak{Y}(\Psi,\iota)}$;

(jj) $\mathrm{Ran}\,\iota$ *is an ideal of* $\mathfrak{Y}$;

(jjj) *if* $\mathrm{Ran}\,\iota \subseteq \mathfrak{Y}(\Psi, \iota)$, *then we have a bounded linear map*

$$\Psi(\cdot)|_{\mathfrak{Y}(\Psi,\iota)}\colon \mathfrak{z}_0 \to \mathrm{Der}(\mathfrak{Y}(\Psi, \iota)).$$

PROOF (j) Let $y_1, y_2 \in \mathfrak{Y}(\Psi, \iota)$ and $\gamma \in \mathfrak{z}_0$ arbitrary. According to the definition of $\mathfrak{Y}(\Psi, \iota)$ in Proposition 9.1, there exist $v_1, v_2 \in \mathfrak{V}$ (depending on γ) such that $\Psi(\gamma)y_j = \iota(v_j)$ for $j = 1, 2$. Consequently we have

$$\begin{aligned}\Psi(\gamma)[y_1, y_2] &= [\Psi(\gamma)y_1, y_2][y_1, \Psi(\gamma)y_2] = [\iota(v_1), y_2] + [y_1, \iota(v_2)] \\ &= -((\mathrm{ad}_{\mathfrak{Y}} y_2 \circ \iota)v_1 + ((\mathrm{ad}_{\mathfrak{Y}} y_1) \circ \iota)v_2 \\ &= -(\iota \circ \rho(y_2))v_1 + (\iota \circ \rho(y_1))v_2 = \iota\big(-\rho(y_2)v_1 + \rho(y_1)v_2\big),\end{aligned}$$

whence $\Psi(\gamma)[y_1, y_2] \in \operatorname{Ran} \iota$ and

$$\iota^{-1}(\Psi(\gamma)[y_1, y_2]) = -(\rho(y_2))(\iota^{-1}(\Psi(\gamma)y_1)) + (\rho(y_1))(\iota^{-1}(\Psi(\gamma)y_2)). \quad (9.4)$$

Then $[y_1, y_2] \in \mathfrak{Y}(\Psi, \iota)$ and

$$\begin{aligned}&\|[y_1, y_2]\|_{\mathfrak{Y}(\Psi,\iota)} \\ &= \|[y_1, y_2]\|_{\mathfrak{Y}} + \|\iota^{-1}(\Psi(\cdot)[y_1, y_2])\|_{\mathcal{B}(\mathfrak{z}_0,\mathfrak{V})} \\ &= \|[y_1, y_2]\|_{\mathfrak{Y}} + \| - (\rho(y_2))(\iota^{-1}(\Psi(\cdot)y_1)) + (\rho(y_1))(\iota^{-1}(\Psi(\cdot)y_2))\|_{\mathcal{B}(\mathfrak{z}_0,\mathfrak{V})}\end{aligned}$$

by (9.4). Consequently

$$\begin{aligned}\|[y_1, y_2]\|_{\mathfrak{Y}(\Psi,\iota)} \leq& M \cdot \|y_1\|_{\mathfrak{Y}} \cdot \|y_2\|_{\mathfrak{Y}} + \|\rho\| \cdot \|y_2\|_{\mathfrak{Y}} \cdot \|\iota^{-1}(\Psi(\cdot)y_1)\|_{\mathcal{B}(\mathfrak{z}_0,\mathfrak{V})} \\ &+ \|\rho\| \cdot \|y_1\|_{\mathfrak{Y}} \cdot \|\iota^{-1}(\Psi(\cdot)y_2)\|_{\mathcal{B}(\mathfrak{z}_0,\mathfrak{V})} \\ \leq& C(\|y_1\|_{\mathfrak{Y}} + \|\iota^{-1}(\Psi(\cdot)y_1)\|_{\mathcal{B}(\mathfrak{z}_0,\mathfrak{V})}) \\ &\times (\|y_2\|_{\mathfrak{Y}} + \|\iota^{-1}(\Psi(\cdot)y_2)\|_{\mathcal{B}(\mathfrak{z}_0,\mathfrak{V})}) \\ =& C\|y_1\|_{\mathfrak{Y}(\Psi,\iota)} \cdot \|y_2\|_{\mathfrak{Y}(\Psi,\iota)},\end{aligned}$$

where $C = \max\{M, \|\rho\|\}$ and $M = \|\mathrm{ad}_{\mathfrak{Y}}\|$, $\mathrm{ad}_{\mathfrak{Y}} \colon \mathfrak{Y} \to \mathcal{B}(\mathfrak{Y})$ being the adjoint representation of $\mathfrak{Y}$.

(jj) See the commutative diagram in the statement.

(jjj) It follows by either Proposition 9.1(v) or Lemma A.15 that the linear map $\Psi(\cdot)|_{\mathfrak{Y}(\Psi,\iota)} \colon \mathfrak{z}_0 \to \mathrm{Der}(\mathfrak{Y}(\Psi, \iota))$ is continuous. ∎

We now give a name to the Banach-Lie algebra $\mathfrak{Y}(\Psi, \iota)$ arising in Proposition 9.3 above.

DEFINITION 9.4 Let $\mathbb{K} \in \{\mathbb{R}, \mathbb{C}\}$ and $\mathfrak{Y}$ a Banach-Lie algebra over $\mathbb{K}$. Also consider the Banach spaces $\mathfrak{V}$ over $\mathbb{K}$ and $\mathfrak{z}_0$ over $\mathbb{R}$, and $\iota \in \mathcal{B}(\mathfrak{V}, \mathfrak{Y})$ with $\operatorname{Ker} \iota = \{0\}$ and $\operatorname{Ran} \iota \trianglelefteq \mathfrak{Y}$. Then Lemma A.16 shows that there exists a bounded representation $\rho \colon \mathfrak{Y} \to \mathcal{B}(\mathfrak{V})$ such that $\iota \circ \rho(y) = (\mathrm{ad}_{\mathfrak{Y}} y) \circ \iota$ for all $y \in \mathfrak{Y}$.

The *pseudo-restricted algebra* associated to a bounded linear map

$$\Psi \colon \mathfrak{z}_0 \to \mathrm{Der}(\mathfrak{Y})$$

is the Banach-Lie algebra (over $\mathbb{K}$)

$$\mathfrak{Y}(\Psi, \iota) := \bigcap_{\gamma \in \mathfrak{z}_0} \Psi(\gamma)^{-1}(\operatorname{Ran} \iota)$$

endowed with the norm defined by

$$(\forall y \in \mathfrak{Y}(\Psi, \iota)) \qquad \|y\|_{\mathfrak{Y}(\Psi,\iota)} = \|y\|_{\mathfrak{Y}} + \|\iota^{-1}(\Psi(\cdot)y)\|_{\mathcal{B}(\mathfrak{z}_0,\mathfrak{V})}$$

(cf. Propositions 9.1 and 9.3 above). When no confusion can arise, we write simply $\mathfrak{Y}(\Psi)$ instead of $\mathfrak{Y}(\Psi, \iota)$.

A *pseudo-restricted group* associated with $\iota\colon \mathfrak{V} \to \mathfrak{Y}$ and $\Psi\colon \mathfrak{z}_0 \to \operatorname{Der}(\mathfrak{Y})$ as above is a Banach-Lie group G whose Lie algebra is isomorphic to the pseudo-restricted algebra $\mathfrak{Y}(\Psi, \iota)$.

EXAMPLE 9.5 Let $\mathcal{H}_\pm$ be two complex separable infinite-dimensional Hilbert spaces and consider their orthogonal sum $\mathcal{H} = \mathcal{H}_+ \oplus \mathcal{H}_-$. According to this orthogonal decomposition, we write each element of $\mathcal{B}(\mathcal{H})$ as a matrix $\begin{pmatrix} A & B \\ C & D \end{pmatrix}$ with $A \in \mathcal{B}(\mathcal{H}_+)$, $D \in \mathcal{B}(\mathcal{H}_-)$, $B \in \mathcal{B}(\mathcal{H}_-, \mathcal{H}_+)$, and $C \in \mathcal{B}(\mathcal{H}_+, \mathcal{H}_-)$. Next consider the operator

$$J = \begin{pmatrix} \mathrm{id}_{\mathcal{H}_+} & 0 \\ 0 & -\mathrm{id}_{\mathcal{H}_-} \end{pmatrix} \in \mathcal{B}(\mathcal{H}).$$

Now consider the special case of the setting of Definition 9.4 defined by $\mathfrak{Y} = \mathcal{B}(\mathcal{H})$, $\mathfrak{V} = \mathcal{C}_2(\mathcal{H})$ (the Hilbert-Schmidt operators), $\iota\colon \mathcal{C}_2(\mathcal{H}) \hookrightarrow \mathcal{B}(\mathcal{H})$ the inclusion map, and $\mathfrak{z}_0 = \mathbb{R}$. Define

$$\Psi\colon \mathbb{R} \to \operatorname{Der}(\mathfrak{Y}), \quad \Psi(t) = t \cdot \mathrm{ad}_{\mathfrak{Y}} J = t[J, \cdot\,].$$

Then it is easy to see that

$$\begin{aligned} \mathfrak{Y}(\Psi, \iota) &= \left\{ T = \begin{pmatrix} A & B \\ C & D \end{pmatrix} \in \mathcal{B}(\mathcal{H}) \mid B^*B \in \mathcal{C}_1(\mathcal{H}_-), C^*C \in \mathcal{C}_1(\mathcal{H}_+) \right\} \\ &= \left\{ T = \begin{pmatrix} A & B \\ C & D \end{pmatrix} \in \mathcal{B}(\mathcal{H}) \mid [J, T] = \begin{pmatrix} 0 & B \\ C & 0 \end{pmatrix} \in \mathcal{C}_2(\mathcal{H}) \right\} \end{aligned}$$

and

$$\|T\|_{\mathfrak{Y}(\Psi,\iota)} = \|T\| + \|[J, T]\|_{\mathcal{C}_2(\mathcal{H})}$$

for all $T \in \mathfrak{Y}(\Psi, \iota)$.

The pseudo-restricted Banach-Lie algebra $\mathfrak{Y}(\Psi, \iota)$ constructed in this example is usually called the *restricted algebra*.

REMARK 9.6 Pseudo-restricted groups always exist in the special case when in Definition 9.4 we have $\mathbb{K} = \mathbb{R}$, $\mathfrak{Y}$ an involutive Banach-Lie algebra, $\mathfrak{V} = \mathfrak{Y}^\#$, and $\iota \in \mathcal{C}_0^+(\mathfrak{Y})$ an injective equivariant monotone operator.

In fact, in this case we have $\mathcal{C}_0^+(\mathfrak{Y}) \neq \emptyset$, hence $\mathfrak{Y}$ is enlargible according to Corollary 8.36. On the other hand, the inclusion map $\mathfrak{Y}(\Psi, \iota) \hookrightarrow \mathfrak{Y}$ is a continuous injective homomorphism of Banach-Lie algebras, hence $\mathfrak{Y}(\Psi, \iota)$ is in turn enlargible as a consequence of Theorem 3.32.

Cocycles of pseudo-restricted algebras

Our next aim (Theorem 9.10) is to show that under natural hypotheses, the pseudo-restricted algebras are equipped with certain continuous 2-cocycles.

LEMMA 9.7 *Let $\mathfrak{z}_0$ be a real Banach space, $\mathfrak{Y}$ and $\mathfrak{V}$ Banach spaces over $\mathbb{K} \in \{\mathbb{R}, \mathbb{C}\}$, $\iota \in \mathcal{B}(\mathfrak{V}, \mathfrak{Y})$ with* $\operatorname{Ker} \iota = \{0\}$, *and $\Psi \colon \mathfrak{z}_0 \to \mathcal{B}(\mathfrak{Y})$ bounded linear. Then the linear operator (cf.* Proposition 9.1*)*

$$\mathfrak{Y}(\Psi, \iota) \to \mathcal{B}(\mathfrak{z}_0, \mathfrak{V}), \quad y \mapsto \iota^{-1}(\Psi(\cdot)y)$$

is bounded.

PROOF Define

$$A \colon \mathfrak{Y} \times \mathfrak{z}_0 \to \mathfrak{V}, \quad (y, \gamma) \mapsto \Psi(\gamma)y$$

and note that, denoting by $j \colon \mathfrak{Y}(\Psi, \iota) \to \mathfrak{Y}$ the inclusion map, we have

$$A\big((\operatorname{Ran} j) \times \mathfrak{z}_0\big) \subseteq \operatorname{Ran} \iota.$$

Since $\mathfrak{Y}(\Psi, \iota)$ is a Banach space (see Proposition 9.1), it then follows by Lemma A.15 that the bilinear map

$$\mathfrak{Y}(\Psi, \iota) \times \mathfrak{z}_0 \to \mathfrak{V}, \quad (y, \gamma) \mapsto \iota^{-1}(\Psi(\gamma)y)$$

is bounded. This implies that for each $y \in \mathfrak{Y}(\psi, \iota)$ we have $\iota^{-1}(\Psi(\cdot)y) \in \mathcal{B}(\mathfrak{z}_0, \mathfrak{V})$ and moreover the linear operator referred to in the statement is bounded. ∎

We now get a first idea of the aforementioned 2-cocycles of pseudo-restricted algebras.

REMARK 9.8 Let $\mathfrak{g}$ be a canonically involutive Banach-Lie algebra and $\iota \in \operatorname{Hom}_{\mathfrak{g}}^*(\mathfrak{g}^\#, \mathfrak{g})$ with $\operatorname{Ker} \iota = \{0\}$. Also consider a real Banach space $\mathfrak{z}_0$ and $\Psi \colon \mathfrak{z}_0 \to \operatorname{Der}(\mathfrak{g})$ bounded linear. Then Lemma 9.7 applied with $\mathfrak{V} = \mathfrak{g}^\#$ shows that the real Banach-Lie algebra $\mathfrak{g}(\Psi, \iota)$ (see Definition 9.4) comes endowed with a *bounded* bilinear map

$$\omega_{\Psi,\iota} \colon \mathfrak{g}(\Psi, \iota) \times \mathfrak{g}(\Psi, \iota) \to \mathfrak{z}_0^\#, \quad (y_1, y_2) \mapsto \langle \iota^{-1}(\Psi(\cdot)y_1), y_2 \rangle.$$

It turns out that under natural hypotheses, we have actually

$$\omega_{\Psi,\iota} \in Z^2_c\big(\mathfrak{g}(\Psi,\iota),\mathfrak{z}_0^{\#}\big)$$

(see Theorem 9.10 below).

REMARK 9.9 We now explain the connection between the bilinear map $\omega_{\Psi,\iota}$ in the above Remark 9.8 and the bilinear form ω dealt with in Lemma 2.8 in [Be05a]. To this end we use the notation of Remark 9.8 and note that

$$\mathfrak{g}(\Psi) := \mathfrak{g}(\Psi,\iota) = \bigcap_{z_0\in\mathfrak{z}_0} \mathfrak{g}(\Psi(z_0))$$

as subalgebras of $\mathfrak{g}$, where, for each $z_0 \in \mathfrak{z}_0$,

$$\mathfrak{g}(\Psi(z_0)) = \Psi(z_0)^{-1}(\operatorname{Ran}\iota)$$

is the subalgebra of $\mathfrak{g}$ considered in the hypothesis 3° before Lemma 2.8 in [Be05a]. For every $z_0 \in \mathfrak{z}_0$, the inclusion map

$$j_{z_0}\colon \mathfrak{g}(\Psi) \to \mathfrak{g}(\Psi(z_0))$$

is continuous and, if we denote by

$$\omega^{z_0}_{\Psi,\iota}\colon \mathfrak{g}(\Psi(z_0)) \times \mathfrak{g}(\Psi(z_0)) \to \mathbb{R}, \quad (y_1,y_2) \mapsto \langle \iota^{-1}(\Psi(z_0)y_1), y_2\rangle$$

the bilinear form dealt with in the above-mentioned Lemma 2.8 in [Be05a], then we have the commutative diagram

$$\begin{array}{ccc}
\mathfrak{g}(\Psi)\times\mathfrak{g}(\Psi) & \xrightarrow{\;j_{z_0}\times j_{z_0}\;} & \mathfrak{g}(\Psi(z_0))\times\mathfrak{g}(\Psi(z_0)) \\
{\scriptstyle\omega_{\Psi,\iota}}\Big\downarrow & & \Big\downarrow{\scriptstyle\omega^{z_0}_{\Psi,\iota}} \\
\mathfrak{z}_0^{\#} & \xrightarrow{\;\operatorname{ev}_{z_0}\;} & \mathbb{R}
\end{array}$$

where $\operatorname{ev}_{z_0}\colon \mathfrak{z}_0^{\#} \to \mathbb{R}$ is given by $f \mapsto f(z_0)$. This fact easily shows that

$$\omega_{\Psi,\iota} \in Z^2_c(\mathfrak{g}(\Psi),\mathfrak{z}_0^{\#}) \iff (\forall z_0 \in \mathfrak{z}_0) \quad \omega^{z_0}_{\Psi,\iota}(\mathfrak{g}(\Psi(z_0)),\mathbb{R}).$$

THEOREM 9.10 *Let $\mathfrak{g}$ be a canonically involutive Banach-Lie algebra with the complexification $\mathfrak{A}$, $\mathfrak{z}_0$ a real Banach space, and $\iota \in \operatorname{Hom}^*_{\mathfrak{g}}(\mathfrak{g}^{\#},\mathfrak{g})$ with $\operatorname{Ker}\iota = \{0\}$ and $\iota^{\#} = \iota$. Assume that $\Psi\colon \mathfrak{z}_0 \to \operatorname{Der}(\mathfrak{g})$ has the property that $\mathrm{i}\Psi \in \mathcal{B}(\mathfrak{z}_0,\mathcal{B}(\mathfrak{A}))$ is hermitian and*

$$(\forall z_0 \in \mathfrak{z}_0) \qquad \Psi(z_0)\circ\iota = -\iota\circ\Psi(z_0)^{\#}.$$

If moreover the Banach space underlying $\mathfrak{g}$ is reflexive, then

$$\omega_{\Psi,\iota} \in Z^2_c\big(\mathfrak{g}(\Psi,\iota),\mathfrak{z}_0^{\#}\big).$$

PROOF According to Remark 9.9, it suffices to show that

$$\omega_{\Psi,\iota}^{z_0} \in Z_c^2(\mathfrak{g}(\Psi(z_0)), \mathbb{R})$$

for each $z_0 \in \mathfrak{z}_0$. But this follows by Proposition 2.9 in [Be05a]. ∎

9.2 Complex polarizations

In this section we obtain a first theorem that supplies complex polarizations of pseudo-restricted algebras (see Theorem 9.12 below). To prove that theorem, we need the following auxiliary result.

LEMMA 9.11 *Let $\mathfrak{g}$ be a canonically involutive Banach-Lie algebra, $\mathfrak{z}_0$ a real Banach space, $\iota \in \mathrm{Hom}_{\mathfrak{g}}^*(\mathfrak{g}^\#, \mathfrak{g})$ with $\mathrm{Ker}\,\iota = \{0\}$, $\Psi \in \mathcal{B}(\mathfrak{z}_0, \mathrm{Der}(\mathfrak{g}))$, and $\alpha \in \mathcal{B}(\mathfrak{g})$ with*

$$[\alpha, \mathrm{Ran}\,\Psi] = \{0\}$$

and

$$\alpha \circ \iota = \varepsilon \cdot \iota \circ \alpha^\#$$

for some $\varepsilon \in \{-1, 1\}$. Then $\mathfrak{g}(\Psi, \iota)$ is invariant to α and

$$(\forall a_1, a_2 \in \mathfrak{g}(\Psi, \iota)) \qquad \omega_{\Psi,\iota}(\alpha(a_1), a_2) = \varepsilon \cdot \omega_{\Psi,\iota}(a_1, \alpha(a_2)).$$

PROOF For $j = 1, 2$, let $a_j \in \mathfrak{g}(\Psi, \iota)$ and denote $V_j := \iota^{-1}(\Psi(\cdot)a_j) \in \mathcal{B}(\mathfrak{z}_0, \mathfrak{g}^\#)$. Then $\iota \circ V_j = \Psi(\cdot)a_j$, hence

$$\Psi(\cdot)(\alpha(a_j)) = \alpha(\Psi(\cdot)a_j) = \alpha \circ \iota \circ V_j = \varepsilon \cdot \iota \circ \alpha^\# \circ V_j.$$

In particular, we have $\alpha(a_j) \in \mathfrak{g}(\Psi, \iota)$.

Moreover

$$\begin{aligned}\omega_{\Psi,\iota}(\alpha(a_1), a_2) &= \langle \iota^{-1}(\Psi(\cdot)\alpha(a_1)), a_2\rangle = \varepsilon \cdot \langle \alpha^\# \circ V_1, a_2\rangle = \varepsilon \cdot \langle V_1, \alpha(a_2)\rangle \\ &= \varepsilon \cdot \langle \iota^{-1}(\Psi(\cdot)a_1), \alpha(a_2)\rangle = \varepsilon \cdot \omega_{\Psi,\iota}(a_1, \alpha(a_2)),\end{aligned}$$

and the proof ends. ∎

Now we can prove one of our main results concerning complex polarizations of Banach-Lie algebras.

THEOREM 9.12 *Let $\mathfrak{g}$ be a canonically involutive Banach-Lie algebra with the complexification $\mathfrak{A}$, $\mathfrak{z}$ a finite-dimensional real vector space, and $\iota \in$*

$\mathrm{Hom}^*_{\mathfrak{g}}(\mathfrak{g}^\#, \mathfrak{g})$ *with* $\mathrm{Ker}\,\iota = \{0\}$ *and* $\iota^\# = \iota$. *Assume that* $\Psi_1 \colon \mathfrak{z}^\# \to \mathrm{Der}(\mathfrak{g})$ *has the property that the map*

$$\Psi_0 := \mathrm{i}\Psi_1 \in \mathcal{B}(\mathfrak{z}^\#, \mathcal{B}(\mathfrak{A}))$$

is hermitian, $[\mathrm{Ran}\,\Psi_1, \mathrm{Ran}\,\Psi_1] = \{0\}$, *and*

$$(\forall \gamma \in \mathfrak{z}^\#) \qquad \Psi_1(\gamma) \circ \iota = -\iota \circ \Psi_1(\gamma)^\#.$$

Also assume that the Banach space underlying $\mathfrak{g}$ *is reflexive and there exists a closed subsemigroup* S *of* $(\mathfrak{z}, +)$ *such that* $(-S) \cap S = \emptyset$ *and*

$$\sigma_W(\Psi_0) \subseteq (-S) \cup \{0\} \cup S.$$

Then the pseudo-restricted algebra

$$\widetilde{\mathfrak{g}} := \mathfrak{g}(\Psi_1, \iota)$$

and its complexification

$$\widetilde{\mathfrak{A}} = \mathfrak{A}(\Psi_0, \iota)$$

have the following properties:

(i) $\widetilde{\omega} := \omega_{\Psi_1, \iota} \in Z^2_c(\widetilde{\mathfrak{g}}, \mathfrak{z})$;

(ii) *for each* $\gamma \in \mathfrak{z}^\#$, $\widetilde{\mathfrak{g}}$ *is invariant to* $\Psi_1(\gamma)$ *and*

$$\widetilde{\Psi}_0 := \Psi_0(\cdot)|_{\widetilde{\mathfrak{A}}} \colon \mathfrak{z}^\# \to \mathcal{B}(\widetilde{\mathfrak{A}})$$

is a hermitian map;

(iii) *the subspace*

$$\mathfrak{p} := \widetilde{\mathfrak{A}}_{\widetilde{\Psi}_0}\big((-S) \cup \{0\}\big)$$

is a complex polarization of $\widetilde{\mathfrak{g}}$ *in* $\widetilde{\omega}$.

PROOF The property (i) follows by Theorem 9.10, after remarking that $(\mathfrak{z}^\#)^\# \simeq \mathfrak{z}$ as a consequence of the fact that $\mathfrak{z}$ is finite dimensional.

Next, the fact that $\widetilde{\mathfrak{g}}$ is invariant to $\Psi_1(\gamma)$ for all $\gamma \in \mathfrak{z}^\#$ follows by either Proposition 9.1(v) or Lemma 9.11 (the latter argument using the fact that we have $[\mathrm{Ran}\,\Psi_1, \mathrm{Ran}\,\Psi_1] = \{0\}$). To see that $\widetilde{\Psi}_0 \in \mathcal{B}(\mathfrak{z}^\#, \mathcal{B}(\widetilde{\mathfrak{A}}))$ is hermitian, first use Proposition 9.1(vi) to deduce that there exists a linear isometry

$$\psi \colon \widetilde{\mathfrak{A}} \to \mathfrak{A} \times \mathcal{B}(\mathfrak{z}, \widetilde{\mathfrak{A}})$$

such that for each $\gamma \in \mathfrak{z}^\#$ the diagram

$$\begin{array}{ccc} \widetilde{\mathfrak{A}} & \xrightarrow{\ \psi\ } & \mathfrak{A} \times \mathcal{B}(\mathfrak{z}, \mathfrak{A}^\#) \\ {\scriptstyle \widetilde{\Psi}_0(\gamma)}\big\downarrow & & \big\downarrow{\scriptstyle \Psi_0(\gamma) \times M_{\Psi_0(\gamma)^\#}} \\ \widetilde{\mathfrak{A}} & \xrightarrow{\ \psi\ } & \mathfrak{A} \times \mathcal{B}(\mathfrak{z}, \mathfrak{A}^\#) \end{array}$$

is commutative, where we denote as usual

$$M_{\Psi_0(\gamma)^\#}\colon \mathcal{B}(\mathfrak{z}^\#,\mathfrak{A}^\#)\to\mathcal{B}(\mathfrak{z},\mathfrak{A}^\#),\quad \chi\mapsto\Psi_0(\gamma)^\#\circ\chi.$$

Since $\Psi_0(\gamma)$ is a hermitian operator, it easily follows that $\Psi_0(\gamma)\times M_{\psi_0(\gamma)^\#}$ is hermitian (use Proposition 9.1(vii) for $f(z)=e^{itz}$ with $t\in\mathbb{R}$), and then the above commutative diagram and the fact that ψ is isometry show that the operator $\Psi_0(\gamma)$ is hermitian as well. Thus the desired assertion (ii) is completely proved.

To prove (ii), first use Lemma 9.11 once again to deduce that

$$(\forall\gamma\in\mathfrak{z}^\#)(\forall a,b\in\widetilde{\mathfrak{A}})\qquad \widetilde{\omega}(\widetilde{\Psi}_0(\gamma)a,b)=-\widetilde{\omega}(a,\widetilde{\Psi}_0(\gamma)b).$$

On the other hand, the above commutative diagram implies by Proposition 5.18(b) that

$$\sigma_W(\widetilde{\Psi}_0)\subseteq\sigma_W(\Psi_0(\cdot)\times M_{\Psi_0(\cdot)^\#})=\sigma_W(\Psi_0)\subseteq(-S)\cup\{0\}\cup S.$$

But

$$\begin{aligned}\widetilde{\mathfrak{A}}_{\widetilde{\Psi}_0}(\{0\})&=\bigcap_{\gamma\in\mathfrak{z}^\#}\operatorname{Ker}\widetilde{\Psi}_0(\gamma) &&\text{(by (5.14) in Example 5.25)}\\ &=\bigcap_{\gamma\in\mathfrak{z}^\#}\operatorname{Ker}\Psi_0(\gamma) &&\text{(see Remark 9.2(a))},\end{aligned}$$

hence

$$\widetilde{\mathfrak{A}}_{\widetilde{\Psi}_0}(\{0\})=\bigcap_{\gamma\in\mathfrak{z}^\#}\operatorname{Ker}\Psi_1(\gamma),\tag{9.5}$$

whence

$$\widetilde{\omega}(\widetilde{\mathfrak{A}}_{\widetilde{\Psi}_0}(\{0\}),\widetilde{\mathfrak{A}}_{\widetilde{\Psi}_0}(\{0\}))=\{0\}.$$

(See the definition of $\widetilde{\omega}=\omega_{\Psi_1,\iota}$ in Remark 9.8.) Now Proposition 5.21 shows that

$$\widetilde{\omega}(\mathfrak{p},\mathfrak{p})=\{0\},$$

with $\mathfrak{p}=\widetilde{\mathfrak{A}}_{\widetilde{\Psi}_0}((-S)\cup\{0\})$ as in the statement.

Next, Corollary 6.3 implies that $\mathfrak{p}$ is a complex subalgebra of $\widetilde{\mathfrak{A}}$, $\mathfrak{p}$ is complemented in $\widetilde{\mathfrak{A}}$, $\mathfrak{p}+\overline{\mathfrak{p}}=\widetilde{\mathfrak{A}}$, $\mathfrak{p}\cap\overline{\mathfrak{p}}=\mathfrak{h}+\mathrm{i}\mathfrak{h}$ and $[\mathfrak{h},\mathfrak{p}]\subseteq\mathfrak{p}$, where

$$\mathfrak{h}:=\widetilde{\mathfrak{g}}\cap\widetilde{\mathfrak{A}}_{\widetilde{\Psi}_0}(\{0\})=\bigcap_{\gamma\in\mathfrak{z}^\#}\operatorname{Ker}\Psi_1(\gamma),$$

by (9.5), and $a\mapsto\bar{a}$ denotes the involution of $\widetilde{\mathfrak{A}}$ whose fixed points are the elements of $\widetilde{\mathfrak{g}}$.

To complete the proof of the fact that $\mathfrak{p}$ is a complex polarization of $\widetilde{\mathfrak{g}}$ in $\widetilde{\omega}$, it only remained to show that

$$\mathfrak{h}=\{x\in\widetilde{\mathfrak{g}}\mid\widetilde{\omega}(x,\widetilde{\mathfrak{g}})=\{0\}\}.$$

To this end, recall that

$$\widetilde{\omega} = \omega_{\Psi_1,\iota} \colon \widetilde{\mathfrak{g}} \times \widetilde{\mathfrak{g}} \to \mathfrak{z}, \quad (x, y) \mapsto \langle \iota^{-1}(\Psi_1(\cdot)x), y\rangle$$

(cf. Remark 9.8). Hence

$$\bigcap_{\gamma \in \mathfrak{z}^{\#}} \operatorname{Ker} \Psi_1(\gamma) \subseteq \{x \in \widetilde{\mathfrak{g}} \mid \widetilde{\omega}(x, \widetilde{\mathfrak{g}}) = \{0\}\}.$$

To prove the converse inclusion, let $x \in \widetilde{\mathfrak{g}}$ such that

$$(\forall y \in \widetilde{\mathfrak{g}}) \qquad \langle \iota^{-1}(\Psi_1(\cdot)x), y\rangle = 0.$$

We have $\operatorname{Ran} \iota \subseteq \mathfrak{g}(\Psi_1, \iota) = \widetilde{\mathfrak{g}}$ (see Proposition 9.1(ii)), and $\operatorname{Ran} \iota$ is dense in $\mathfrak{g}$ because $\iota^{\#} = \iota$ and $\operatorname{Ker} \iota = \{0\}$. Hence $\widetilde{\mathfrak{g}}$ is in turn dense in $\mathfrak{g}$, and then the above equality implies that $\iota^{-1}(\Psi_1(\cdot)x) = 0$, whence $\Psi_1(\cdot)x = 0$. Consequently $x \in \bigcap_{\gamma \in \mathfrak{z}^{\#}} \operatorname{Ker} \Psi_1(\gamma)$, and the proof is finished. ∎

9.3 Kähler polarizations

Now we are going to refine Theorem 9.12, in the sense that we require polarizations satisfying the positivity condition described in Definition 6.10.

The proof of the density assertion in the following statement turns out to be rather involved, thus reflecting the fact that the topology of the pseudo-restricted algebra $\mathfrak{A}(\Psi_0, \iota)$ is in general stronger than the one inherited from the norm topology of $\mathfrak{A}$.

PROPOSITION 9.13 *Let $\mathfrak{A}$ be a complex Banach-Lie algebra and pick $\iota \in \mathcal{B}(\mathfrak{A}^{\#}, \mathfrak{A})$ with $\operatorname{Ker} \iota = \{0\}$ and $\operatorname{Ran} \iota$ a dense ideal of $\mathfrak{A}$. Also let $\mathfrak{z}$ be a finite-dimensional real vector space, and $\Psi_0 \in \mathcal{B}(\mathfrak{z}^{\#}, \operatorname{Der}(\mathfrak{A}))$ hermitian with $[\operatorname{Ran} \Psi_0, \operatorname{Ran} \Psi_0] = \{0\}$. If*

$$(\forall \gamma \in \mathfrak{z}^{\#}) \qquad \Psi_0(\gamma) \circ \iota = -\iota \circ \Psi_0(\gamma)^{\#}$$

and there exists a closed convex set C in $\mathfrak{z}$ with $(-C) \cap C = \emptyset$ and

$$\sigma_W(\Psi_0) \subseteq (-C) \cup \{0\} \cup C,$$

then $\operatorname{Ran} \iota$ is a dense ideal of the pseudo-restricted algebra $\mathfrak{A}(\Psi_0, \iota)$.

PROOF First note that since $[\operatorname{Ran} \Psi_0, \operatorname{Ran} \Psi_0] = \{0\}$, it follows that the pseudo-restricted algebra $\widetilde{\mathfrak{A}} := \mathfrak{A}(\Psi_0, \iota)$ is invariant to $\Psi_0(\gamma)$ for all $\gamma \in \mathfrak{z}^{\#}$, according to Lemma 9.11. Then the implication (iii) $\Rightarrow$ (ii) in Proposition 9.1 shows that $\operatorname{Ran}\iota \subseteq \widetilde{\mathfrak{A}}$. Since $\operatorname{Ran}\iota \trianglelefteq \mathfrak{A}$, we have also $\operatorname{Ran}\iota \trianglelefteq \widetilde{\mathfrak{A}}$. It only remains to prove that $\operatorname{Ran}\iota$ is dense in $\widetilde{\mathfrak{A}}$.

To this end, first recall from the proof of Theorem 9.12(ii) that the map

$$\widetilde{\Psi}_0 := \Psi_0(\cdot)|_{\widetilde{\mathfrak{A}}} \in \mathcal{B}(\mathfrak{z}^{\#}, \operatorname{Der}(\widetilde{\mathfrak{A}}))$$

is hermitian. The hypothesis also implies that $[\operatorname{Ran}\widetilde{\Psi}_0, \operatorname{Ran}\widetilde{\Psi}_0] = \{0\}$. Then, according to Example 5.25, there exists a multiplicative map $\widetilde{\Psi} \in \mathcal{E}'(\mathfrak{z}, \mathcal{B}(\widetilde{\mathfrak{A}}))$ such that $\widetilde{\Psi}|_{\mathfrak{z}^{\#}} = \widetilde{\Psi}_0$. For the same reason, there exists a multiplicative map $\Psi \in \mathcal{E}'(\mathfrak{z}, \mathcal{B}(\mathfrak{A}))$ with $\Psi|_{\mathfrak{z}^{\#}} = \Psi_0$.

If $j \colon \widetilde{\mathfrak{A}} \to \mathfrak{A}$ denotes the inclusion map, then $\operatorname{Ker} j = \{0\}$ and $\operatorname{Ran} j = \widetilde{\mathfrak{A}}$ is dense in $\mathfrak{A}$ (because we have already seen that $\operatorname{Ran}\iota \subseteq \widetilde{\mathfrak{A}}$, and $\operatorname{Ran}\iota$ is dense in $\mathfrak{A}$ by hypothesis). We then get by Proposition 5.18(a)–(b) that

$$\operatorname{supp}\Psi_N^{\#} = \operatorname{supp}\widetilde{\Psi} = \operatorname{supp}\Psi \subseteq (-C) \cup \{0\} \cup C, \tag{9.6}$$

where $\Psi_N^{\#} \in \mathcal{E}'(\mathfrak{z}, \mathcal{B}(\mathfrak{A}^{\#}))$ is defined by $f \mapsto \Psi(f \circ N)^{\#}$, with $N \colon \mathfrak{z} \to \mathfrak{z}$, $z \mapsto -z$.

Now, since C is a closed convex subset of $\mathfrak{z}$ with $C \cap (-C) = \emptyset$, there exists $\gamma_0 \in \mathfrak{z}^{\#}$ such that $\inf\limits_{C} \gamma_0 =: c > 0$. Pick $c_0 \in \mathbb{R}$ with $0 < c_0 < c$. Then

$$\begin{aligned}
\widetilde{\mathfrak{A}}_{\widetilde{\Psi}}((-C) \cup C) &= \widetilde{\mathfrak{A}}_{\widetilde{\Psi}}(\gamma_0^{-1}((-\infty, -c_0] \cup [c_0, \infty))) \\
&= \widetilde{\mathfrak{A}}_{\widetilde{\Psi}(\gamma_0)}((-\infty, -c_0] \cup [c_0, \infty)) \\
&= \widetilde{\mathfrak{A}}_{\widetilde{\Psi}(\gamma_0)}((-\infty, -c_0]) + \widetilde{\mathfrak{A}}_{\widetilde{\Psi}(\gamma_0)}([c_0, \infty)) \\
&\subseteq \operatorname{Ran}(\widetilde{\Psi}(\gamma_0)^2).
\end{aligned}$$

But for each $y \in \widetilde{\mathfrak{A}}$ we have $\Psi(\gamma_0) y = \iota(v)$ for some $v \in \mathfrak{A}^{\#}$, hence

$$\widetilde{\Psi}(\gamma_0)^2 y = (\widetilde{\Psi}(\gamma_0) \circ \iota) v = -(\iota \circ \widetilde{\Psi}(\gamma_0)^{\#}) v \in \operatorname{Ran}\iota.$$

In other words,

$$\operatorname{Ran}(\widetilde{\Psi}(\gamma_0)^2) \subseteq \operatorname{Ran}\iota,$$

and then

$$\widetilde{\mathfrak{A}}_{\widetilde{\Psi}}((-C) \cup C) \subseteq \operatorname{Ran}\iota. \tag{9.7}$$

On the other hand,

$$\iota(\mathfrak{A}^{\#}_{\Psi_N^{\#}}(\pm C)) \subseteq \mathfrak{A}_{\Psi}(\mp C) \quad \text{and} \quad \iota(\mathfrak{A}^{\#}_{\Psi_N^{\#}}(\{0\})) \subseteq \mathfrak{A}_{\Psi}(\{0\}) \tag{9.8}$$

by Remark 5.19, because we have by the hypothesis that $\Psi_0(\gamma) \circ \iota = \iota \circ \Psi_0(\gamma \circ N)^{\#}$ for all $\gamma \in \mathfrak{z}^{\#} \cup \{1\}$. Since

$$\begin{aligned}
\mathfrak{A}^{\#} &= \mathfrak{A}^{\#}_{\Psi_N^{\#}}(-C) \dotplus \mathfrak{A}^{\#}_{\Psi_N^{\#}}(\{0\}) \dotplus \mathfrak{A}^{\#}_{\Psi_N^{\#}}(C), \\
\mathfrak{A} &= \mathfrak{A}_{\Psi}(C) \dotplus \mathfrak{A}_{\Psi}(\{0\}) \dotplus \mathfrak{A}_{\Psi}(-C)
\end{aligned}$$

(by the above relation (9.6) and Theorem 5.14) and $\iota\colon \mathfrak{A}^{\#} \to \mathfrak{A}$ has dense range, it follows by (9.8) in particular that $\iota\big(\mathfrak{A}^{\#}_{\Psi^{\#}_{N}}(\{0\})\big)$ is dense in $\mathfrak{A}_{\Psi}(\{0\})$. Then Remark 9.2(a) shows that $\iota\big(\mathfrak{A}^{\#}_{\Psi^{\#}_{N}}(\{0\})\big)$ is dense also in $\widetilde{\mathfrak{A}}_{\widetilde{\Psi}}(\{0\})$. Now, since

$$\widetilde{\mathfrak{A}} = \widetilde{\mathfrak{A}}_{\widetilde{\Psi}}((-C)\cup C) + \widetilde{\mathfrak{A}}_{\widetilde{\Psi}}(\{0\})$$

(again by the above relation (9.6) and Theorem 5.14), it follows that $\operatorname{Ran}\iota$ is dense in $\widetilde{\mathfrak{A}}$, and the proof is finished. □

REMARK 9.14 The proof of the above Proposition 9.13 actually shows that

$$\iota\big(\mathfrak{A}^{\#}_{\Psi^{\#}_{N}}(\pm C)\big) = \widetilde{\mathfrak{A}}_{\widetilde{\Psi}}(\pm C) \text{ and } \widetilde{\mathfrak{A}} = \iota\big(\mathfrak{A}^{\#}_{\Psi^{\#}_{N}}(\mathfrak{z}\setminus\{0\})\big) + \widetilde{\mathfrak{A}}_{\widetilde{\Psi}}(\{0\}).$$

See also Remark 2.11 in [Be05a].

Now we are sufficiently equipped to construct complex polarizations satisfying the positivity condition (C5) in Definition 6.10. (See the property (iii) in the statement of the following theorem.)

THEOREM 9.15 *Let $\mathfrak{g}$ be a canonically involutive Banach-Lie algebra with the complexification $\mathfrak{A}$, $\iota \in \mathcal{C}_0^+(\mathfrak{g})$ and $\mathfrak{z}$ a finite-dimensional real vector space. Assume that*

$$\Psi_1\colon \mathfrak{z}^{\#} \to \operatorname{Der}(\mathfrak{g})$$

has the property that, for some spectral measure E on $\mathfrak{z}$ with values in $\mathcal{B}(\mathfrak{A})$, we have

$$(\forall \gamma\in\mathfrak{z}^{\#}) \qquad \mathrm{i}\Psi_1(\gamma) = \int_{\mathfrak{z}} \gamma(z)dE(z) \quad \text{and} \quad \Psi_1(\gamma)\circ\iota = -\iota\circ\Psi_1(\gamma)^{\#}.$$

Also assume that the Banach space underlying to $\mathfrak{g}$ is reflexive and there exists a compact convex set C in $\mathfrak{z}$ such that $(-C)\cap C=\emptyset$ and

$$\operatorname{supp} E \subseteq (-C)\cup\{0\}\cup C.$$

Then the pseudo-restricted algebra $\widetilde{\mathfrak{g}} := \mathfrak{g}(\Psi_1,\iota)$ and its complexification $\widetilde{\mathfrak{A}}$ have the following properties:

(i) $\widetilde{\omega} := \omega_{\Psi_1,\iota} \in Z^2_c(\widetilde{\mathfrak{g}},\mathfrak{z})$;

(ii) *for each Borel subset δ of $\mathfrak{z}$, $\widetilde{\mathfrak{A}}$ is invariant to $E(\delta)$ and*

$$\widetilde{E}(\cdot) = E(\cdot)|_{\widetilde{\mathfrak{A}}}$$

is a spectral measure on $\mathfrak{z}$ with values in $\mathcal{B}(\widetilde{\mathfrak{A}})$; and

(iii) *the subspace*

$$\mathfrak{p} := \widetilde{E}((-C) \cup \{0\})\widetilde{\mathfrak{A}}$$

is a complex polarization of $\widetilde{\mathfrak{g}}$ *in* $\widetilde{\omega}$ *and, for each* $y \in \mathfrak{p} \setminus \widetilde{E}(\{0\})\widetilde{\mathfrak{A}}$, *we have*

$$\mathrm{i}\widetilde{\omega}(y, \bar{y}) \in -\mathbb{R}_+^* \cdot C,$$

where $y \mapsto \bar{y}$ *stands for the involution of* $\mathfrak{A}$ *whose set of fixed points is* $\mathfrak{g}$.

PROOF The remark made in the end of Example 5.26 shows that for some equivalent norm on $\mathfrak{A}$, the map $\Psi_0 := \mathrm{i}\Psi_1 \in \mathcal{B}(\mathfrak{z}^{\#}, \mathcal{B}(\mathfrak{A}))$ is hermitian, hence Theorem 9.12 applies and we deduce that the assertion (i) holds.

The assertion (ii) easily follows by Proposition 5.27, taking into account the commutative diagram used in the proof of Theorem 9.12 above.

The fact that $\mathfrak{p}$ is a complex polarization of $\widetilde{\mathfrak{g}}$ in $\widetilde{\omega}$ follows by the similar assertion in Theorem 9.12.

To prove the last assertion from (iii), we consider the spectral measures

$$E^{\#}, E_N^{\#} \colon \mathrm{Bor}(\mathfrak{z}) \to \mathcal{B}(\mathfrak{A}^{\#}), \quad E^{\#}(\delta) := E(\delta)^{\#}, \quad E_N^{\#}(\delta) := E(-\delta)^{\#}.$$

Then the hypothesis together with Remark 5.28 imply that

$$(\forall \delta \in \mathrm{Bor}(\mathfrak{z})) \qquad E(\delta) \circ \iota = \iota \circ E_N^{\#}(\delta) = \widetilde{E}(\delta) \circ \iota.$$

Next note that by Remark 9.14,

$$\iota\big(\mathrm{Ran}\, E_N^{\#}(\pm C)\big) = \mathrm{Ran}\, \widetilde{E}(\pm C).$$

Thus, in order to prove that

$$(\forall y \in \mathfrak{p} \setminus \widetilde{E}(\{0\})\widetilde{\mathfrak{A}}) \qquad \mathrm{i}\widetilde{\omega}(y, \bar{y}) \in -\mathbb{R}_+^* \cdot C, \tag{9.9}$$

it suffices to prove that

$$(\forall v \in \mathrm{Ran}\, E_N^{\#}(-C) \setminus \{0\}) \qquad \mathrm{i}\widetilde{\omega}(\iota(v), \overline{\iota(v)}) \in -\mathbb{R}_+^* \cdot C.$$

Let us fix $v \in \mathrm{Ran}\, E_N^{\#}(-C) \setminus \{0\}$ and note that since $\iota(v) \in \mathrm{Ran}\, E(-C)$, we have

$$(\forall \gamma \in \mathfrak{z}^{\#}) \qquad \mathrm{i}\Psi_1(\gamma)(\iota(v)) = \int\limits_{-C} \gamma(z) dE(z)(\iota(v)).$$

But

$$\begin{aligned}\widetilde{\omega}(\iota(v), \overline{\iota(v)}) &= \langle \iota^{-1}(\Psi_1(\cdot)\iota(v)), \overline{\iota(v)} \rangle = \langle -\Psi_1(\cdot)^{\#} v, \overline{\iota(v)} \rangle = \langle \bar{v}, \iota(-\Psi_1(\cdot)^{\#} v) \rangle \\ &= \langle \bar{v}, \Psi_1(\cdot)\iota(v) \rangle,\end{aligned}$$

where the third equality follows since $\iota^{\#} = \iota$. Hence for every $\gamma \in \mathfrak{z}^{\#}$ (extended to a complex-linear functional on the complexification of $\mathfrak{z}$) we have

$$\begin{aligned}\gamma(\mathrm{i}\widetilde{\omega}(\iota(v), \overline{\iota(v)})) &= \langle \bar{v}, \mathrm{i}\Psi_1(\gamma)\iota(v)\rangle = \int\limits_{-C} \gamma(z)\langle \bar{v}, E(dz)\iota(v)\rangle \\ &= \int\limits_{-C} \gamma(z)\langle \bar{v}, E(dz)^2\iota(v)\rangle = \int\limits_{-C} \gamma(z)\langle E^{\#}(dz)\bar{v}, E(dz)\iota(v)\rangle \\ &= \int\limits_{-C} \gamma(z)\langle E^{\#}(dz)\bar{v}, \iota E_N^{\#}(dz)v\rangle.\end{aligned}$$

Thus

$$\mathrm{i}\widetilde{\omega}(\iota(v), \overline{\iota(v)}) = \int\limits_{-C} z\langle E^{\#}(dz)\bar{v}, \iota E_N^{\#}(dz)v\rangle. \tag{9.10}$$

But for each Borel subset δ of $-C$ we have $E^{\#}(\delta)\bar{v} = \overline{E_N^{\#}(\delta)v}$ by Remark 5.29, hence

$$\langle E^{\#}(\delta)\bar{v}, \iota E_N^{\#}(\delta)v\rangle = \langle \overline{E_N^{\#}(\delta)v}, \iota(E_N^{\#}(\delta)v)\rangle \geq \frac{1}{\|\iota\|} \cdot \|\iota(\overline{E_N^{\#}(\delta)v})\|^2,$$

according to inequality (7.8) in Remark 7.25. Consequently, we have in (9.10) an integral with respect to a non-negative measure. Since $-C$ is a closed convex set and $E_N^{\#}(-C)v = v$, it then follows by (9.10) that

$$\mathrm{i}\widetilde{\omega}(\iota(v), \overline{\iota(v)}) \in \langle E^{\#}(-C)\bar{v}, \iota E_N^{\#}(-C)v\rangle \cdot (-C) = \langle \bar{v}, \iota(v)\rangle \cdot (-C) \subseteq \mathbb{R}_+^* \cdot (-C)$$

by inequality (7.8) in Remark 7.25. This finishes the proof of (9.9). ∎

We now think it worthwhile to show how the preceding constructions look in the situation of L^*-algebras, where the equivariant monotone operator ι is surjective (compare Proposition 8.22).

EXAMPLE 9.16 Let $\mathfrak{g}$ be a compact (i.e., canonically involutive) L^*-algebra with the scalar product denoted by $(\cdot \mid \cdot)$. The complexification of $\mathfrak{g}$ will be a complex L^*-algebra $\mathfrak{A}$, whose scalar product is denoted again by $(\cdot \mid \cdot)$. We also consider a finite-dimensional real vector space $\mathfrak{z}$.

We perform the usual identification of $\mathfrak{g}$ to its topological dual $\mathfrak{g}^{\#}$, by means of the scalar product $(\cdot \mid \cdot)$. In other words, we set $\mathfrak{g}^{\#} = \mathfrak{g}$ with the duality pairing $\langle \cdot, \cdot \rangle \colon \mathfrak{g} \times \mathfrak{g} \to \mathbb{R}$ given by $(\cdot \mid \cdot)$. Then the operator $\iota = \mathrm{id}_{\mathfrak{g}}$ has all of the properties needed in Theorem 9.12.

Next consider $\Psi_1 \colon \mathfrak{z}^{\#} \to \mathrm{Der}(\mathfrak{g})$ such that $[\mathrm{Ran}\,\Psi_1, \mathrm{Ran}\,\Psi_1] = \{0\}$ and

$$(\forall \gamma \in \mathfrak{z}^{\#}) \qquad \Psi_1(\gamma)^{\#} = -\Psi_1(\gamma).$$

In other words, $\operatorname{Ran}\Psi_1$ is a commuting set of skew-symmetric derivations of $\mathfrak{g}$. Then, for

$$\Psi_0 := \mathrm{i}\Psi_1(\cdot) \in \mathcal{B}(\mathfrak{z}^{\#}, \mathcal{B}(\mathfrak{A})),$$

$\operatorname{Ran}\Psi_0$ will be a commuting set of self-adjoint operators on the complex Hilbert space $\mathfrak{A}$, hence Ψ_0 is automatically a hermitian map. In fact, there exists a spectral measure E on $\mathfrak{z}$ with values orthogonal projections on $\mathfrak{A}$ such that

$$(\forall \gamma \in \mathfrak{z}^{\#}) \qquad \mathrm{i}\Psi_1(\gamma) = \int\limits_{\mathfrak{z}} \gamma(z) dE(z)$$

(see e.g., Theorem 10.3 in Chapter IV in [Va82]).

On the other hand, since ι is surjective, we have $\mathfrak{g}(\Psi_1, \iota) = \mathfrak{g}$ (with an equivalent norm). Then

$$\omega_{\Psi_1,\iota} \colon \mathfrak{g} \times \mathfrak{g} \to \mathfrak{z}, \qquad (x, y) \mapsto (\Psi_1(\cdot)x \mid y),$$

and we have $\omega_{\Psi_1,\iota} \in Z^2_c(\mathfrak{g}, \mathfrak{z})$ either by Theorem 9.10 or by direct checking, using the fact that $\mathfrak{g}$ is a canonically involutive L^*-algebra and the values of Ψ_1 are skew-symmetric derivations of $\mathfrak{g}$.

Now assume that there exists a compact convex subset C of $\mathfrak{z}$ such that the sets $-C, \{0\}, C$ are pairwise disjoint and

$$\operatorname{supp} E \subseteq (-C) \cup \{0\} \cup C.$$

It then follows by Theorem 9.15 that

$$\mathfrak{p} := \operatorname{Ran} E\big((-C) \cup \{0\}\big)$$

is a weakly Kähler polarization of $\mathfrak{g}$ in the 2-cocycle $\omega_{\Psi_1,\iota}$.

Next let $\gamma_0 \in \mathfrak{z}^{\#}$ such that $\gamma_0(C) \subseteq (0, \infty)$, and denote

$$A_0 := \Psi_0(\gamma_0) \in \mathcal{B}(\mathfrak{A}).$$

Then A_0 is a self-adjoint operator on $\mathfrak{A}$ and, denoting by E_{A_0} the spectral measure of A_0, we easily get

$$\sigma(A_0) = \big(-\gamma_0(C \cap \operatorname{supp} E)\big) \cup \big(\{0\} \cap \operatorname{supp} E\big) \cup \big(\gamma_0(C \cap \operatorname{supp} E)\big),$$

and

$$\operatorname{Ran} E_{A_0}(\pm\gamma_0(C)) = \operatorname{Ran} E(\pm C) \quad \text{and} \quad \operatorname{Ker} A_0 = \bigcap_{\gamma \in \mathfrak{z}^{\#}} \operatorname{Ker} \Psi_0(\gamma).$$

On the other hand, the preceding arguments with $\mathfrak{z}$ replaced by $\mathbb{R} \cdot \gamma_0$ show that $\mathfrak{p}$ is a weakly Kähler polarization also for the cocycle $\omega^{\gamma_0}_{\Psi_1,\iota} \in Z^2_c(\mathfrak{g}, \mathbb{R})$ defined by

$$\omega^{\gamma_0}_{\Psi_1,\iota} \colon \mathfrak{g} \times \mathfrak{g} \to \mathbb{R}, (x, y) \mapsto (\Psi_1(\gamma_0)x \mid y).$$

We now show that $\mathfrak{p}$ is actually a *strongly* Kähler polarization of $\mathfrak{g}$ in the cocycle $\omega_{\Psi_1,\iota}^{\gamma_0} \in Z_c^2(\mathfrak{g},\mathbb{R})$. To this end, denote

$$\mathfrak{h} := \{x \in \mathfrak{g} \mid \omega_{\Psi_1,\iota}^{\gamma_0}(x,\mathfrak{g}) = \{0\}\} = \bigcap_{\gamma\in\mathfrak{z}^\#} \operatorname{Ker}\Psi_1(\gamma) = (\operatorname{Ker} A_0)\cap\mathfrak{g} = \operatorname{Ker}\Psi_1(\gamma_0)$$

and perform the identification $\mathfrak{g}/\mathfrak{h} \simeq \mathfrak{h}^\perp \subseteq \mathfrak{g}$. Since $\operatorname{Ker} A_0 = \mathfrak{h} + \mathrm{i}\mathfrak{h}$, it follows that the orthogonal complement of $\mathfrak{h}$ in $\mathfrak{A}$ is $\operatorname{Ran} A_0 = \operatorname{Ran} E_{A_0}\big((-C)\cup C\big)$, hence

$$(\forall x \in \mathfrak{h}^\perp \subseteq \mathfrak{g}) \qquad \|\Psi_1(\gamma_0)x\|^2 \geq c\cdot\|x\|^2,$$

where $c := \inf\gamma_0(C) > 0$. It then follows that the operator $\Psi_1(\gamma_0)|_{\mathfrak{h}^\perp}$ has closed range. On the other hand $\operatorname{Ran}\big(\Psi_1(\gamma_0)|_{\mathfrak{h}^\perp}\big) = \operatorname{Ran}\Psi_1(\gamma_0)$ and $\mathfrak{h}^\perp$ is the closure of $\operatorname{Ran}\Psi_1(\gamma_0)$, hence $\Psi_1(\gamma_0)|_{\mathfrak{h}^\perp}\colon \mathfrak{h}^\perp \to \mathfrak{h}^\perp$ is an invertible operator. And this fact implies that

$$\mathfrak{h}^\perp \to (\mathfrak{h}^\perp)^\#, \quad x \mapsto \big(\Psi_1(\gamma_0)x \mid \cdot\big)$$

is a topological isomorphism. In other words, $\mathfrak{p}$ is a strongly Kähler polarization of $\mathfrak{g}$ in $\omega_{\Psi_1,\iota}^{\gamma_0} \in Z_c^2(\mathfrak{g}.\mathbb{R})$.

We now formulate a problem concerning the preceding constructions of complex polarizations.

PROBLEM 9.17 Is it possible to extend Theorems 9.12 and 9.15 (or at least Example 9.16) to infinite-dimensional real Banach spaces $\mathfrak{z}$? Note that no additional condition is imposed upon the real Banach space $\mathfrak{z}_0$ in Theorem 9.10 and Definitions 6.8 and 6.10, hence the preceding question makes sense. A positive answer to it would be a first step towards enlarging the class of infinite-dimensional Kähler homogeneous spaces constructed in this book. One of the main technical difficulties raised by the aforementioned question seems to be the dropping of the assumption $\dim\mathfrak{z} < \infty$ needed in Chapter 5.

9.4 Admissible pairs of operator ideals

Although the Schatten ideals have been widely used in order to construct Banach-Lie groups and related objects (see e.g., [dlH72] and [Ne04]), it seems that other ideals of compact operators have not yet been much studied from the point of view of the Lie theory.

We try in this section to fill that gap, describing in Definition 9.22 some axioms for pairs of operator ideals which, on one side, lead to symplectic homogeneous spaces (see Theorem 9.29) and Kähler homogeneous spaces (Theorem 9.33), and on the other side are fulfilled by operator ideals as the Lorentz ones (see Example 9.25).

DEFINITION 9.18 By *Banach ideal* we mean a two-sided ideal $\mathfrak{I}$ of $\mathcal{B}(\mathcal{H})$ equipped with a norm $\|\cdot\|_{\mathfrak{I}}$ satisfying $\|T\| \le \|T\|_{\mathfrak{I}} = \|T^*\|_{\mathfrak{I}}$ and $\|ATB\|_{\mathfrak{I}} \le \|A\|\,\|T\|_{\mathfrak{I}}\,\|B\|$ whenever $A, B \in \mathcal{B}(\mathcal{H})$.

DEFINITION 9.19 Let $\widehat{c}$ be the vector space of all sequences of real numbers $\{\xi_j\}_{j\ge 1}$ such that $\xi_j = 0$ for all but finitely many indices. A *symmetric norming function* is a function $\Phi\colon \widehat{c} \to \mathbb{R}$ satisfying the following conditions:

I) $\Phi(\xi) > 0$ whenever $0 \ne \xi \in \widehat{c}$,

II) $\Phi(\alpha\xi) = |\alpha|\Phi(\xi)$ whenever $\alpha \in \mathbb{R}$ and $\xi \in \widehat{c}$,

III) $\Phi(\xi + \eta) \le \Phi(\xi) + \Phi(\eta)$ whenever $\xi, \eta \in \widehat{c}$,

IV) $\Phi((1, 0, 0, \dots)) = 1$, and

V) $\Phi(\{\xi_j\}_{j\ge 1}) = \Phi(\{\xi_{\pi(j)}\}_{j\ge 1})$ whenever $\{\xi_j\}_{j\ge 1} \in \widehat{c}$ and $\pi\colon \{1, 2, \dots\} \to \{1, 2, \dots\}$ is bijective.

Any symmetric norming function Φ gives rise to two Banach ideals $\mathfrak{S}_\Phi$ and $\mathfrak{S}_\Phi^{(0)}$ as follows. For every bounded sequence of real numbers $\xi = \{\xi_j\}_{j\ge 1}$ define

$$\Phi(\xi) := \sup_{n\ge 1} \Phi(\xi_1, \xi_2, \dots, \xi_n, 0, 0, \dots) \in [0, \infty].$$

For all $T \in \mathcal{B}(\mathcal{H})$ denote

$$\|T\|_\Phi := \Phi(\{s_j(T)\}_{j\ge 1}) \in [0, \infty],$$

where $s_j(T) = \inf\{\|T - F\| \mid F \in \mathcal{B}(\mathcal{H}), \operatorname{rank} F < j\}$ whenever $j \ge 1$. With this notation we can define

$$\begin{aligned} \mathfrak{S}_\Phi &= \{T \in \mathcal{B}(\mathcal{H}) \mid \|T\|_\Phi < \infty\}, \\ \mathfrak{S}_\Phi^{(0)} &= \overline{\mathfrak{F}}^{\|\cdot\|_\Phi} \quad (\subseteq \mathfrak{S}_\Phi), \end{aligned}$$

that is, $\mathfrak{S}_\Phi^{(0)}$ is the $\|\cdot\|_\Phi$-closure of the finite-rank operators $\mathfrak{F}$ in $\mathfrak{S}_\Phi$. Then $\|\cdot\|_\Phi$ is a norm making $\mathfrak{S}_\Phi$ and $\mathfrak{S}_\Phi^{(0)}$ into Banach ideals (see §4 in Chapter III in [GK69]).

REMARK 9.20 Every separable Banach ideal equals $\mathfrak{S}_\Phi^{(0)}$ for some symmetric norming function Φ (see Theorem 6.2 in Chapter III in [GK69]).

REMARK 9.21 For every symmetric norming function $\Phi\colon \widehat{c} \to \mathbb{R}$ there exists a unique symmetric norming function $\Phi^*\colon \widehat{c} \to \mathbb{R}$ such that

$$\Phi^*(\eta) = \sup\left\{ \frac{1}{\Phi(\xi)} \sum_{j=1}^{\infty} \xi_j \eta_j \;\middle|\; \xi = \{\xi_j\}_{j\geq 1} \in \widehat{c} \text{ and } \xi_1 \geq \xi_2 \geq \cdots \geq 0 \right\}$$

whenever $\eta = \{\eta_j\}_{j\geq 1} \in \widehat{c}$ and $\eta_1 \geq \eta_2 \geq \cdots \geq 0$. The function Φ^* is said to be *adjoint* to Φ and we always have $(\Phi^*)^* = \Phi$ (see Theorem 11.1 in Chapter III in [GK69]). For instance, if $1 \leq p, q \leq \infty$, $1/p + 1/q = 1$, $\Phi_p(\xi) = \|\xi\|_{\ell^p}$ and $\Phi_q(\xi) = \|\xi\|_{\ell^q}$ whenever $\xi \in \widehat{c}$, then $(\Phi_p)^* = \Phi_q$. If Φ is any symmetric norming function, then the topological dual of the Banach space $\mathfrak{S}_\Phi^{(0)}$ is isometrically isomorphic to $\mathfrak{S}_{\Phi^*}$ by means of the duality pairing

$$\mathfrak{S}_{\Phi^*} \times \mathfrak{S}_\Phi^{(0)} \to \mathbb{C}, \quad (T, S) \mapsto \operatorname{Tr}(TS)$$

(see Theorems 12.2 and 12.4 in Chapter III in [GK69]).

The following definition describes an appropriate framework in order to construct a wide class of infinite-dimensional symplectic homogeneous spaces (see Theorem 9.29).

DEFINITION 9.22 Let $\mathcal{H}$ be a complex Hilbert space with the scalar product denoted by $(\cdot \mid \cdot)$. An *admissible pair of ideals* of $\mathcal{B}(\mathcal{H})$ is a pair $(\mathfrak{J}_0, \mathfrak{J}_1)$ of two-sided self-adjoint ideals of $\mathcal{B}(\mathcal{H})$ satisfying the following conditions.

(a) The ideal $\mathfrak{J}_0$ is equipped with a norm $\|\cdot\|_{\mathfrak{J}_0}$ making it into a reflexive Banach space.

(b) For all $A, B \in \mathcal{B}(\mathcal{H})$ and $T \in \mathfrak{J}_0$ we have

$$\|T\| \leq \|T\|_{\mathfrak{J}_0} = \|T^*\|_{\mathfrak{J}_0} \quad \text{and} \quad \|ATB\|_{\mathfrak{J}_0} \leq \|A\| \cdot \|T\|_{\mathfrak{J}_0} \cdot \|B\|.$$

For all $x, y \in \mathcal{H}$ we have $x(\cdot \mid y) \in \mathfrak{J}_0$ and

$$\|x(\cdot \mid y)\|_{\mathfrak{J}_0} = \|x\| \cdot \|y\|.$$

(c) We have $\mathfrak{J}_0 \cdot \mathfrak{J}_1 + \mathfrak{J}_1 \cdot \mathfrak{J}_0 \subseteq \mathcal{C}_1(\mathcal{H})$ and $\operatorname{Tr}(TK) = \operatorname{Tr}(KT)$ whenever $T \in \mathfrak{J}_0$ and $K \in \mathfrak{J}_1$. Moreover, the bilinear functional

$$\mathfrak{J}_1 \times \mathfrak{J}_0 \to \mathbb{C}, \qquad (K, T) \mapsto \operatorname{Tr}(KT),$$

induces a vector space isomorphism of $\mathfrak{J}_1$ onto the topological dual of the Banach space $(\mathfrak{J}_0, \|\cdot\|_{\mathfrak{J}_0})$. We denote by $\|\cdot\|_{\mathfrak{J}_1}$ the norm which $\mathfrak{J}_1$ gets in this way.

(d) We have $\mathfrak{J}_1 \subseteq \mathfrak{J}_0$.

In these conditions, we denote

$$\mathfrak{u}_{\mathfrak{J}_k} = \{T \in \mathfrak{J}_k \mid T^* = -T\}$$

for $k = 0, 1$.

REMARK 9.23

(a) In Definition 9.22, $\mathfrak{J}_1$ is in particular the adjoint to $\mathfrak{J}_0$ in the sense of Section 9.1.1. (see also 10.3.6) in [Pit78]. We also note that the inclusion map $\mathfrak{J}_1 \hookrightarrow \mathfrak{J}_0$ is continuous as an easy consequence of the closed graph theorem.

(b) It is worth mentioning that some of the conditions in Definition 9.22 actually are consequences of the others. Thus, every two-sided ideal of $\mathcal{B}(\mathcal{H})$ is self-adjoint and contains all the finite-rank operators, provided it does not reduce to $\{0\}$ (see §1 in Chapter III in [GK69]). Moreover, if $\mathfrak{J}_0$ and $\mathfrak{J}_1$ are two-sided ideals of $\mathcal{B}(\mathcal{H})$ and $\mathfrak{J}_0 \cdot \mathfrak{J}_1 + \mathfrak{J}_1 \cdot \mathfrak{J}_0 \subseteq \mathcal{C}_1(\mathcal{H})$, then it follows by Theorems 1.1 and 8.2 in Chapter III in [GK69] that $\mathrm{Tr}\,(TK) = \mathrm{Tr}\,(KT)$ for all $T \in \mathfrak{J}_0$ and $K \in \mathfrak{J}_1$.

EXAMPLE 9.24 The simplest example of an admissible pair of ideals of $\mathcal{B}(\mathcal{H})$ is a pair of Schatten ideals $\big(\mathcal{C}_{p_0}(\mathcal{H}), \mathcal{C}_{p_1}(\mathcal{H})\big)$ with $2 \le p_0 < \infty$ and $\frac{1}{p_0} + \frac{1}{p_1} = 1$.

EXAMPLE 9.25 More sophisticated examples of admissible pairs of ideals can be constructed using Remark 9.21. By way of illustration, we now briefly consider the duality theory of Lorentz ideals (see [GL74], [Co88]). More specifically, denote as in Example 7.39 by $\mathcal{C}_\infty(\mathcal{H})$ the set of compact operators on $\mathcal{H}$. For $T \in \mathcal{C}_\infty(\mathcal{H})$, let $\{s_n(T)\}_{n\ge 1}$ be the sequence of singular numbers of T, i.e., the non-increasing sequence of eigenvalues of $|T| = (T^*T)^{1/2}$ listed with multiplicity. (One can prove that the sequence of singular numbers $\{s_n(T)\}_{n\ge 1}$ is the one used in Definition 9.19; see e.g., [GK69] for details in this connection.) For all $p, q \in (0, \infty)$ we define the corresponding *Lorentz ideal* by

$$\mathcal{C}_{q,p} = \Big\{T \in \mathcal{C}_\infty(\mathcal{H}) \mid \|T\|_{q,p} := \Big(\sum_{n=1}^{\infty} \frac{1}{n^{1-\frac{p}{q}}} s_n(T)^p\Big)^{1/p} < \infty\Big\}.$$

(In particular, if $p = q$, then $\mathcal{C}_{q,p}$ coincides with the Schatten ideal $\mathcal{C}_p(\mathcal{H})$.) *If*

$$2 \le p_0 \le q_0 < \infty \text{ and } \frac{1}{p_1} + \frac{1}{p_0} = \frac{1}{q_1} + \frac{1}{q_0} = 1,$$

then $(\mathcal{C}_{q_0,p_0}, \mathcal{C}_{q_1,p_1})$ *is an admissible pair of ideals of* $\mathcal{B}(\mathcal{H})$*, and one can take just* $\|\cdot\|_{q_0,p_0}$ *in the role of the needed norm on* $\mathcal{C}_{q_0,p_0}$. Indeed, property (a) in Definition 9.22 follows by Theorem 10 (and its proof) and Corollary 13 in [GL74]. The checking of property (b) is straightforward. Property (c) follows by Lemma 9 and Theorem 4 (and its proof) in [GL74]. Finally, to prove that property (d) in Definition 9.22 holds, note that $p_0 \ge p_1$ since $2 \le p_0$, and $\frac{p_0}{q_0} \le \frac{p_1}{q_1}$ since $p_0 \le q_0$, whence

$$(\forall T \in \mathcal{C}_\infty(\mathcal{H})) \quad (\forall n \ge 1) \qquad \frac{1}{n^{1-\frac{p_0}{q_0}}} s_n(T)^{p_0} \le \frac{1}{n^{1-\frac{p_1}{q_1}}} s_n(T)^{p_1} \cdot \|T\|^{p_0 - p_1},$$

which implies at once that $\mathcal{C}_{q_1,p_1} \subseteq \mathcal{C}_{q_0,p_0}$.

REMARK 9.26 If $(\mathfrak{J}_0, \mathfrak{J}_1)$ is an admissible pair of ideals of $\mathcal{B}(\mathcal{H})$, then the following assertions hold.

(i) Endowed with the restriction of $\|\cdot\|_{\mathfrak{J}_0}$, the set $\mathfrak{u}_{\mathfrak{J}_0}$ (see the last part of Definition 9.22) is a real Banach-Lie algebra whose underlying Banach space is reflexive.

(ii) The topological dual of $\mathfrak{u}_{\mathfrak{J}_0}$ can be identified with $\mathfrak{u}_{\mathfrak{J}_1}$ (endowed with the restriction of $\|\cdot\|_{\mathfrak{J}_1}$) by means of the pairing

$$\mathfrak{u}_{\mathfrak{J}_1} \times \mathfrak{u}_{\mathfrak{J}_0} \to \mathbb{R}, \quad (K, T) \mapsto \langle K, T \rangle := -\mathrm{Tr}\,(KT).$$

(iii) We have $\mathfrak{u}_{\mathfrak{J}_1} \subseteq \mathfrak{u}_{\mathfrak{J}_0}$ and, if we identify $\mathfrak{u}_{\mathfrak{J}_0}^{\#} = \mathfrak{u}_{\mathfrak{J}_1}$ according to (ii), then the inclusion map $\iota\colon \mathfrak{u}_{\mathfrak{J}_1} \to \mathfrak{u}_{\mathfrak{J}_0}$ is an equivariant monotone operator for the canonically involutive Banach-Lie algebra $\mathfrak{u}_{\mathfrak{J}_0}$.

Note that the assertion (iii) extends the assertion $\iota \in \mathcal{C}^+(\mathfrak{u}_p)$ from Example 8.12 and can be proved by the reasoning from the final of that example.

DEFINITION 9.27 For every two-sided self-adjoint ideal $\mathfrak{J}$ of $\mathcal{B}(\mathcal{H})$, we define

$$\mathrm{GL}_{\mathfrak{J}} := \mathrm{GL}(\mathcal{H}) \cap (\mathrm{id}_{\mathcal{H}} + \mathfrak{J})$$

and

$$\mathrm{U}_{\mathfrak{J}} := \mathrm{U}(\mathcal{H}) \cap (\mathrm{id}_{\mathcal{H}} + \mathfrak{J}),$$

where $\mathrm{GL}(\mathcal{H})$ denotes the group of all invertible operators on $\mathcal{H}$, and $\mathrm{U}(\mathcal{H})$ stands for the group of the unitary ones.

PROPOSITION 9.28 *If $\mathfrak{J}$ is a two-sided self-adjoint ideal of $\mathcal{B}(\mathcal{H})$, endowed with a complete norm $\|\cdot\|_{\mathfrak{J}}$ stronger than the operator norm, then the following assertions hold.*

(i) *The set $\mathrm{GL}_{\mathfrak{J}}$ has a structure of a complex Banach-Lie group with the Lie algebra $\mathfrak{J}$, such that the inclusion map $\mathrm{GL}_{\mathfrak{J}} \hookrightarrow \mathrm{GL}(\mathcal{H})$ is a homomorphism of Banach-Lie groups.*

(ii) *The set* $\mathrm{U}_{\mathfrak{J}}$ *is a real Banach-Lie subgroup of* $\mathrm{GL}_{\mathfrak{J}}$ *with the Lie algebra*

$$\mathfrak{u}_{\mathfrak{J}} = \{T \in \mathfrak{J} \mid T^* = -T\}.$$

PROOF First note that as a straightforward consequence of the hypotheses and of Lemma A.15, there exists a constant $M > 0$ such that

$$(\forall T, S \in \mathfrak{J}) \qquad \|TS\|_{\mathfrak{J}} \leq M \cdot \|T\|_{\mathfrak{J}} \cdot \|S\|_{\mathfrak{J}},$$

hence $\mathfrak{J}$ is in fact a Banach algebra. Now the desired assertion (i) follows from the fact that if $\mathfrak{J} \neq \mathcal{B}(\mathcal{H})$, then $\mathrm{GL}_{\mathfrak{J}}$ is a Banach-Lie subgroup (of complex codimension 1) of the group of invertible elements in the unitalization of $\mathfrak{J}$. (See e.g., Examples IV.15 (c)-(d) in [Ne04].)

The assertion (ii) follows by (i), using Proposition 4.8 and the fact that $\mathrm{U}(\mathcal{H})$ is a real Banach-Lie subgroup of $\mathrm{GL}(\mathcal{H})$ (cf. also Remark 4.14 for the C^*-algebra $\mathcal{B}(\mathcal{H})$). Note that $\{T \in \mathfrak{J} \mid T^* = T\}$ is a *closed* complement to $\mathfrak{u}_{\mathfrak{J}}$ in $\mathfrak{J}$, in view of the fact that by the closed graph theorem, the mapping $T \mapsto T^*$ is continuous on $\mathfrak{J}$. ∎

9.5 Some Kähler homogeneous spaces

We recall that for a subset $\mathcal{S}$ of $\mathcal{B}(\mathcal{H})$, the symbol $\mathcal{S}'$ stands for the *commutant* of $\mathcal{S}$ (that is, the set of all elements of $\mathcal{B}(\mathcal{H})$ which commute with each element of $\mathcal{S}$). Our next aim is to describe a class of symplectic homogeneous spaces associated to admissible pairs of ideals of $\mathcal{B}(\mathcal{H})$.

THEOREM 9.29 *Let* $(\mathfrak{J}_0, \mathfrak{J}_1)$ *be an admissible pair of ideals of* $\mathcal{B}(\mathcal{H})$ *and denote by* $\iota\colon \mathfrak{u}_{\mathfrak{J}_1} \to \mathfrak{u}_{\mathfrak{J}_0}$ *the inclusion map. Next let* $n \geq 1$ *and* $A = (A_1, \dots, A_n)$ *be a commuting* n*-tuple of bounded self-adjoint operators on* $\mathcal{H}$*, and define*

$$\Psi_1\colon \mathbb{R}^n \to \mathrm{Der}(\mathfrak{u}_{\mathfrak{J}_0}), \qquad (t_1, \dots, t_n) \mapsto \mathrm{ad}_{\mathfrak{u}_{\mathfrak{J}_0}}\big(i(t_1A_1 + \cdots + t_nA_n)\big),$$

and then

$$\mathrm{U}_{\mathfrak{J}_0,\mathfrak{J}_1}(A) := \{T \in \mathrm{U}_{\mathfrak{J}_0} \mid [A_j, T] \in \mathfrak{J}_1 \text{ for } j = 1, \dots, n\}$$

and

$$H := \mathrm{U}_{\mathfrak{J}_0,\mathfrak{J}_1}(A) \cap \{A_1, \dots, A_n\}'.$$

Then the following assertions hold.

(i) *The set* $\mathrm{U}_{\mathfrak{J}_0,\mathfrak{J}_1}(A)$ *has a structure of a real Banach-Lie group whose Lie algebra is the pseudo-restricted algebra* $\mathfrak{u}_{\mathfrak{J}_0}(\Psi_1,\iota)$, *such that the inclusion map*

$$\mathrm{U}_{\mathfrak{J}_0,\mathfrak{J}_1}(A) \hookrightarrow \mathrm{U}_{\mathfrak{J}_0}$$

is a homomorphism of Banach-Lie groups.

(ii) *The set* H *is a Banach-Lie subgroup of* $\mathrm{U}_{\mathfrak{J}_0,\mathfrak{J}_1}(A)$ *with the Lie algebra*

$$\mathfrak{h} := \mathfrak{u}_{\mathfrak{J}_0} \cap \{A_1,\dots,A_n\}'.$$

(iii) *The cocycle* $\omega_{\Psi_1,\iota} \in Z^2_c(\mathfrak{u}_{\mathfrak{J}_0}(\Psi_1,\iota),\mathbb{R}^n)$ *induces an* $\mathbb{R}^n$*-valued smooth 2-form* Ω *on* $\mathrm{U}_{\mathfrak{J}_0,\mathfrak{J}_1}(A)/H$ *such that the pair* $(\mathrm{U}_{\mathfrak{J}_0,\mathfrak{J}_1}(A)/H,\Omega)$ *is a weakly symplectic homogeneous space of* $\mathrm{U}_{\mathfrak{J}_0,\mathfrak{J}_1}(A)$ *of type* $\mathbb{R}^n$.

PROOF (i) Let us denote

$$\mathrm{G}_{\mathfrak{J}_0,\mathfrak{J}_1}(A) = \{T \in \mathrm{GL}_{\mathfrak{J}_0} \mid [A_j,T] \in \mathfrak{J}_1 \text{ for } j=1,\dots,n\}.$$

Since $\mathfrak{J}_1$ is a two-sided ideal of $\mathcal{B}(\mathcal{H})$, it follows at once that $\mathrm{G}_{\mathfrak{J}_0,\mathfrak{J}_1}(A)$ is an abstract subgroup of $\mathrm{GL}_{\mathfrak{J}_0}$. Actually, defining (as usual)

$$\Psi_0\colon \mathbb{R}^n \to \mathrm{Der}(\mathfrak{J}_0), \qquad (t_1,\dots,t_n) \mapsto \mathrm{ad}_{\mathfrak{J}_0}(-(t_1A_1+\cdots+t_nA_n)),$$

and denoting also by ι the inclusion map $\mathfrak{J}_1 \hookrightarrow \mathfrak{J}_0$, it is easy to see that $\mathrm{G}_{\mathfrak{J}_0,\mathfrak{J}_1}(A)$ is just the group of invertible elements of the complex associative unital Banach algebra

$$\begin{aligned}\mathfrak{J}_0(\Psi_0,\iota) &= \{T \in \mathfrak{J}_0 \mid (\forall t \in \mathbb{R}^n \quad \Psi_0(t)T \in \mathfrak{J}_1\} \\ &= \{T \in \mathfrak{J}_0 \mid [A_j,T] \in \mathfrak{J}_1 \text{ for } j=1,\dots,n\}\end{aligned}$$

with the norm defined by

$$\|T\|_{\mathfrak{J}_0(\Psi_0,\iota)} = \|T\|_{\mathfrak{J}_0} + \sup\{\|\Psi_0(t)T\|_{\mathfrak{J}_1} \mid t \in \mathbb{R}^n,\ \|t\|_2 \le 1\}.$$

(Compare Propositions 9.1 and 9.3.) Consequently, $\mathrm{G}_{\mathfrak{J}_0,\mathfrak{J}_1}(A)$ has a structure of a complex Banach-Lie group with the Lie algebra $\mathfrak{J}_0(\Psi_0,\iota)$. The inclusion map

$$\theta\colon \mathrm{G}_{\mathfrak{J}_0,\mathfrak{J}_1}(A) \to \mathrm{GL}_{\mathfrak{J}_0}$$

is a homomorphism of Banach-Lie groups, as an easy consequence of the fact that the inclusion map $\mathfrak{J}_0(\Psi_0,\iota) \hookrightarrow \mathfrak{J}_0$ is continuous.

Now the assertion (i) follows by the remark that

$$\mathrm{U}_{\mathfrak{J}_0,\mathfrak{J}_1}(A) = \theta^{-1}(\mathrm{U}_{\mathfrak{J}_0}),$$

using Propositions 9.28(ii) and 4.8. Note that since all of $A_1,\dots,A_n$ are self-adjoint operators, it follows that $\mathfrak{J}_0(\Psi_0,\iota)$ is closed under the taking of

Hilbert space adjoints, whence $i\mathfrak{u}_{\mathfrak{J}_0}(\Psi_1,\iota)$ is a direct complement to $\mathfrak{u}_{\mathfrak{J}_0}(\Psi_1,\iota)$ in $\mathfrak{J}_0(\psi_0,\iota)$. (Compare the final of the proof of Proposition 9.28(ii).)

(ii) First note that $\mathrm{U}(\mathcal{H})\cap\{A_1,\dots,A_n\}'$ is the group of unitary elements in the unital C^*-algebra $\{A_1,\dots,A_n\}'$, hence it has a natural structure of a real Banach-Lie group (see Remark 4.14). Since the topology of this Banach-Lie group is just the operator norm topology, $\mathrm{U}(\mathcal{H})\cap\{A_1,\dots,A_n\}'$ is in fact a Banach-Lie subgroup of $\mathrm{U}(\mathcal{H})$. (One can use an argument from the proof of Theorem 4.33 to get a direct complement of the Lie algebra of $\mathrm{U}(\mathcal{H})\cap\{A_1,\dots,A_n\}'$ in the Lie algebra of $\mathrm{U}(\mathcal{H})$.)

On the other hand, the inclusion map

$$\eta\colon \mathrm{U}_{\mathfrak{J}_0,\mathfrak{J}_1}(A)\to\mathrm{U}(\mathcal{H})$$

is a homomorphism of Banach-Lie groups (by the already proved assertion (i) and Proposition 9.28) and

$$\mathrm{U}_{\mathfrak{J}_0,\mathfrak{J}_1}(A)\cap\{A_1,\dots,A_n\}=\eta^{-1}(\mathrm{U}(\mathcal{H})\cap\{A_1,\dots,A_n\}'),$$

hence, using Proposition 4.8 once again, we deduce that $H=\mathrm{U}_{\mathfrak{J}_0,\mathfrak{J}_1}(A)\cap\{A_1,\dots,A_n\}$ is a Banach-Lie subgroup of $\mathrm{U}_{\mathfrak{J}_0,\mathfrak{J}_1}(A)$. (Since the Lie algebra of H clearly coincides with $\mathfrak{h}=\mathfrak{u}_{\mathfrak{J}_0}\cap\{A_1,\dots,A_n\}'$, the fact that it is complemented in $\mathfrak{u}_{\mathfrak{J}_0}(\Psi_1,\iota)$ follows by Remark 9.2(b).)

(iii) Note that $\omega_{\Psi_1,\iota}\in Z^2_c(\mathfrak{u}_{\mathfrak{J}_0}(\Psi_1,\iota),\mathbb{R}^n)$ by Theorem 9.10. Moreover, we have

$$\mathfrak{h}=\{Y\in\mathfrak{u}_{\mathfrak{J}_0}(\Psi_1,\iota)\mid\omega_{\psi_1,\iota}(Y,\cdot)=\{0\}\},$$

as a consequence of the fact that $\mathfrak{u}_{\mathfrak{J}_0}(\Psi_1,\iota)$ is dense in $\mathfrak{u}_{\mathfrak{J}_0}$. (The latter fact follows because $\operatorname{Ran}\iota$ is dense in $\mathfrak{u}_{\mathfrak{J}_0}$ and $\operatorname{Ran}\iota\subseteq\mathfrak{u}_{\mathfrak{J}_0}(\Psi_1,\iota)$.)

Now the desired assertion follows by Theorem 4.30. ∎

REMARK 9.30 In Theorem 9.29, if $\mathfrak{J}_0=\mathfrak{J}_1=\mathcal{C}_2(\mathcal{H})$ (the Hilbert-Schmidt class), $n=1$, and the spectrum of $A=A_1$ is a finite set, then $(U_{\mathfrak{J}_0,\mathfrak{J}_1}(A)/H,\Omega)$ is a *strongly* symplectic homogeneous space of $U_{\mathfrak{J}_0,\mathfrak{J}_1}(A)$ of type $\mathbb{R}$. See Corollary 9.34 below.

LEMMA 9.31 *Assume that $\mathfrak{J}$ is a two-sided self-adjoint ideal of $\mathcal{B}(\mathcal{H})$, endowed with a complete norm $\|\cdot\|_{\mathfrak{J}}$ stronger than the operator norm, such that*

$$(\forall A,B\in\mathcal{B}(\mathcal{H}))\ (\forall T\in\mathfrak{J})\qquad \|ATB\|_{\mathfrak{J}}\le\|A\|\cdot\|T\|_{\mathfrak{J}}\|B\|$$

and

$$(\forall x,y\in\mathcal{H})\qquad x(\cdot\mid y)\in\mathfrak{J}\text{ and }\|x(\cdot\mid y)\|_{\mathfrak{J}}=\|x\|\cdot\|y\|.$$

If $A_1,\dots,A_n\in\mathcal{B}(\mathcal{H})$ are commuting self-adjoint operators, then the map

$$\Psi_0\colon\mathbb{R}^n\to\mathcal{B}(\mathfrak{J}),\quad(t_1,\dots,t_n)\mapsto\operatorname{ad}_{\mathfrak{J}}(t_1A_1+\cdots+t_nA_n)$$

is hermitian and

$$\sigma_W(\Psi_0) = \{(t_1 - s_1, \dots, t_n - s_n) \mid t_j, s_j \in \sigma(A_j) \text{ for } j = 1, \dots, n\}.$$

PROOF In the statement, we identify as usual $(\mathbb{R}^n)^\#$ to $\mathbb{R}^n$ by means of the usual scalar product of $\mathbb{R}^n$.

The fact that Ψ_0 is a hermitian map is a consequence of the well-known fact that

$$(\forall A, B \in \mathcal{B}(\mathcal{H})) \qquad \big(\exp(\mathrm{ad}_{\mathcal{B}(\mathcal{H})} A)\big)B = e^A B e^{-A}.$$

Next recall that, as mentioned in Example 5.25, $\sigma_W(\Psi_0)$ agrees with (i.e., equals) the Taylor joint spectrum of the commuting n-tuple

$$(\mathrm{ad}_{\mathfrak{J}} A_1, \dots, \mathrm{ad}_{\mathfrak{J}} A_n)$$

of bounded linear operators on $\mathfrak{J}$. But the latter spectrum can be computed by means of Korollar 4.3 and Satz 4.1 in [Es87], and this leads to the desired conclusion. ∎

In the proof of Theorem 9.33 we need the following elementary fact.

LEMMA 9.32 *For every finite subset F of $\mathbb{R}^n$, there exists a compact convex subset C of $\mathbb{R}^n$ such that $F \subseteq (-C) \cup \{0\} \cup C$ and $(-C) \cap C = \emptyset$.*

PROOF Consider

$$L = \{0\} \cup \{(t_1, \dots, t_n) \in \mathbb{R}^n \mid \text{either } t_1 > 0, \text{ or } t_1 = 0 \text{ and } t_2 > 0, \dots, \\ \text{or } t_1 = \dots = t_{n-1} = 0 \text{ and } t_n > 0\}.$$

Then L is a convex cone in $\mathbb{R}^n$ such that

$$(-L) \cap L = \{0\} \text{ and } (-L) \cup L = \mathbb{R}^n.$$

(In fact, L is the convex cone which defines the lexicographic ordering on $\mathbb{R}^n$.) Also note that the convex hull of every finite subset of $L \setminus \{0\}$ is a compact convex set contained in $L \setminus \{0\}$.

Now the desired conclusion follows taking C to be the convex hull of the set $F \cap (L \setminus \{0\})$. ∎

THEOREM 9.33 *In the setting of* Theorem 9.29, *if the spectra of all the operators $A_1, \dots, A_n$ are finite, then there exists a complex structure I on $U_{\mathfrak{J}_0,\mathfrak{J}_1}(A)/H$ such that the triple $(U_{\mathfrak{J}_0,\mathfrak{J}_1}(A)/H, \Omega, I)$ is a weakly Kähler homogeneous space of $U_{\mathfrak{J}_0,\mathfrak{J}_1}(A)$ of type $\mathbb{R}^n$.*

PROOF First note that the map

$$\Psi_0 \colon \mathbb{R}^n \to \mathrm{Der}(\mathfrak{J}_0), \quad (t_1, \dots, t_n) \mapsto \mathrm{ad}_{\mathfrak{J}}(-(t_1 A_1 + \cdots + t_n A_n))$$

is hermitian and $\sigma_W(\Psi_0)$ is a finite subset of $\mathbb{R}^n$ by Lemma 9.31. It then follows by Lemma 9.32 that there exists a compact convex subset C of $\mathbb{R}^n$ such that

$$(-C) \cap C = \emptyset \text{ and } \sigma_W(\Psi_0) \subseteq (-C) \cup \{0\} \cup C.$$

We now proceed to check the hypotheses of Theorem 9.15. Since all of the commuting operators $A_1, \dots, A_n$ are diagonalizable with finite spectra, the existence of the needed spectral measure E is well known. (See e.g., Proposition 4.3 in [Hl90].) We have $\mathrm{supp}\, E = \sigma_W(\Psi_0)$ (see Example 5.26), hence

$$\mathrm{supp}\, E \subseteq (-C) \cup \{0\} \cup C.$$

Next note that, for all $A \in \mathcal{B}(\mathcal{H})$, $K \in \mathfrak{J}_1$ and $T \in \mathfrak{J}_0$, we have

$$\begin{aligned}
\langle (\mathrm{ad} A)^{\#} K, T \rangle &= \langle K (\mathrm{ad} A) T \rangle \\
&= \mathrm{Tr}\,(KAT - KTA) \\
&= \mathrm{Tr}\,(KAT - AKT) \qquad \text{(since } KT \in \mathfrak{J}_1 \cdot \mathfrak{J}_0 \subseteq \mathcal{C}_1(\mathcal{H})) \\
&= \langle -(\mathrm{ad} A) K, T \rangle
\end{aligned}$$

with respect to the duality pairing $\mathfrak{J}_1 \times \mathfrak{J}_0 \to \mathbb{C}$, $(K, T) \mapsto \mathrm{Tr}\,(KT)$. (See Definition 9.22.) Consequently,

$$(\forall t \in \mathbb{R}^n) \qquad \Psi_1(t) \circ \iota = -\iota \circ \Psi_1(t)^{\#}.$$

It then follows by Theorem 9.15 that there exists a weakly Kähler polarization $\mathfrak{p}$ of $\mathfrak{u}_{\mathfrak{J}_0}(\Psi_1, \iota)$ in the cocycle $\omega_{\Psi_1, \iota} \in Z^2_c(\mathfrak{u}_{\mathfrak{J}_0}(\Psi_1, \iota), \mathbb{R}^n)$. Now the desired assertion follows by Theorem 6.12. ∎

We now draw from Theorem 9.33 the following consequence.

COROLLARY 9.34 *In the setting of* Theorem 9.29, *if* $\mathfrak{J}_0 = \mathfrak{J}_1 = \mathcal{C}_2(\mathcal{H})$ *(the Hilbert-Schmidt ideal) and all of the operators* $A_1, \dots, A_n$ *have finite spectra, then there exists a complex structure* I *on the manifold* $\mathrm{U}_{\mathcal{C}_2(\mathcal{H})}/H$ *such that, for each* $\gamma_0 \in (\mathbb{R}^n)^{\#}$ *with*

$$\mathrm{Ker}\, \gamma_0 \cap \{(t_1 - s_1, \dots, t_n - s_n) \mid t_j, s_j \in \sigma(A_j) \text{ for } j = 1, \dots, n\} = \{0\},$$

the triple $(\mathrm{U}_{\mathcal{C}_2(\mathcal{H})}/H, \gamma_0 \circ \Omega, I)$ *is a strongly Kähler homogeneous space of* $\mathrm{U}_{\mathcal{C}_2(\mathcal{H})}$ *of type* $\mathbb{R}$.

PROOF Use Theorem 9.33 and Example 9.16. ∎

As mentioned in the Introduction of this book, the examples of infinite-dimensional Kähler homogeneous spaces constructed in the present chapter point to several research directions in various fields. We would like to conclude by mentioning a couple of problems concerning the aforementioned homogeneous spaces *themselves*.

PROBLEM 9.35 In the framework of Theorem 9.29, for the canonically involutive Banach-Lie algebra $\mathfrak{u}_{\mathfrak{J}_0}$ we have $\iota \in \mathcal{C}_0^+(\mathfrak{u}_{\mathfrak{J}_0})$ (see assertion (iii) in Remark 9.26), hence $\Phi(\iota) \in L^*\mathrm{Id}(\mathfrak{u}_{\mathfrak{J}_0})$ by Theorem 8.13. Thus, there exist a canonically involutive L^*-algebra $\mathfrak{X}_{\mathfrak{J}_0,\mathfrak{J}_1}$ and $\varphi \in \mathrm{Hom}^*(\mathfrak{X}_{\mathfrak{J}_0,\mathfrak{J}_1}, \mathfrak{u}_{\mathfrak{J}_0})$ such that $\Phi(\iota) = \mathrm{Ran}\,\varphi \trianglelefteq \mathfrak{u}_{\mathfrak{J}_0}$.

Now let $X_{\mathfrak{J}_0,\mathfrak{J}_1}$ be a connected and simply connected Banach-Lie group whose Lie algebra is $\mathfrak{X}_{\mathfrak{J}_0,\mathfrak{J}_1}$, and integrate φ to a homomorphism of Banach-Lie groups

$$\widetilde{\varphi}\colon X_{\mathfrak{J}_0,\mathfrak{J}_1} \to U_{\mathfrak{J}_0}.$$

Under the additional assumption that $U_{\mathfrak{J}_0,\mathfrak{J}_1}(A)$ is a Banach-Lie subgroup of $U_{\mathfrak{J}_0}$ (which is the case e.g., when all of the operators $A_1, \dots, A_n$ have finite spectra), it follows by Proposition 4.8 that

$$X_{\mathfrak{J}_0,\mathfrak{J}_1}(A) := \widetilde{\varphi}^{-1}(U_{\mathfrak{J}_0,\mathfrak{J}_1}(A))$$

is a Banach-Lie subgroup of (the L^*-group) $X_{\mathfrak{J}_0,\mathfrak{J}_1}$, and $\widetilde{\varphi}^{-1}(H)$ is a Banach-Lie subgroup of $X_{\mathfrak{J}_0,\mathfrak{J}_1}(A)$.

Now, the map $\widetilde{\varphi}$ induces a smooth map

$$X_{\mathfrak{J}_0,\mathfrak{J}_1}(A)/\widetilde{\varphi}^{-1}(H) \to U_{\mathfrak{J}_0,\mathfrak{J}_1}(A)/H,$$

and the problem is: Study the similar features which the latter mapping imposes to various differential geometric structures (respectively, topological properties) of the homogeneous spaces $X_{\mathfrak{J}_0,\mathfrak{J}_1}(A)/\widetilde{\varphi}^{-1}(H)$ and $U_{\mathfrak{J}_0,\mathfrak{J}_1}(A)/H$.

The answer to this problem is important in order to understand the role played by the special case considered in Corollary 10.17 ($\mathfrak{J}_0 = \mathfrak{J}_1 = \mathcal{C}_2(\mathcal{H})$) among the situations corresponding to arbitrary admissible pairs $(\mathfrak{J}_0, \mathfrak{J}_1)$, to which Theorem 9.33 applies.

PROBLEM 9.36 Is it possible to point out properties of the operator ideals $\mathfrak{J}_0$ and $\mathfrak{J}_1$ that are reflected by special complex geometric properties of the Kähler homogeneous spaces in Theorem 9.33? And a similar question can be asked in connection with the symplectic geometry of the weakly symplectic homogeneous spaces constructed in Theorem 9.29.

Notes

The terminology of pseudo-restricted groups and algebras claims its origins from the restricted groups and algebras (Example 9.5) dealt with in the theory of loop groups. (See e.g., Section 6.2 in [PS90].) That terminology was introduced in [Be03]. In the same paper were announced some of the main results of the present chapter, namely Theorems 9.10, 9.15, 9.29, and 9.33, as well as Example 9.16.

We note that Theorem 9.15 in the present chapter is a generalization of Theorem 2.10 in [Be05a]. In the case when $\dim \mathfrak{z}_0 = 1$, the result of Proposition 9.1 above reduces to Proposition 7.25 in [Be05a]. Similarly, Proposition 9.3 in the present chapter extends Proposition 7.29 in [Be05a]. Corollary 9.34 might be thought of as extending Theorem VII.6 in [Ne04]. (See Remark 9.30 above.)

Our basic references for the theory of operator ideals are [GK69], [Pit78], and [DFWW01]. We refer to [BR04] for some constructions of Kähler homogeneous spaces that are closely related to the ones from the present chapter. We refer to [dlH72] for more open problems on the relationship between the Schatten ideals and infinite-dimensional Lie groups. Our Definition 9.22 is somehow related to the project proposed at page 95 in [dlH72] concerning a generalization of the notion of L^*-algebra.

Appendices

Appendix A

Differential Calculus and Smooth Manifolds

Abstract. In this appendix we first collect some basic elements of differential calculus in topological vector spaces, and particularly in locally convex spaces. We discuss real and complex analytic mappings on open subsets of locally convex spaces. As an important example, we prove that the inversion mapping of a continuous inverse algebra is analytic. Then we introduce the notion of smooth manifold modeled on a locally convex space, as well as the closely related notions of tangent vector, tangent bundle and vector field. The main result of this appendix (Theorem A.67) shows that the set of all vector fields on a smooth locally convex manifold has a natural structure of Lie algebra.

A.1 Locally convex spaces and algebras

The main aim of this section is to expose some basic facts related to complexifications of topological vector spaces and to continuous inverse algebras.

DEFINITION A.1 A *topological vector space* over $\mathbb{K} \in \{\mathbb{R}, \mathbb{C}\}$ is a vector space X over $\mathbb{K}$ equipped with a Hausdorff topology such that both the vector addition $X \times X \to X$, $(x, y) \mapsto x+y$, and the scalar multiplication $\mathbb{K} \times X \to X$, $(\lambda, x) \mapsto \lambda x$, are continuous mappings.

DEFINITION A.2 A topological vector space X is said to be *locally convex* if each point of X has a basis of convex neighborhoods.

For the following statement, we recall that a *seminorm* on a vector space X over $\mathbb{K} \in \{\mathbb{R}, \mathbb{C}\}$ is a function $p\colon X \to [0, \infty)$ such that for all $x, y \in X$ and $\alpha \in \mathbb{K}$ we have $p(x + y) \le p(x) + p(y)$ and $p(\alpha x) = |\alpha| p(x)$.

THEOREM A.3 *Let X be a topological vector space over $\mathbb{K} \in \{\mathbb{R}, \mathbb{C}\}$. Then X is locally convex if and only if there exists a family of seminorms $\{p_i\}_{i\in I}$ defining the topology of X in the sense that if for $n \geq 1$, $i_1, \dots, i_n \in I$, and $\varepsilon > 0$ we denote*

$$V_{i_1,\dots,i_n;\varepsilon} := \{x \in X \mid \max_{1\leq k\leq n} |p_{i_k}(x)| < \varepsilon\},$$

then $\mathcal{V} := \{V_{i_1,\dots,i_n;\varepsilon} \mid n \geq 1, i_1, \dots, i_n \in I, \varepsilon > 0\}$ is a basis of neighborhoods of $0 \in X$.

PROOF See e.g., §1 in Chapter II in [Sf66] for the connection between convex sets and seminorms in a topological vector space. ∎

DEFINITION A.4 Let X be a topological vector space. We say that a sequence $\{x_n\}_{n\in\mathbb{N}}$ in X is *convergent* to $x \in X$ if for every neighborhood V of x there exists $n_V \in \mathbb{N}$ such that for all $n \geq n_V$ we have $x_n \in V$. With the same notation, we say that $\{x_n\}_{n\in\mathbb{N}}$ is a *Cauchy sequence* if for every neighborhood W of $0 \in X$ there exists $m_V \in \mathbb{N}$ such that $x_n - x_m \in W$ whenever $n, m \geq m_V$.

Finally, we say that the topological vector space X is *sequentially complete* if every Cauchy sequence in X is convergent.

We note that every Banach space is in particular a sequentially complete locally convex space. The next definition singles out a more general class of topological vector spaces of the latter type.

DEFINITION A.5 We say that a locally convex space X over $\mathbb{K} \in \{\mathbb{R}, \mathbb{C}\}$ is a *Fréchet space* if it is sequentially complete and its topology can be defined by a countable family of seminorms (see Theorem A.3).

We now state a version of the Hahn-Banach theorem.

THEOREM A.6 *If X is a locally convex space over $\mathbb{K} \in \{\mathbb{R}, \mathbb{C}\}$, then for all $x, y \in X$, with $x \neq y$, there exists a continuous linear functional $l\colon X \to \mathbb{K}$ such that $l(x) \neq l(y)$.*

PROOF See e.g., Theorem 9.2 in Chapter II in [Sf66]. ∎

PROPOSITION A.7 *Let X be a sequentially complete locally convex space and $f\colon [0,1] \to X$ a continuous function. Then there exists the Riemann integral $\int_0^1 f(t)\mathrm{d}t \in X$.*

PROOF The conclusion means that there exists $x \in X$ (to be denoted $\int_0^1 f(t)\mathrm{d}t$) such that for every neighborhood V of $0 \in V$ there exists $\delta > 0$ such that

$$(t_1 - t_0)f(\xi_1) + \cdots + (t_n - t_{n-1})f(\xi_n) \in x + V \tag{A.1}$$

whenever $0 = t_0 \leq \xi_1 \leq t_1 \leq \xi_2 \leq t_2 \leq \cdots \leq t_{n-1} \leq \xi_n \leq t_n = 1$ and $\sup\limits_{1\leq i\leq n} (t_n - t_{n-1}) < \delta$.

Since X is sequentially complete, it suffices (just as in the case $X = \mathbb{R}$) to show that for every sequence of subdivisions of $[0,1]$ with the mesh tending to 0, and for arbitrary choices of ξ's, the sequence of the corresponding Riemann sums (as in the left-hand side of (A.1)) is a Cauchy sequence.

Since X is a locally convex space, the latter property follows because of the fact that $f\colon [0,1] \to X$ is uniformly continuous in the following sense: for every neighborhood V of $0 \in X$ there exists $\varepsilon > 0$ such that, for all $s, t \in [0,1]$ with $|s - t| < \varepsilon$, we have $f(s) - f(t) \in V$. ∎

Complexifications and continuous inverse algebras

DEFINITION A.8 Let X be a real topological vector space. The *complexification* of X is the complex topological vector space $X_{\mathbb{C}} := X \times X$ equipped with the product topology, with the componentwise vector addition

$$X_{\mathbb{C}} \times X_{\mathbb{C}} \to X_{\mathbb{C}}, \quad (x_1, x_2) + (y_1, y_2) = (x_1 + y_1, x_2 + y_2),$$

and with the scalar multiplication defined by

$$\mathbb{C} \times X_{\mathbb{C}} \to X_{\mathbb{C}}, \quad (a + \mathrm{i}b) \cdot (x_1, x_2) = (ax_1 - bx_2, ax_2 + bx_1).$$

We usually perform the identifications

$$X \simeq X \times \{0\} \hookrightarrow X_{\mathbb{C}} \text{ and } \mathrm{i}X \simeq \{0\} \times X \hookrightarrow X_{\mathbb{C}},$$

and thus

$$X_{\mathbb{C}} = X \dot{+} \mathrm{i}X,$$

thinking of X as a real vector subspace of $X_{\mathbb{C}}$. In particular we write $x_1 + \mathrm{i}x_2$ instead of (x_1, x_2) whenever $x_1, x_2 \in X$.

EXAMPLE A.9

(a) If X is a real Banach space with the norm $\|\cdot\|_X$, then the complexification $X_{\mathbb{C}}(= X \times X)$ is a complex Banach space with the norm $\|\cdot\|_{X_{\mathbb{C}}}$ defined by

$$(\forall x_1, x_2 \in X) \qquad \|(x_1, x_2)\|_{X_{\mathbb{C}}} := \sup_{t\in[0,2\pi]} \|(\cos t)x_1 + (\sin t)x_2\|_X.$$

(b) If H is a real Hilbert space with the scalar product $\langle\cdot,\cdot\rangle_H$, then the complexification $H_{\mathbb{C}}(= H \times H)$ is a complex Hilbert space with the scalar product $\langle\cdot,\cdot\rangle_{H_{\mathbb{C}}}$ defined by

$$\langle(x_1,x_2),(y_1,y_2)\rangle_{H_{\mathbb{C}}} := \langle x_1,y_1\rangle_H + \langle x_2,y_2\rangle_H + \mathrm{i}(\langle x_2,y_1\rangle_H - \langle x_1,y_2\rangle_H)$$

whenever $x_1,x_2,y_1,y_2 \in H$.

EXAMPLE A.10 Let X and Y be real Banach spaces and denote by $\mathcal{B}_{\mathbb{R}}(X,Y)$ the real Banach space of all bounded $\mathbb{R}$-linear operators from X into Y. Also denote by $\mathcal{B}_{\mathbb{C}}(X_{\mathbb{C}},Y_{\mathbb{C}})$ the complex Banach space of all bounded $\mathbb{C}$-linear operators from $X_{\mathbb{C}}$ into $Y_{\mathbb{C}}$. Then $\mathcal{B}_{\mathbb{C}}(X_{\mathbb{C}},Y_{\mathbb{C}})$ is the complexification of $\mathcal{B}_{\mathbb{R}}(X,Y)$, the natural embedding

$$\mathcal{B}_{\mathbb{R}}(X,Y) \hookrightarrow \mathcal{B}_{\mathbb{C}}(X_{\mathbb{C}},Y_{\mathbb{C}})$$

being the one which associates to each $T \in \mathcal{B}_{\mathbb{R}}(X,Y)$ the operator

$$X_{\mathbb{C}} = X \times X \to Y_{\mathbb{C}} = Y \times Y, \quad (x_1,x_2) \mapsto (Tx_1,Tx_2).$$

NOTATION A.11 For a Banach space X over $\mathbb{K} \in \{\mathbb{R},\mathbb{C}\}$, we always denote by

$$X^{\#} := \{l\colon X \to \mathbb{K} \mid l \text{ is } \mathbb{K}\text{-linear continuous}\}$$

its *topological dual* and by

$$\langle\cdot,\cdot\rangle\colon X^{\#} \times X \to \mathbb{K}$$

the natural pairing. If Y is another Banach space over $\mathbb{K}$, then $\mathcal{B}(X,Y)$ stands for the set of all bounded $\mathbb{K}$-linear operators from X into Y. For $T \in \mathcal{B}(X,Y)$, we denote by $T^{\#} \in \mathcal{B}(Y^{\#},X^{\#})$ the operator dual to T, by $\operatorname{Ran} T$ the range of T, and by $\operatorname{Ker} T$ the kernel of T.

We denote $\mathcal{B}(X) = \mathcal{B}(X,X)$. If $\mathbb{K} = \mathbb{C}$ and $T \in \mathcal{B}(X)$, then $\sigma(T)$ denotes the spectrum of T. In the case $\mathbb{K} = \mathbb{R}$, we always endow the complexification $X_{\mathbb{C}} = X \oplus \mathrm{i}X$ of X with a norm such that

$$(\forall x,y \in X) \quad \|x+\mathrm{i}y\|_{X_{\mathbb{C}}} = \|x-\mathrm{i}y\|_{X_{\mathbb{C}}}.$$

(See Example A.9.) In the same case of a real Banach space X, for $T \in \mathcal{B}(X)$ we denote by $\sigma(T)$ the spectrum of the $\mathbb{C}$-linear extension of T to $X_{\mathbb{C}}$ and note that the norm of that $\mathbb{C}$-linear extension of T equals $\|T\|$.

DEFINITION A.12 A *topological algebra* A is a topological vector space equipped with a continuous bilinear mapping

$$A \times A \to A, \quad (a,b) \mapsto a \cdot b,$$

called the *multiplication* of A. If the topological vector space underlying A is locally convex, Fréchet, Banach, or Hilbert, then we say that A is a *locally convex, Fréchet, Banach,* or *Hilbert algebra*, respectively.

We say that the topological algebra A is *associative* if for all $a, b, c \in A$ we have $(a \cdot b) \cdot c = a \cdot (b \cdot c)$, and that A is *unital* if there exists an element $\mathbf{1} \in A$ (called the *unit element* of A) such that $a \cdot \mathbf{1} = \mathbf{1} \cdot a = a$ for all $a \in A$. If A is both unital and associative, then an element $x \in A$ is said to be *invertible* if there exists $x_0 \in A$ such that $x \cdot x_0 = x_0 \cdot x = \mathbf{1}$. In this case, x_0 is uniquely determined by x and we denote $x_0 =: x^{-1}$. The invertible elements of A are sometimes called the *units of* A. We denote

$$A^\times := \{x \in A \mid x \text{ is invertible}\}.$$

Note that $\mathbf{1} \in A^\times$ and $A^\times$ is a group with respect to the multiplication inherited from A.

If A is a topological algebra, then a *subalgebra* of A is a vector subspace A_0 of A such that $a \cdot b \in A_0$ whenever $a, b \in A_0$. If moreover A is unital and the unit of A belongs to A_0, then we say that A_0 is a *unital subalgebra* of A.

If A is a real topological algebra, then the *complexification* $A_{\mathbb{C}}$ of the topological vector space underlying A has a natural structure of complex topological algebra with the multiplication defined by

$$(a_1 + \mathrm{i}a_2) \cdot (b_1 + \mathrm{i}b_2) = (a_1 \cdot b_1 - a_2 \cdot b_2) + \mathrm{i}(a_1 \cdot b_2 + a_2 \cdot b_1)$$

whenever $a_1, a_2, b_1, b_2 \in A$. If A is associative or unital, then so is $A_{\mathbb{C}}$.

DEFINITION A.13 Let $\mathfrak{g}$ be a Banach algebra over $\mathbb{K} \in \{\mathbb{R}, \mathbb{C}\}$ with the multiplication $\mathfrak{g} \times \mathfrak{g} \to \mathfrak{g}$, $(a, b) \mapsto ab$. For every $a \in \mathfrak{g}$ we define $L_a, R_a \in \mathcal{B}(\mathfrak{g})$ by $L_a(b) = ab$ and $R_a(b) = ba$, respectively, for all $b \in \mathfrak{g}$. The *annihilator* of $\mathfrak{g}$ is defined by

$$\mathrm{Ann}(\mathfrak{g}) = \bigcap_{a \in \mathfrak{g}} (\mathrm{Ker}\, L_a \cap \mathrm{Ker}\, R_a).$$

(In the case when $\mathfrak{g}$ is a Lie algebra, we call it a *Banach-Lie algebra*, denote $L_a = \mathrm{ad}_{\mathfrak{g}} a$ and $ab = [a, b]$ for all $a, b \in \mathfrak{g}$, and define the *center* $\mathcal{Z}(\mathfrak{g})$ of $\mathfrak{g}$ by $\mathcal{Z}(\mathfrak{g}) = \mathrm{Ann}(\mathfrak{g})$.) Moreover, we denote by $\mathrm{Der}(\mathfrak{g})$ the set of all bounded *derivations* of $\mathfrak{g}$ into itself, that is, the operators $D \in \mathcal{B}(\mathfrak{g})$ such that

$$(\forall a, b \in \mathfrak{g}) \qquad D(a \cdot b) = (Da) \cdot b + a \cdot (Db).$$

Then $\mathrm{Der}(\mathfrak{g})$ is a closed subalgebra of the Banach-Lie algebra $\mathcal{B}(\mathfrak{g})$.

We say that $\mathfrak{g}$ is an *involutive Banach algebra* if it is moreover endowed with a continuous map

$$\mathfrak{g} \to \mathfrak{g}, \quad a \mapsto a^*,$$

such that for every $a, b \in \mathfrak{g}$ and $\lambda \in \mathbb{K}$ we have

$$(a^*)^* = a, \quad (ab)^* = b^* a^*, \quad (a + b)^* = a^* + b^*, \quad \text{and} \quad (\lambda a)^* = \bar{\lambda} a^*,$$

where $\bar{\lambda} = \lambda$ if $\mathbb{K} = \mathbb{R}$, and $\bar{\lambda}$ is the complex-conjugate of λ if $\mathbb{K} = \mathbb{C}$.

We say that $\mathfrak{I}$ is an *ideal* of $\mathfrak{g}$, and denote $\mathfrak{I} \trianglelefteq \mathfrak{g}$, if $\mathfrak{I}$ is a linear subspace of $\mathfrak{g}$ which is invariant to both L_a and R_a for all $a \in \mathfrak{g}$. The involutive Banach algebra $\mathfrak{g}$ is said to be *topologically simple* if its only closed ideals which are invariant to the involution $a \mapsto a^*$ are $\{0\}$ and $\mathfrak{g}$.

If $\mathfrak{h}$ is another Banach algebra over $\mathbb{K}$, then $\mathrm{Hom}(\mathfrak{h}, \mathfrak{g})$ denotes the set of all continuous homomorphisms from $\mathfrak{h}$ into $\mathfrak{g}$. When moreover $\mathfrak{h}$ is an involutive Banach algebra, we denote by $\mathrm{Hom}^*(\mathfrak{h}, \mathfrak{g})$ the set of all $*$-homomorphisms from $\mathfrak{h}$ into $\mathfrak{g}$. The set of all invertible maps in $\mathrm{Hom}(\mathfrak{g}, \mathfrak{g})$ and $\mathrm{Hom}^*(\mathfrak{g}, \mathfrak{g})$ is denoted by $\mathrm{Aut}(\mathfrak{g})$ and $\mathrm{Aut}^*(\mathfrak{g})$, respectively.

REMARK A.14

(a) If $\mathfrak{g}$ is a real Banach algebra such that $ab = -ba$ for all $a, b \in \mathfrak{g}$, then it can be made into an involutive Banach algebra, defining $a^* = -a$ for each $a \in \mathfrak{g}$. This can be done e.g., when $\mathfrak{g}$ is a real Banach-Lie algebra, and the real involutive Banach-Lie algebra obtained in this way is called a *canonically involutive Banach-Lie algebra.*

(b) The forgetful functor from the category of the abelian canonically involutive Banach-Lie algebras (with the continuous $*$-homomorphisms as morphisms) into the category of real Banach spaces (with the bounded linear operators as morphisms) is an equivalence of categories.

We collect in the next two lemmas some useful consequences of the closed graph theorem.

LEMMA A.15 *Let X, Y, and Z be Banach spaces over $\mathbb{K} \in \{\mathbb{R}, \mathbb{C}\}$, and $A\colon X \times Y \to Z$ a continuous bilinear map. Suppose that X_1, Y_1, and Z_1 are Banach spaces over $\mathbb{K}$ with continuous injective linear maps $\eta_X\colon X_1 \to X$, $\eta_Y\colon Y_1 \to Y$ and $\eta_Z\colon Z_1 \to Z$. If $A(\mathrm{Ran}\,\eta_X \times \mathrm{Ran}\,\eta_Y) \subseteq \mathrm{Ran}\,\eta_Z$, then there exists a unique continuous bilinear map*

$$A_1\colon X_1 \times Y_1 \to Z_1$$

with $\eta_Z \circ A = A \circ (\eta_X \times \eta_Y)$.

PROOF See e.g., Lemma A.2 in the Appendix of [Ne02b]. ∎

LEMMA A.16 *Let $\mathfrak{g}$ and $\mathfrak{X}$ be Banach-Lie algebras over $\mathbb{K} \in \{\mathbb{R}, \mathbb{C}\}$ and consider $\varphi \in \mathrm{Hom}(\mathfrak{X}, \mathfrak{g})$ injective such that $\mathrm{Ran}\,\varphi \trianglelefteq \mathfrak{g}$. Then there exists a homomorphism of Banach-Lie algebras $\rho\colon \mathfrak{g} \to \mathrm{Der}(\mathfrak{X})$ such that for each*

$a \in \mathfrak{g}$ the diagram

$$\begin{array}{ccc} \mathfrak{X} & \xrightarrow{\varphi} & \mathfrak{g} \\ {\scriptstyle \rho(a)}\downarrow & & \downarrow{\scriptstyle \mathrm{ad}_{\mathfrak{g}} a} \\ \mathfrak{X} & \xrightarrow{\varphi} & \mathfrak{g} \end{array}$$

is commutative.

PROOF See e.g., Lemma A.3 in the Appendix of [Ne02b]. ∎

DEFINITION A.17 Let A be a unital associative topological algebra. We say that A is a *continuous inverse algebra* if $A^\times$ is an open subset of A and the inversion mapping

$$\eta\colon A^\times \to A^\times, \quad x \mapsto x^{-1},$$

is continuous.

LEMMA A.18 *Let A be a unital associative topological algebra. If there exists an open subset W of A such that $\mathbf{1} \in W \subseteq A^\times$ and the inversion mapping $\eta|_W \colon W \to A^\times$ is continuous, then A is a continuous inverse algebra.*

PROOF Fix $a \in A^\times$. Since a is invertible, it is easy to see that the mapping

$$L_a \colon A \to A, \quad b \mapsto ab,$$

is a homeomorphism and $L_a(A^\times) = A^\times$, hence $L_a(W) = aW$ is an open subset of $A^\times$ containing $L_a(\mathbf{1}) = a$. Since $a \in A^\times$ is arbitrary, it follows that $A^\times$ is open in A. Moreover, for all $c \in aW$ we have

$$\eta(c) = c^{-1} = \big(a(a^{-1}c)\big)^{-1} = (a^{-1}c)^{-1}a^{-1} = \eta(a^{-1}c)a^{-1}.$$

Since $a^{-1}c \in W$ and $\eta|_W$ is continuous, it then follows that η is continuous on the open neighborhood aW of a. Since $a \in A^\times$ is arbitrary, it follows that the inversion mapping $\eta\colon A^\times \to A^\times$ is continuous throughout on $A^\times$. ∎

PROPOSITION A.19 *If A is a real continuous inverse algebra, then the complexification $A_{\mathbb{C}}$ is a complex continuous inverse algebra.*

PROOF Denote as usual by $\eta\colon A^\times \to A^\times$ the inversion mapping, and define

$$\psi\colon A^\times \times A \to A, \quad \psi(a,b) = \mathbf{1} + (a^{-1}b)^2.$$

Then $\psi(a,b) = \mathbf{1} + \big(\eta(a)\cdot b\big)^2$, hence ψ is a continuous mapping. Since $A^\times$ is open in A and $\psi(\mathbf{1},0) = \mathbf{1} \in A^\times$, it then follows that we can find an open neighborhood U of $\mathbf{1} \in A^\times$ and an open neighborhood V of $0 \in A$ such that $\psi(U \times V) \subseteq A^\times$. Hence

$$(\forall a \in U)\,(\forall b \in V) \quad \mathbf{1} + (a^{-1}b)^2 \in A^\times.$$

For arbitrary $a \in U$ and $b \in V$, we have

$$a + ib = a(\mathbf{1} + \mathrm{i}\underbrace{a^{-1}b}_{=:c}) = a(\mathbf{1} + \mathrm{i}c).$$

On the other hand, we have $(\mathbf{1} + \mathrm{i}c)(\mathbf{1} - \mathrm{i}c) = \mathbf{1} + c^2 \in A^\times$, whence $\mathbf{1} + \mathrm{i}c$ is invertible, in fact $(\mathbf{1} + \mathrm{i}c)^{-1} = (\mathbf{1} - \mathrm{i}c)(\mathbf{1} + c^2)^{-1}$. Since a is also invertible, it then follows from the above equation that $a + ib$ is also invertible and

$$\begin{aligned}(a + \mathrm{i}b)^{-1} &= (\mathbf{1} + \mathrm{i}c)^{-1}a^{-1} = (\mathbf{1} - \mathrm{i}c)(\mathbf{1} + c^2)^{-1}a^{-1} \\ &= (\mathbf{1} - \mathrm{i}a^{-1}b)(\mathbf{1} + (a^{-1}b)^2)^{-1}a^{-1} \\ &= (\mathbf{1} - \mathrm{i}\eta(a)\cdot b)\cdot \eta(\mathbf{1} + (\eta(a)\cdot b)^2)\cdot \eta(a).\end{aligned}$$

Hence $U + \mathrm{i}V \subseteq (A_{\mathbb{C}})^\times$ and, moreover, the inversion mapping

$$U + \mathrm{i}V \to A_{\mathbb{C}}, \quad z \mapsto z^{-1},$$

is continuous since η is a continuous mapping. Since $W := U + \mathrm{i}V$ is an open neighborhood of $\mathbf{1} \in A_{\mathbb{C}}$, the desired conclusion follows by Lemma A.18. ∎

REMARK A.20 Every unital associative Banach algebra is a continuous inverse algebra.

REMARK A.21 If B is a continuous inverse algebra and A is a unital subalgebra of B such that $A^\times = A \cap B^\times$, then A is in turn a continuous inverse algebra.

A.2 Differential calculus and infinite-dimensional holomorphy

DEFINITION A.22 Let X be a topological vector space over $\mathbb{K} \in \{\mathbb{R}, \mathbb{C}\}$, and D an open subset of $\mathbb{K}$. We say that a continuous mapping $f\colon D \to X$ is of *class* $\mathcal{C}^1$ if the limit

$$\dot{f}(t_0) := \lim_{t \to t_0} \frac{f(t) - f(t_0)}{t - t_0}$$

exists in X for all $t_0 \in D$, and the mapping $\dot{f}\colon D \to X$ is continuous. Denote $f_1 := \dot{f}\colon D \to X$ and suppose that we have already defined $f_n\colon D \to X$ for some $n \geq 1$. If the mapping $f_{n+1} := \dot{f}_n\colon D \to X$ is defined and continuous, then we say that f is *of class* $\mathcal{C}^n$. Finally, we say that f is *smooth* if it is of class $\mathcal{C}^n$ for all $n \geq 1$.

DEFINITION A.23 Let X be a real locally convex space, I an interval in $\mathbb{R}$, and $a, b \in I$. We say that a continuous function $f\colon I \to X$ is *weakly integrable from a to b* if there exists $x_0 \in X$ such that for every continuous linear functional $l\colon X \to \mathbb{R}$ we have

$$l(x_0) = \int_a^b (l \circ f)(t)\mathrm{d}t.$$

In this case we denote x_0 by $\int_a^b f(t)\mathrm{d}t$ and call it the *weak integral of f from a to b*.

REMARK A.24 In the setting of Definition A.23, the vector x_0 is unique whenever it exists, since we are working only with Hausdorff spaces.

REMARK A.25 In the setting of Definition A.23, if X is moreover sequentially complete, then f is always weakly integrable from a to b.

THEOREM A.26 *Let X be a locally convex space, I an open interval in $\mathbb{R}$, and $a, b \in I$. If $f\colon I \to X$ is of class $\mathcal{C}^1$, then the continuous function $\dot{f}\colon I \to X$ is weakly integrable from a to b and $\int_a^b \dot{f}(t)\mathrm{d}t = f(b) - f(a)$.*

PROOF See Theorem 1.5 in [Gl02a]. ∎

DEFINITION A.27 Let X and Y be real topological vector spaces, U an open subset of X, and $f\colon U \to Y$ a continuous mapping. We say that f is *of class* $\mathcal{C}^1$ if the limit

$$f'_x(h) := \lim_{t \to 0} \frac{f(x + th) - f(x)}{t}$$

exists in Y for all $x \in U$ and $h \in X$, and the mapping

$$df\colon U \times X \to Y, \qquad df(x, h) := f'_x(h)$$

is continuous.

Now suppose that $n \geq 1$ and

$$d^n f\colon U \times \underbrace{X \times \cdots \times X}_{n \text{ times}} \to Y, \quad (x; h_1, \ldots, h_n) \mapsto f_x^{(n)}(h_1, \ldots, h_n)$$

was already defined and is continuous. If, for all $x \in U$ and $h_1, \dots, h_n, h_{n+1} \in X$, the limit

$$d^{n+1}f(x; h_1, \dots, h_n, h_{n+1}) := f_x^{(n+1)}(h_1, \dots, h_n, h_{n+1})$$
$$:= \lim_{t \to 0} \frac{f_{x+th_{n+1}}^{(n)}(h_1, \dots, h_n) - f_x^{(n)}(h_1, \dots, h_n)}{t}$$

exists in X and the mapping $d^{n+1}f \colon U \times X^{n+1} \to Y$ is continuous, then we say that f is *of class* C^{n+1}. Furthermore, we say that f is *smooth* or *of class* $\mathcal{C}^\infty(U, V)$ if it is of class $\mathcal{C}^n$ for all $n \geq 1$.

For $n = 1, 2, \dots, \infty$, we denote by $\mathcal{C}^n(U, Y)$ the set of all mappings $U \to Y$ of class $\mathcal{C}^n$.

Finally, if $X_1, \dots, X_n$ are real topological vector spaces, U is an open subset of $X_1 \times \cdots \times X_n$, and $g \colon U \to Y$ is a continuous mapping, then for every point $x = (x_1, \dots, x_n) \in U$ and $j \in \{1, \dots, n\}$ we define by

$$\partial_j g(x) := (g_j^x)'_{x_j} \colon X_j \to Y$$

the *j-th partial derivative of first order of* g *at* x (whenever it exists), where the mapping $g_j^x \colon U_j \to Y$ is defined by $g_j^x(z) := g(x_1, \dots, x_{j-1}, z, x_{j+1}, \dots, x_n)$ for all z in the open subset $U_j := \{\tilde{z} \in X_j \mid (x_1, \dots, x_{j-1}, \tilde{z}, x_{j+1}, \dots, x_n) \in U\}$ of X_j.

PROPOSITION A.28 *Let X and Y be real locally convex spaces, U an open subset of X, and $f \colon U \to Y$ a mapping of class $\mathcal{C}^1$. Then for all $x \in U$ the mapping $f'_x \colon X \to Y$ is linear and continuous.*

If moreover f is of class $\mathcal{C}^n$ with $n \geq 2$, then for all $x \in U$ the mapping

$$f_x^{(n)} \colon \underbrace{X \times \cdots \times X}_{n \text{ times}} \to Y$$

is symmetric, continuous and n-linear.

PROOF See Lemma 1.9 and Proposition 1.13 in [Gl02a]. ∎

REMARK A.29 In the setting of Proposition A.28, if moreover $X = \mathbb{R}$, then for all $x \in U$ and $h \in X = \mathbb{R}$ we have $f'_x(h) = h \cdot \dot{f}(x)$, where $\dot{f}(x) \in Y$ is introduced in Definition A.22. Similarly, if moreover f is of class $\mathcal{C}^n$, then for all $x \in U$ and $h_1, \dots, h_n \in X = \mathbb{R}$ we have $f_x^{(n)}(h_1, \dots, h_n) = h_1 \cdots h_n \cdot f_n(x)$, where $f_n(x) \in Y$ is introduced in Definition A.22.

PROPOSITION A.30 *Let X and Y be real locally convex spaces, U an open subset of X, and $f \colon U \to Y$ a mapping of class $\mathcal{C}^1$. If for all $x \in U$ we have $f'_x = 0$, then f is constant on each connected component of U.*

PROOF See Proposition 1.11 in [Gl02a]. ∎

PROPOSITION A.31 *Let X, Y, and Z be real locally convex spaces, U an open subset of X, V an open subset of Y, and $U \xrightarrow{f} V \xrightarrow{g} Z$ mappings of class $\mathcal{C}^n$, where $n \geq 1$. Then $g \circ f \colon U \to Z$ is of class $\mathcal{C}^n$ and for all $x \in U$ we have $(g \circ f)'_x = g'_{f(x)} \circ f'_x$.*

PROOF See Propositions 1.15 and 1.12 in [Gl02a]. ∎

We now state a version of Taylor's formula suitable for our purposes.

THEOREM A.32 *Let X and Y be real locally convex spaces, U an open subset of X, and $f \colon U \to Y$ a mapping of class $\mathcal{C}^{n+1}$, where $n \geq 0$. If $x \in U$ and $h \in X$ have the property that $x + th \in U$ whenever $0 \leq t \leq 1$, then*

$$
\begin{aligned}
f(x+h) = & f(x) + f'_x(h) + \frac{1}{2!} f''_x(h,h) + \cdots + \frac{1}{n!} f^{(n)}_x(h,\dots,h) \\
& + \int_0^1 \frac{(1-t)^n}{n!} f^{(n+1)}_{x+th}(h,\dots,h)\mathrm{d}t.
\end{aligned}
$$

PROOF See Proposition A2.17 in [Gl02a]. ∎

COROLLARY A.33 *Let X and Y be real Banach spaces, U an open subset of X, and $f \colon U \to Y$ a mapping of class $\mathcal{C}^{n+1}$, where $n \geq 0$. If $x \in U$ and we denote $V_x := \{h \in X \mid (\forall t \in [0,1])\ x + th \in U\}$, then V_x is an open neighborhood of $0 \in X$ and the function $\theta \colon V_x \to Y$ defined for all $h \in V_x$ by*

$$
f(x+h) = f(x) + f'_x(h) + \frac{1}{2!} f''_x(h,h) + \cdots + \frac{1}{n!} f^{(n)}_x(h,\dots,h) + \theta(h)
$$

has the property

$$
\lim_{h \to 0} \frac{\|\theta(h)\|}{\|h\|^n} = 0.
$$

PROOF See either Corollary 4.4 in Chapter I in [La01], or Theorem 6 in Chapter 1 in [Nel69]. ∎

The converse to Corollary A.33 holds under the following form.

THEOREM A.34 *Let X and Y be real Banach spaces, U an open subset of X, and $f \colon X \to Y$. Suppose that for some positive integer n there exist,* *for $j = 0, 1, \ldots, n$, the continuous mappings*

$$a_j : U \to \mathcal{B}^j(X, Y)$$

(where $\mathcal{B}^j(X, Y)$ stands for the real Banach space of symmetric multilinear mappings $X^j = X \times \cdots \times X \to Y$) such that the following assertion holds: For all $x \in U$ there exist $r > 0$ and $\theta : B_X(0, r) \to Y$ such that $B_X(x, r) \subseteq U$ and

$$(\forall h \in B_X(0,r)) \quad f(x+h) = a_0(x) + a_1(x)(h) + \cdots + \frac{1}{n!} a_n(x)(h, \ldots, h) + \theta(h)$$

and $\lim_{h \to 0} \|\theta(h)\| / \|h\|^n = 0$. Then f is of class $\mathcal{C}^n$ and $a_j = f^{(j)}$ for $j = 0, 1, \ldots, n$.

PROOF See Theorem 3 in Chapter 1 in [Nel69]. ∎

Analytic mappings

DEFINITION A.35 Let E and F be complex locally convex spaces, W an open subset of E, and $g : W \to F$. We say that g is *complex analytic* if it is of class $\mathcal{C}^1$ (when we view both E and F as real vector spaces) and for each $x \in W$ the mapping $g'_x : E \to F$ is $\mathbb{C}$-linear. (Recall from Proposition A.28 that g'_x is always $\mathbb{R}$-linear.) The complex analytic mappings are sometimes called *holomorphic.*

If X and Y are real locally convex spaces and U is an open subset of X, then a mapping $f : U \to Y$ is said to be *real analytic* if there exist an open subset U_1 of the complexification $X_{\mathbb{C}}$ of X and a complex analytic mapping $f_1 : U_1 \to Y_{\mathbb{C}}$ such that $U \subseteq U_1$ and $f_1|_U = f$.

PROPOSITION A.36 *Every real or complex analytic mapping is smooth.*

PROOF See Proposition 2.4 in [Gl02a]. ∎

THEOREM A.37 *Let X and Y be real (respectively complex) locally convex spaces, U an open subset of X, and $f : U \to Y$ a smooth mapping. Then f is real (respectively complex) analytic if and only if for every $x \in U$ there exists a neighborhood V of $0 \in X$ such that $x + V \subseteq U$ and for all $h \in V$ we have*

$$f(x+h) = \sum_{n=0}^{\infty} \frac{1}{n!} f_x^{(n)}(h, \ldots, h),$$

where $f_x^{(0)} := f(x)$.

PROOF See Lemma 2.5 and Definition 2.1 in [Gl02a]. ∎

PROPOSITION A.38 *Compositions of real or complex analytic mappings are real or complex analytic, respectively.*

PROOF See Propositions 2.7 and 2.8 in [Gl02a]. ∎

We now recall the definition of Fréchet differentiability.

DEFINITION A.39 Let X and Y be real Banach spaces, U an open subset of X, and $f\colon U \to Y$. We say that f is *Fréchet differentiable* if for every $x_0 \in U$ there exists $T \in \mathcal{B}(X,Y)$ such that

$$\lim_{x\to x_0} \frac{\|f(x)-f(x_0)-T(x-x_0)\|}{\|x-x_0\|} = 0.$$

In this case T is uniquely determined by x_0 and we denote

$$df(x_0) := f'_{x_0} := T.$$

We say that f is *Fréchet smooth* if

$$df\colon U \to \mathcal{B}(X, \mathcal{B}(X,Y))$$

is Fréchet differentiable and moreover

$$d^2 f := d(df)\colon U \to \mathcal{B}(X, \mathcal{B}(X, \mathcal{B}(X,Y)))$$

is Fréchet differentiable ... and moreover

$$d^k f = d(d^{k-1} f)\colon U \to \underbrace{\mathcal{B}(X, \dots, \mathcal{B}(X}_{k+1 \text{ times}}, Y)\dots)$$

is Fréchet differentiable ... ad infinitum.

REMARK A.40 It is clear that the notations f'_x, df etc. introduced in Definition A.39 agree with the ones in Definition A.27. Moreover, a mapping is Fréchet smooth if and only if it is smooth in the sense of Definition A.27. (For a proof of this fact, see Theorem I.7 in [Ne01a].)

THEOREM A.41 *Let X and Y be real Banach spaces, U an open subset of X, and for all $a, b \in X$ denote $D_{a,b} = \{t \in \mathbb{R} \mid a + tb \in U\}$. Then a smooth mapping $f\colon U \to Y$ is real analytic if and only if for all $a, b \in X$ the function*

$$D_{a,b} \to Y, \quad t \mapsto f(a+tb),$$

is real analytic.

PROOF See Theorem 7.5 in [BS71b]. □

THEOREM A.42 *Let E and F be complex Banach spaces, V an open subset of E and, for each $n \geq 1$, let $f_n \colon V \to F$ be a holomorphic mapping such that $\sup_{x \in V} \|f_n(x)\| < \infty$. If moreover $\lim_{m,n \to \infty} \left(\sup_{x \in V} \|f_n(x) - f_m(x)\| \right) = 0$, then there exists a holomorphic mapping $f \colon V \to F$ such that $\lim_{n \to \infty} \left(\sup_{x \in V} \|f_n(x) - f(x)\| \right) = 0$.*

PROOF This is an easy application of Proposition 6.2 in [BS71b]. □

PROPOSITION A.43 *Let E and F be complex Banach spaces, V an open subset of E, $r > 0$, and $g \colon B_{\mathbb{C}}(0,r) \times V \to F$ a holomorphic mapping such that*

$$\sup\{\|g(t,x)\| \mid t \in B_{\mathbb{C}}(0,r), x \in V\} < \infty.$$

Define

$$f \colon B_{\mathbb{C}}(0,r) \times V \to F, \quad f(t,x) = \int_0^t g(s,x) \mathrm{d}s.$$

Then f is holomorphic.

PROOF For all $t \in B_{\mathbb{C}}(0,r)$ and $x \in V$ we have $f(t,x) = \lim_{n \to \infty} f_n(t,x)$, where

$$f_n(t,x) = \frac{t}{n} \sum_{j=1}^{n} g(jt/n, x).$$

Then reason as in the proof of Proposition 6.3 in [BS71b], since $f_n \colon B_{\mathbb{C}}(0,r) \times V \to F$ is holomorphic for all $n \geq 1$. □

The following statement concerns the notion of continuous inverse algebra as introduced in Definition A.17.

LEMMA A.44 *If A is a continuous inverse locally convex algebra, then the inversion mapping*

$$\eta \colon A^{\times} \to A^{\times}, \quad x \mapsto x^{-1},$$

is smooth and $d\eta \colon A^{\times} \times A \to A$ is given by $(x,y) \mapsto -x^{-1}yx^{-1}$.

PROOF For all $x, y \in A^\times$ we have

$$y^{-1} - x^{-1} = x^{-1}(x-y)y^{-1} = y^{-1}(x-y)x^{-1}.$$

Since $A^\times$ is an open subset of A, it follows that for all $x \in A^\times$ and $y \in A$ we have $x + ty \in A^\times$ whenever $|t|$ is small enough. For such t, we have by the above equation $\eta(x+ty) - \eta(x) = x^{-1}(-ty)(x+ty)^{-1}$. Since η is continuous, we get

$$\begin{aligned}\eta'_x(y) &= \lim_{t\to 0} \frac{\eta(x+ty)-\eta(x)}{t} = \lim_{t\to 0}\bigl(-\eta(x)\cdot y\cdot \eta(x+ty)\bigr)\\ &= -\eta(x)\cdot y\cdot\eta(x) = -x^{-1}yx^{-1}.\end{aligned}$$

Thus, if we consider the mapping $\tau\colon A\times A\times A \to A$, $(a,b,c)\mapsto abc$, and the natural projections $\mathrm{pr}_{A^\times}\colon A^\times \times A \to A^\times$ and $\mathrm{pr}_A\colon A^\times\times A\to A$, we get the following formula for the differential of η:

$$d\eta\colon A^\times \times A \to A,\quad d\eta = -\tau\circ\bigl((\eta\circ \mathrm{pr}_{A^\times})\times \mathrm{pr}_A \times (\eta\circ\mathrm{pr}_A)\bigr).$$

Since all of the mappings η, $\mathrm{pr}_{A^\times}$, pr_A, and τ are continuous, we deduce that $d\eta\colon A^\times\times A\to A$ is continuous, hence η is of class $\mathcal{C}^1$. Then using the chain rule (Proposition A.31) and the above formula for $d\eta$, we can prove by induction that η is of class $\mathcal{C}^k$ for $k = 1, 2, \ldots$, hence η is smooth. ∎

PROPOSITION A.45 *If A is a complex continuous inverse locally convex algebra, then the inversion mapping $\eta\colon A^\times \to A^\times$ is complex analytic.*

PROOF We have seen in Lemma A.44 that for each $x\in A^\times$ we have

$$\eta'_x\colon A\to A,\quad \eta'_x(y) = -x^{-1}yx^{-1},$$

hence η'_x is clearly $\mathbb{C}$-linear, and this is just the condition required in Definition A.35. ∎

PROPOSITION A.46 *If A is a real continuous inverse locally convex algebra, then the inversion mapping $\eta\colon A^\times\to A^\times$ is real analytic.*

PROOF First recall from Lemma A.44 that η is smooth. On the other hand, it follows by Proposition A.19 that the complexification $A_{\mathbb{C}}$ of A is a complex continuous inverse algebra. Hence the inversion mapping of $A_{\mathbb{C}}$,

$$\eta_{\mathbb{C}}\colon (A_{\mathbb{C}})^\times \to (A_{\mathbb{C}})^\times,\quad z\mapsto z^{-1},$$

is a complex analytic mapping according to Proposition A.45.

Since $A^\times \subseteq (A_{\mathbb{C}})^\times$ and $\eta_{\mathbb{C}}|_A = \eta$, it then follows that η is real analytic. ∎

Other examples of analytic mappings are provided by the following remark and proposition (needed in the proof of Theorem 4.13).

REMARK A.47

(a) Let X be a real Banach space and

$$\Theta\colon \mathbb{C} \to \mathbb{C}, \quad \Theta(z) = \sum_{n=0}^{\infty} a_n z^n,$$

an entire function with $a_n \in \mathbb{R}$ for all $n \geq 0$. Then for every $T \in \mathcal{B}(X)$ the series $\sum\limits_{n=0}^{\infty} a_n T^n$ is convergent in the real Banach space $\mathcal{B}(E)$, and the mapping

$$\Theta\colon \mathcal{B}(X) \to \mathcal{B}(X), \quad \Theta(T) := \sum_{n=0}^{\infty} a_n T^n,$$

is real analytic.

(b) An assertion similar to (a) holds when X is replaced by a complex Banach space and the entire series Θ has arbitrary coefficients.

PROPOSITION A.48 *In a complex associative unital Banach algebra A consider the subsets $\mathcal{U} = \left\{a \in A \mid \sigma(a) \subseteq \mathbb{R} + \mathrm{i}(-\frac{\pi}{n}, \frac{\pi}{n})\right\}$ and $\mathcal{V} = \left\{a \in A^{\times} \mid (\forall z \in \sigma(a))|\arg z| < \frac{\pi}{n}\right\}$. Then both $\mathcal{U}$ and $\mathcal{V}$ are open subsets of A and the mapping*

$$\exp\colon \mathcal{U} \to \mathcal{V}, \quad a \mapsto \exp a = \sum_{j=0}^{\infty} \frac{a^j}{j!}$$

is a complex analytic bijection with the inverse defined by

$$\log\colon \mathcal{V} \to \mathcal{U}, \quad a \mapsto \log a = \sum_{j=0}^{\infty} \frac{(-1)^n}{n+1}(a - \mathbf{1})^{n+1}.$$

PROOF See Lemma 2.11 in [Up85]. ∎

A.3 Smooth manifolds

DEFINITION A.49 A topological space X is called *regular* if it is Hausdorff and for every $x \in X$ and every neighborhood U of x there exists another neighborhood V of x such that $\overline{V} \subseteq U$. In other words, each point of x has a basis of closed neighborhoods.

REMARK A.50 In order for the Hausdorff topological space X to be regular, it suffices that it is of one of the following types:

(a) X is locally compact;

(b) X is a topological group; or

(c) X is a locally convex topological vector space.

DEFINITION A.51 A *smooth manifold* modeled on a locally convex topological vector space V is a regular topological space M equipped with a maximal family of homeomorphisms $\{\varphi_\alpha \colon V_\alpha \to M_\alpha\}_{\alpha \in A}$ satisfying the following conditions.

(i) For every $\alpha \in A$, V_α is an open subset of V and M_α is an open subset of M.

(ii) We have $M = \bigcup\limits_{\alpha \in A} M_\alpha$.

(iii) If $\alpha, \beta \in A$ and $M_\alpha \cap M_\beta \neq \emptyset$, then the corresponding change of coordinate function

$$\varphi_\beta^{-1} \circ \varphi_\alpha|_{\varphi_\alpha^{-1}(M_\alpha \cap M_\beta)} \colon \varphi_\alpha^{-1}(M_\alpha \cap M_\beta) \to \varphi_\beta^{-1}(M_\alpha \cap M_\beta)$$

is smooth. Note that both $\varphi_\alpha^{-1}(M_\alpha \cap M_\beta)$ and $\varphi_\beta^{-1}(M_\alpha \cap M_\beta)$ are open subsets of the locally convex topological vector space V.

In this case, the maps $\varphi_\alpha \colon V_\alpha \to M_\alpha$ will be called *local coordinate systems*, while the maps $\varphi_\alpha^{-1} \colon V_\alpha \to M_\alpha$ are called *local coordinate charts.*

A smooth manifold modeled on a locally convex, Fréchet, Banach, or Hilbert space will be called, respectivelly, a *locally convex, Fréchet, Banach*, or *Hilbert manifold.*

DEFINITION A.52 Let M be a locally convex smooth manifold modeled on V with the family of local coordinate systems $\{\varphi_\alpha \colon V_\alpha \to M_\alpha\}_{\alpha \in A}$, and $\widehat{M}$ a locally convex smooth manifold modeled on $\widehat{V}$ with the family of local coordinate systems $\{\widehat{\varphi}_{\widehat{\alpha}} \colon \widehat{V}_{\widehat{\alpha}} \to \widehat{M}_{\widehat{\alpha}}\}_{\widehat{\alpha} \in \widehat{A}}$.

Then a continuous function $f \colon M \to \widehat{M}$ is *smooth* if for every $x \in M$ there exist $\alpha \in A$ and $\widehat{\beta} \in \widehat{A}$ such that $x \in M_\alpha$ and the map

$$\widehat{\varphi}_{\widehat{\beta}}^{-1} \circ f \circ \varphi_\alpha|_{\varphi_\alpha^{-1}(M_\alpha \cap f^{-1}(\widehat{M}_{\widehat{\beta}}))} \colon \varphi_\alpha^{-1}(M_\alpha \cap f^{-1}(\widehat{M}_{\widehat{\beta}})) \to \widehat{V}_{\widehat{\beta}}$$

is smooth. Note that $\varphi_\alpha^{-1}(M_\alpha \cap f^{-1}(\widehat{M}_{\widehat{\beta}}))$ is always an open (maybe empty) subset of the model space V. We will denote the set of all smooth mappings from M into $\widehat{M}$ by $\mathcal{C}^\infty(M, \widehat{M})$.

The mapping $f\colon M \to \widehat{M}$ is called a *diffeomorphism* if it is bijective and both f and f^{-1} are smooth.

REMARK A.53 If M, N and P are smooth manifolds, and $f\colon M \to N$ and $g\colon N \to P$ are smooth mappings, then $g \circ f\colon M \to P$ is in turn a smooth mapping.

This shows that there exists a category **Man** whose objects are the smooth manifolds and whose morphisms are the smooth mappings.

REMARK A.54 Real (or complex) *analytic manifolds* and real (or complex) analytic mappings on smooth manifolds can be defined by replacing the word "smooth" by "real (or complex) analytic" in Definitions A.51 and A.52.

DEFINITION A.55 Let M be a smooth manifold modeled on V, with the local coordinate systems $\{\varphi_\alpha\colon V_\alpha \to M_\alpha\}_{\alpha\in A}$. Fix a point $x_0 \in M$. A *tangent vector* at x_0 is an equivalence class of parameterized paths through x_0 in the following sense. Let I_1 and I_2 be open intervals in $\mathbb{R}$, with $t_0 \in I_1 \cap I_2$, and $p_1\colon I_1 \to M$, $p_2\colon I_2 \to M$ smooth mappings (*smooth paths*) with $p_1(t_0) = p_2(t_0) = x_0$. We say that p_1 and p_2 are *equivalent* at t_0 if there exists $\alpha \in A$ such that $x_0 \in M_\alpha$ and moreover the two smooth mappings

$$\mathbb{R} \supseteq p_i^{-1}(M_\alpha) \ni t \mapsto \varphi_\alpha^{-1}(p_i(t)) \in V \qquad (i = 1, 2)$$

have the same derivative at t_0. If $p\colon I \to M$ is a smooth path and $t_0 \in I$, then the equivalence class of p at t_0 is denoted by $\dot{p}(t_0)$ and is called the *velocity vector* of p at t_0.

The set of all such tangent vectors at x_0 is denoted by $T_{x_0}M$ and is called the *tangent space* at x_0. Note that if $x_0 \in M_\alpha$ as above, then there exists a natural bijective mapping

$$\Phi_\alpha\colon V \to T_{x_0}M$$

such that, for each $v \in V$, the tangent vector $\Phi_\alpha(v) \in T_{x_0}M$ is the equivalence class of the path $t \mapsto \varphi_\alpha(v_0 + tv)$, where $v_0 := \varphi_\alpha^{-1}(x_0) \in V_\alpha$. Using the bijection Φ_α we can equip $T_{x_0}M$ with the structure of a topological vector space isomorphic to V.

REMARK A.56 In the setting of Definition A.55, it is not hard to see that the following assertions hold.

(a) The definition of the equivalence relation for paths through x_0 does not depend on the choice of the index α with $x_0 \in M_\alpha$.

(b) The mapping $\Phi_\alpha \colon V \to T_{x_0}M$ is indeed bijective.

(c) If $\beta \in A$ is another index with $x_0 \in M_\beta$, then $\Phi_\beta^{-1} \circ \Phi_\alpha \colon V \to V$ is an isomorphism of topological vector spaces.

(d) The structure of the topological vector space of $T_{x_0}M$ is natural in the sense that it does not depend on the choice of $\alpha \in A$ with $x_0 \in M_\alpha$.

REMARK A.57 Let U be an open subset of a locally convex vector space V, viewed as a smooth manifold with the local coordinate system $\mathrm{id}_U \colon U \hookrightarrow V$. Then $TU = U \times V$.

DEFINITION A.58 Let M be a smooth manifold modeled on the locally convex space V, with the local coordinate systems $\{\varphi_\alpha \colon V_\alpha \to M_\alpha\}_{\alpha \in A}$. The *tangent bundle* of M is defined as the disjoint union

$$TM := \bigcup_{x \in M} T_x M,$$

and its *canonical projection* $p \colon TM \to M$ is defined such that $v \in T_{p(v)}M$ for all $v \in TM$.

For every $\alpha \in A$ we also introduce the mapping

$$\psi_\alpha \colon V_\alpha \times V \to TM$$

such that, for $u \in V_\alpha$ and $v \in V$, the tangent vector $\psi_\alpha(u, v) \in T_{\varphi_\alpha(u)}M$ is by definition the equivalence class of the smooth path $t \mapsto \varphi_\alpha(u + tv)$ through $\varphi_\alpha(u) \in M$. If we denote the image of ψ_α by TM_α, then the tangent bundle TM has a natural structure of a smooth manifold modeled on $V \times V$, with the local coordinate systems $\{\psi_\alpha \colon V_\alpha \times V \to TM_\alpha\}_{\alpha \in A}$.

REMARK A.59 In the setting of Definition A.55, the following assertions hold.

(a) The tangent bundle TM indeed has a structure of a smooth manifold (in particular a topology) as indicated in Definition A.58.

(b) For every $\alpha \in A$ and $u \in V_\alpha$ we have

$$(\forall v \in V) \qquad \Phi_\alpha(v) = \psi_\alpha(u, v),$$

where Φ_α is as in Definition A.58.

DEFINITION A.60 Let M and $\widehat{M}$ be smooth manifolds modeled on the locally convex spaces V and $\widehat{V}$, with the local coordinate systems $\{\varphi_\alpha \colon V_\alpha \to$

$M_\alpha\}_{\alpha\in A}$ and $\{\widehat{\varphi}_{\widehat{\alpha}}\colon \widehat{V}_{\widehat{\alpha}} \to \widehat{M}_{\widehat{\alpha}}\}_{\widehat{\alpha}\in\widehat{A}}$, respectively. If $f\colon M \to \widehat{M}$ is a smooth mapping and $x \in M$, then the *tangent* of f at x is the mapping

$$f'_x\colon T_xM \to T_{f(x)}\widehat{M}$$

defined in the following way. If $v \in T_xM$ and $p\colon I \to M$ is a smooth path such that $0 \in I$, $p(0) = x$ and $\dot{p}(0) = v$, then

$$f'_x(v) := \dot{q}(0) \in T_{f(x)}\widehat{M},$$

where $q := f \circ p\colon I \to \widehat{M}$. (Note that q is a smooth path and $q(0) = f(x)$.)

Then the *tangent* of f is the mapping

$$Tf\colon TM \to T\widehat{M}$$

defined by

$$(\forall x \in M) \qquad Tf|_{T_xM} := T_xf := f'_x.$$

Then Tf is a smooth mapping and for every $x \in M$ the restriction of Tf to T_xM is a continuous linear operator $T_xM \to T_{f(x)}\widehat{M}$.

REMARK A.61 In the setting of Definition A.60, it is not hard to see that the following assertions hold.

(a) For each $x \in M$, the mapping $T_xf\colon T_xM \to T_{f(x)}\widehat{M}$ is correctly defined, and indeed linear and continuous.

(b) The mapping $Tf\colon TM \to T\widehat{M}$ is indeed smooth.

REMARK A.62 If **Man** stands for the category of smooth manifolds (see Remark A.53), it is easy to see that the correspondence

$$T\colon \mathbf{Man} \to \mathbf{Man}$$

which associates to each locally convex smooth manifold its tangent bundle, and to each smooth map its tangent mapping, has all of the properties (i)–(vi) in Remark 2.3.

Vector fields

DEFINITION A.63 Let M be a locally convex smooth manifold with the tangent bundle TM. A *smooth vector field* on M is a smooth map

$$v\colon M \to TM$$

such that $v(x) \in T_xM$ for all $x \in M$.

The set of all vector fields on M clearly has a structure of vector space (with pointwise defined addition and scalar multiplication), and we denote that vector space by $\mathcal{V}(M)$.

DEFINITION A.64 Let M be a locally convex smooth manifold, $v \in \mathcal{V}(M)$, and Y a locally convex topological vector space. We define a linear operator

$$D_v \colon \mathcal{C}^\infty(M,Y) \to \mathcal{C}^\infty(M,Y)$$

in the following way.

Let $\pi \colon Y \times Y \to Y$, $(y_1,y_2) \mapsto y_2$. Observing that the tangent bundle of Y is $TY = Y \times Y$ with the canonical projection $p \colon TY \to Y$, $(y_1,y_2) \mapsto y_1$ (which is different from π!), we define, for all $f \in \mathcal{C}^\infty(M,Y)$, a smooth function $D_v f \in \mathcal{C}^\infty(M,Y)$ by the commutative diagram

$$\begin{array}{ccc} TM & \xrightarrow{Tf} & TY \\ {\scriptstyle v}\uparrow & & \downarrow{\scriptstyle \pi} \\ M & \xrightarrow{D_v f} & Y \end{array}$$

that is,

$$(\forall x \in M) \qquad (Tf \circ v)(x) = \big(f(x),(D_v f)(x)\big) \in Y \times Y = TY.$$

REMARK A.65 In the setting of Definition A.64, it is easy to see that the mapping

$$\mathcal{V}(M) \to \mathrm{End}\,(\mathcal{C}^\infty(M,Y)), \qquad v \mapsto D_v,$$

is linear.

LEMMA A.66 *Let V be a locally convex vector space and V_0 an open subset of V. Then for all $u,w \in \mathcal{V}(V_0)$ there exists a unique vector field $[u,w] \in \mathcal{V}(V_0)$ such that, for each open subset D of V_0 and each locally convex vector space Y we have*

$$(\forall f \in \mathcal{C}^\infty(D,Y)) \qquad D_{[u,w]}f = D_u(D_w f) - D_w(D_u f).$$

PROOF Note that $TV_0 = V_0 \times V$ with the canonical projection $V_0 \times V \to V_0$, $(x,t) \mapsto x$, so that

$$\mathcal{V}(V_0) = \{v \colon V_0 \to V_0 \times V \mid (\exists \widetilde{v} \in \mathcal{C}^\infty(V_0,V)) \quad v(\cdot) = (\cdot,\widetilde{v}(\cdot))\}.$$

Then for $u,w \in \mathcal{V}(V_0)$ fixed, let $\widetilde{u},\widetilde{w} \in \mathcal{C}^\infty(V_0,V)$ with $u(\cdot) = (\cdot,\widetilde{u}(\cdot))$, $w(\cdot) = (\cdot,\widetilde{w}(\cdot))$, and define $[u,w] \in \mathcal{V}(V_0)$ by

$$[u,w] \colon V_0 \to V_0 \times V, \quad [u,w](\cdot) = \big(\cdot,(D_u\widetilde{w} - D_w\widetilde{u})(\cdot)\big). \tag{A.2}$$

Now, for each locally convex space Y, and all open subsets D of V_0 and $f \in \mathcal{C}^\infty(D, Y)$, we clearly have

$$(\forall v \in \mathcal{V}(V_0))\,(\forall x \in D) \qquad (D_v f)(x) = f'_x \widetilde{v}(x), \tag{A.3}$$

whence

$$\begin{aligned}(D_{[u,w]} f)(x) &= f'_x(D_u\widetilde{w} - D_w\widetilde{u})(x) = f'_x((\widetilde{w})'_x\widetilde{u}(x) - (\widetilde{u})'_x\widetilde{w}(x)) \\ &= f'_x(\widetilde{w})'_x\widetilde{u}(x) - f'_x(\widetilde{u})'_x\widetilde{w}(x)\end{aligned}$$

On the other hand,

$$D_u(D_w f)(x) = (D_w f)'_x\widetilde{u}(x) = (f'_\bullet\widetilde{w}(\cdot))'_x\widetilde{u}(x) = f''_x(\widetilde{w}(x), \widetilde{u}(x)) + f'_x(\widetilde{w})'_x\widetilde{u}(x)$$

and similarly

$$D_w(D_u f)(x) = f''_x(\widetilde{u}(x), \widetilde{w}(x)) + f'_x(\widetilde{u})'_x\widetilde{w}(x).$$

Thus $D_{[u,w]} f = D_u(D_w f) - D_w(D_u f)$ by the symmetry property of f''_x (see Proposition A.28).

To prove the uniqueness assertion, apply the property of $[v, w]$ in the special case when $D = V_0$, $Y = V$ and f is the inclusion mapping $V_0 \hookrightarrow V$. Then $f'_x = \mathrm{id}_V$ for all $x \in V_0$, and $f'' = 0$, hence by the above computations we get

$$D_u(D_w f)(x) - D_w(D_u f)(x) = (\widetilde{w})'_x\widetilde{u}(x) - (\widetilde{u})'_x\widetilde{w}(x) = (D_u\widetilde{w})(x) - (D_w\widetilde{u})(x),$$

while

$$(D_{[u,w]} f)(x) = \widetilde{[u, w]}(x).$$

Thus $\widetilde{[u, w]} = D_u\widetilde{w} - D_w\widetilde{u}$, that is, the vector field $[u, w] \in \mathcal{V}(V_0)$ is necessarily given by (A.2), and the proof ends. ∎

THEOREM A.67 *If M is a locally convex smooth manifold, then $\mathcal{V}(M)$ has a unique structure of Lie algebra such that, for each locally convex vector space Y and each open subset U of M, the linear map*

$$\mathcal{V}(M) \to \mathrm{End}\,(\mathcal{C}^\infty(U, Y)), \quad v \mapsto D_v,$$

is a Lie algebra homomorphism.

PROOF Let $\{\varphi_\alpha\colon V_\alpha \to M_\alpha\}_{\alpha \in A}$ as in Definition A.51. Also let $u, w \in \mathcal{V}(M)$ fixed for the moment. Then for each $\alpha \in A$ there clearly exist $\bar{u}_\alpha, \bar{w}_\alpha \in \mathcal{V}(V_\alpha)$ such that the diagrams

$$\begin{array}{ccc} T(V_\alpha) & \xrightarrow{T(\varphi_\alpha)} & T(M_\alpha) \\ {\scriptstyle \bar{u}_\alpha}\uparrow & & \uparrow{\scriptstyle u|_{M_\alpha}} \\ V_\alpha & \xrightarrow{\varphi_\alpha} & M_\alpha \end{array} \quad \text{and} \quad \begin{array}{ccc} T(V_\alpha) & \xrightarrow{T(\varphi_\alpha)} & T(M_\alpha) \\ {\scriptstyle \bar{w}_\alpha}\uparrow & & \uparrow{\scriptstyle w|_{M_\alpha}} \\ V_\alpha & \xrightarrow{\varphi_\alpha} & M_\alpha \end{array}$$

are commutative (since both $\varphi_\alpha\colon V_\alpha \to M_\alpha$ and $T(\varphi_\alpha)\colon T(V_\alpha) \to T(M_\alpha)$ are diffeomorphisms, and $T(\varphi_\alpha)$ maps the fiber $T_x(V_\alpha)$ into $T_{\varphi_\alpha(x)}(M_\alpha)$ for each $x \in V_\alpha$). Now define $[\bar{u}_\alpha, \bar{w}_\alpha] \in \mathcal{V}(V_\alpha)$ by Lemma A.66, and then denote

$$[u,w]_\alpha := T(\varphi_\alpha) \circ [\bar{u}_\alpha, \bar{w}_\alpha] \circ (\varphi_\alpha)^{-1} \in \mathcal{V}(M_\alpha).$$

It then easily follows by the uniqueness assertion in Lemma A.66 that if $M_\alpha \cap M_\beta \neq \emptyset$, then $[u,w]_\alpha|_{M_\alpha \cap M_\beta} = [u,w]_\beta|_{M_\alpha \cap M_\beta}$. Hence there exists a unique vector field $[u,w] \in \mathcal{V}(M)$ such that $[u,w]|_{M_\alpha} = [u,w]_\alpha$ for all $\alpha \in A$.

Now, for Y and U as in the statement, and $f \in \mathcal{C}^\infty(U,Y)$, we easily get by Lemma A.66 that $D_{[u,w]}f = D_u(D_w f) - D_w(D_u f)$ on $U \cap M_\alpha$ for all α, whence $D_{[u,w]}f = D_u(D_w f) - D_w(D_u f)$ on U. In other words, the mapping

$$\mathcal{V}(M) \times \mathcal{V}(M) \to \mathcal{V}(M), \quad (u,w) \mapsto [u,w],$$

has the property that for every open subset U of M we have

$$(\forall v, w \in \mathcal{V}(M)) \qquad D_{[v,w]} = [D_v, D_w] \in \mathrm{End}\,(\mathcal{C}^\infty(U,Y)).$$

This easily implies that $[\cdot,\cdot]$ is a Lie algebra structure on $\mathcal{V}(M)$, in view of the fact that if V stands for the model space of the manifold M, then the linear mapping

$$\mathcal{V}(M) \to \prod_{\alpha \in A} \mathrm{End}\,(\mathcal{C}^\infty(M_\alpha, V)), \qquad v \mapsto (D_{v|_{M_\alpha}})_{\alpha \in A}$$

is injective. (In fact, if $v \in \mathcal{V}(M)$ and $D_{v|_{M_\alpha}} = 0$, then $(D_{v|_{M_\alpha}})(\varphi_\alpha^{-1}) = 0$, whence $\pi \circ T(\varphi_\alpha^{-1}) \circ v|_{M_\alpha} = 0$, where $\pi\colon V \times V \to V$ is the projection on the second coordinate (as in Definition A.64). But the linear operator $T_x(\varphi_\alpha^{-1})$ is invertible for all $x \in M_\alpha$, hence $v|_{M_\alpha} = 0$.)

For the uniqueness assertion, use Lemma A.66 for $U = M_\alpha$ with arbitrary $\alpha \in A$. ∎

PROPOSITION A.68 *Let M and N be two locally convex smooth manifolds, and $\varphi\colon M \to N$ a smooth mapping. If $v_1, v_2 \in \mathcal{V}(M)$ and $w_1, w_2 \in \mathcal{V}(N)$ satisfy $T\varphi \circ v_j = w_j \circ \varphi$ for $j = 1, 2$, then $T\varphi \circ [v_1, v_2] = [w_1, w_2] \circ \varphi$.*

PROOF It suffices to show that the desired equality holds on some neighborhood of an arbitrary point $p \in M$. Thus, replacing M by a suitably small neighborhood of p, and N by an appropriate neighborhood of $\varphi(p) \in N$, we may assume that there exist the locally convex vector spaces V and W such that M is an open subset of V and N is an open subset of W. Then there exist $\widetilde{v}_1, \widetilde{v}_2 \in \mathcal{C}^\infty(M,V)$ and $\widetilde{w}_1, \widetilde{w}_2 \in \mathcal{C}^\infty(N,W)$ such that for $j = 1, 2$ we have

$$\begin{gathered} v_j(\cdot) = (\cdot, \widetilde{v}_j(\cdot))\colon M \to M \times V = TM \\ \text{and} \\ w_j(\cdot) = (\cdot, \widetilde{w}_j(\cdot))\colon N \to N \times W = TN \end{gathered} \tag{A.4}$$

(see the proof of Lemma A.66). Now, using the fact that we have

$$T\varphi\colon M\times V\to N\times W,\quad (m_0,v_0)\mapsto(\varphi(m_0),\varphi'_{m_0}(v_0)),\tag{A.5}$$

along with formula (A.3) in the proof of Lemma A.66, we get for every $m_0\in M$

$$\begin{aligned}(T\varphi\circ[v_1,v_2])(m_0)&=T\varphi(m_0,\widetilde{[v_1,v_2]}(m_0))=(\varphi(m_0),\varphi'_{m_0}(\widetilde{[v_1,v_2]}(m_0)))\\&=(\varphi(m_0),(D_{[v_1,v_2]}\varphi)(m_0)).\end{aligned}$$

We further deduce by Lemma A.66 that

$$(T\varphi\circ[v_1,v_2])(m_0)=(\varphi(m_0),(D_{v_1}D_{v_2}\varphi)(m_0)-(D_{v_2}D_{v_1}\varphi)(m_0)).\tag{A.6}$$

On the other hand, note that the hypothesis $T\varphi\circ v_j=w_j\circ\varphi$ implies by means of (A.4) and (A.5) that

$$(\forall x\in M)\qquad \big(\varphi(x),\varphi'_x(\widetilde{v}_j(x))\big)=\big(\varphi(x),(\widetilde{w}_j\circ\varphi)(x)\big),$$

hence by formula (A.3) in the proof of Lemma A.66 we get for $j=1,2$

$$(\forall x\in M)\qquad (D_{v_j}\varphi)(x)=\varphi'_x(\widetilde{v}_j(x))=(\widetilde{w}_j\circ\varphi)(x).\tag{A.7}$$

Consequently

$$\begin{aligned}D_{v_1}(D_{v_2}\varphi)(m_0)&=D_{v_1}(\widetilde{w}_2\circ\varphi)(m_0)=(\widetilde{w}_2\circ\varphi)'_{m_0}(\widetilde{v}_1(m_0))\\&=(\widetilde{w}_2)'_{\varphi(m_0)}\varphi'_{m_0}\widetilde{v}_1(m_0)=(\widetilde{w}_2)'_{\varphi(m_0)}\widetilde{w}_1(\varphi(m_0))\\&=(D_{w_1}\widetilde{w}_2)(\varphi(m_0)),\end{aligned}$$

where the second equality follows by formula (A.3) while the fourth equality follows by (A.7). Similarly,

$$D_{v_2}(D_{v_1}\varphi)(m_0)=(D_{w_2}\widetilde{w}_1)(\varphi(m_0)).$$

Thus

$$\begin{aligned}D_{v_1}(D_{v_2}\varphi)(m_0)-D_{v_2}(D_{v_1}\varphi)(m_0)&=(D_{w_1}\widetilde{w}_2-D_{w_2}\widetilde{w}_1)(\varphi(m_0))\\&=\widetilde{[w_1,w_2]}(\varphi(m_0)),\end{aligned}$$

by formula (A.2) in the proof of Lemma A.66. Consequently, by (A.6) we get

$$(T\varphi\circ[v_1,v_2])(m_0)=\big(\varphi(m_0),\widetilde{[w_1,w_2]}(\varphi(m_0))\big)=([w_1,w_2]\circ\varphi)(m_0),$$

using again formula (A.2) in the proof of Lemma A.66. ∎

REMARK A.69 Let M be an open subset of the real locally convex space V. For all $u,v\in\mathcal{C}^\infty(M,V)$ define $[u,v]\in\mathcal{C}^\infty(M,V)$ by

$$(\forall x\in M)\quad [u,v](x)=u'_x(v(x))-v'_x(u(x)).$$

Then the bracket $[\cdot,\cdot]$ turns $\mathcal{C}^\infty(M,V)$ into a Lie algebra.

Differential forms

NOTATION A.70 If X and $\mathfrak{z}$ are real Banach spaces and $n \geq 1$ is an integer, we denote

$$\mathcal{B}^n_{\text{skew}}(X, \mathfrak{z}) = \{\psi \colon \underbrace{X \times \cdots \times X}_{n \text{ times}} \to \mathfrak{z} \mid \psi \text{ } n\text{-linear skew-symmetric}\}.$$

Then $\mathcal{B}^n_{\text{skew}}(X, Y)$ is a Banach space with respect to the usual norm of multilinear maps.

DEFINITION A.71 Let M be a Banach manifold, $\mathfrak{z}$ a real Banach space, and $n \geq 1$ an integer. Assume that the manifold M is modeled on the Banach space V and let $\{\varphi_\alpha \colon V_\alpha \to M_\alpha\}_{\alpha \in A}$ as in Definition A.51. Accordingly, for $\alpha \in A$ and $x \in M_\alpha$ let $\Phi_{\alpha,x} \colon V \to T_x M$ be the isomorphism of Banach spaces used in Definition A.55.

A *differential n-form of type* $\mathfrak{z}$ on M is a family $\psi = \{\psi_x\}_{x \in M}$ satisfying the following conditions:

(i) For each $x \in M$ we have $\psi_x \in \mathcal{B}^n_{\text{skew}}(T_x M, \mathfrak{z})$.

(ii) For all $\alpha \in A$ the function

$$V_\alpha \mapsto \mathcal{B}^n_{\text{skew}}(V, \mathfrak{z}), \quad x \mapsto \omega_x \circ (\Phi_{\alpha,x} \times \cdots \times \Phi_{\alpha,x})$$

is smooth.

We usually denote by $\Omega^n(M, \mathfrak{z})$ the set of all differential n-form of type $\mathfrak{z}$ on M. Note that $\Omega^n(M, \mathfrak{z})$ has a natural structure of real vector space.

If moreover N is another Banach manifold and $f \colon N \to M$ is a smooth mapping, then the *pull-back* of ψ by f is the differential n-form of type $\mathfrak{z}$ on N defined by

$$f^*(\psi) = \{\psi_{f(x)} \circ (T_x f \times \cdots \times T_x f)\}_{x \in M}.$$

In the case when $n = 2$ we say that the differential 2-form ψ is *weakly nondegenerate* if we have

$$\{v \in T_x M \mid (\forall w \in T_x M) \quad \psi_x(v, w) = 0\} = \{0\}$$

for all $x \in M$. If moreover the operator

$$T_x M \to \mathcal{B}(T_x M, \mathfrak{z}), \quad v \mapsto \psi_x(v, \cdot)$$

is surjective for all $x \in M$ then we say that ψ is *strongly nondegenerate.*

REMARK A.72 In the setting of Definition A.71, let ψ be a differential n-form of type $\mathfrak{z}$ on M and $v_1, \ldots, v_n \in \mathcal{V}(M)$. Then it is clear that the function defined by

$$\psi(v_1, \ldots, v_n) \colon M \to \mathfrak{z}, \quad x \mapsto \psi_x(v_1(x), \ldots, v_n(x)),$$

is smooth.

PROPOSITION A.73 *Let M be a Banach manifold and $\mathfrak{z}$ a real Banach space. Then for each integer $n \geq 1$ and each $\psi \in \Omega^n(M, \mathfrak{z})$ there exists a unique differential $(n+1)$-form of type $\mathfrak{z}$ on M, denoted by $d\psi \in \Omega^{n+1}(M, \mathfrak{z})$, such that for every open subset M_0 of M and all $v_1, \dots, v_n \in \mathcal{V}(M_0)$ we have*

$$d\psi(v_1, \dots, v_{n+1}) = \sum_{i=1}^{n} (-1)^i D_{v_i}(\psi(v_1, \dots, \widehat{v_i}, \dots, v_n)) + \sum_{1 \leq i < j \leq n} (-1)^{i+j} \psi([v_i, v_j], v_1, \dots, \widehat{v_i}, \dots, \widehat{v_j}, \dots, v_n).$$

Moreover the operator $d\colon \Omega^n(M, \mathfrak{z}) \to \Omega^{n+1}(M, \mathfrak{z})$ is linear, the composition of the operators

$$\Omega^n(M, \mathfrak{z}) \xrightarrow{d} \Omega^{n+1}(M, \mathfrak{z}) \xrightarrow{d} \Omega^{n+2}(M, \mathfrak{z})$$

equals 0 and the diagram

$$\begin{array}{ccc} \Omega^n(N, \mathfrak{z}) & \xrightarrow{d} & \Omega^{n+1}(N, \mathfrak{z}) \\ {\scriptstyle f^*}\downarrow & & \downarrow{\scriptstyle f^*} \\ \Omega^n(M, \mathfrak{z}) & \xrightarrow{d} & \Omega^{n+1}(M, \mathfrak{z}) \end{array}$$

is commutative whenever $f\colon M \to N$ is a smooth mapping from M into another Banach manifold N.

PROOF The proof of Proposition 3.2 in Chapter V of [La01] extends line by line to vector-valued differential forms. □

DEFINITION A.74 Let M be a Banach manifold, $\mathfrak{z}$ a real Banach space, and $n \geq 1$ an integer. The operator

$$d\colon \Omega^n(M, \mathfrak{z}) \to \Omega^{n+1}(M, \mathfrak{z})$$

introduced in Proposition A.73 is called the *exterior derivative.* We say that a differential form $\psi \in \Omega^n(M, \mathfrak{z})$ is *closed* if $d\eta = 0$.

Notes

The first part of the above review of topological vector spaces follows the lines of the corresponding section in the paper of J. Milnor [Mi84]. See also the books [Tr67] and [Sf66].

See Lemma 1.1 in [BS71b] for more details on the result contained in our Proposition A.7. Further information on the complexifications of topological vector spaces can be found in Section 2 of the paper [BS71a]. The results concerning continuous inverse algebras are taken from the paper [Gl02b]. The algebras of that type play an important role in K-theory; see e.g., [Swa77].

We refer to the paper by J. Milnor [Mi84] for a quick review of the differential calculus in locally convex spaces. The detailed proofs of these results can be found in [Gl02a]. See also [Ke74], [Ht82], and [Ne01a].

The book [La01] contains a good exposition of the basic results in differential calculus in the framework of Banach spaces. See also [Nel69].

An introduction to analytic mappings on Banach spaces can be found in Chapter 1 in [Up85]. See e.g., [BS71a] and [BS71b] for analytic mappings on more general topological vector spaces. The analyticity of the inversion mapping in a continuous inverse algebra (Proposition A.46) was proved in [Gl02b].

We refer to the book [Wa71] for an elementary introduction to the theory of finite-dimensional manifolds.

A quick introduction to infinite-dimensional manifolds modeled on locally convex spaces can be found in [Mi84]. For manifolds modeled on Banach spaces, see [La01] and [Up85]. The more special setting of analytic Banach manifolds is developed in [Up85].

Appendix B

Basic Differential Equations of Lie Theory

Abstract. We develop here the equations and formulas describing the most basic level of Lie theory, namely the theory of local Lie groups. One of the main results (Theorem B.12) essentially says that every local Lie group modeled on a Banach space is real analytic.

Throughout this appendix we denote by Y a real Banach space, and for every $r > 0$ we denote by

$$B_Y(y_0, r) = \{y \in Y \mid \|y - y_0\| < r\}$$

the open ball with center at y_0 and radius r. Also, we denote by $\mathrm{GL}(Y)$ the set of all invertible bounded linear operators on Y.

We begin by a couple of general statements on ordinary differential equations.

THEOREM B.1 *Let B be an open subset of Y such that $0 \in B$, J an open interval in $\mathbb{R}$, and $g\colon J \times B \to Y$ a smooth mapping. For $j \in \{1, 2\}$, consider an open interval I_j contained in J and a smooth function $\gamma_j\colon I_j \to Y$ such that for each $t \in I_j$ we have $\dot{\gamma}_j(t) = g(t, \gamma_j(t))$. If there exists $t_0 \in I_1 \cap I_2$ such that $\gamma_1(t_0) = \gamma_2(t_0)$, then $\gamma_1|_{I_1 \cap I_2} = \gamma_2|_{I_1 \cap I_2}$.*

PROOF See Theorem 1.3 in Chapter IV in [La01]. ∎

THEOREM B.2 *Let B be an open subset of Y such that $0 \in B$, and $g\colon B \times Y \to Y$ a smooth mapping. Then there exist $r > 0$ and $\varepsilon \in (0, 1)$ such that the following conditions are fulfilled.*

(a) *We have $B_Y(0, r) \subseteq B$.*

(b) *There exists a unique smooth mapping $\gamma\colon (-\varepsilon, \varepsilon) \times B_Y(0, r) \to B$ such that, for all $v \in B_Y(0, r)$, the mapping $\gamma_v(\cdot) := \gamma(\cdot, v)\colon (-\varepsilon, \varepsilon) \to B$ has the properties*

$$\gamma_v(0) = 0$$

and

$$(\forall t \in (-\varepsilon, \varepsilon)) \qquad \dot{\gamma}_v(t) = g(\gamma_v(t), v).$$

PROOF Define $E := Y \times Y$, $U := B \times Y$, $J = (-1, 1)$, and

$$f\colon J \times U \to E, \qquad f(t, (y, v)) := (g(y, v), 0) \text{ for } t \in J, y \in B, v \in Y.$$

Then use Theorem 1.11 in Chapter IV in [La01] to get $r > 0$, $\varepsilon \in (0, 1)$ and $\alpha\colon J_0 \times U_0 \to U$ such that $J_0 := (-\varepsilon, \varepsilon) \subseteq J$, $0 \in U_0$, U_0 is an open subset of U and α is the unique smooth mapping satisfying the conditions that for all $x \in U_0$ we have $\alpha(0, x) = x$ and the function $\alpha_x(\cdot) := \alpha(\cdot, x)\colon J_0 \to U$ has the property $\dot{\alpha}_x(t) = f(t, \alpha_x(t))$ for each $t \in J_0$.

On the other hand, since U_0 is an open subset of $U(\subseteq Y \times Y)$ and $0 \in U_0$, we can find $r > 0$ such that $B_Y(0, r) \times B_Y(0, r) \subseteq U_0$. Also, since $U = B \times Y$ and α takes values in U, it follows that there exist $\gamma_1\colon J_0 \times U_0 \to B$ and $\beta\colon J_0 \times U_0 \to Y$ such that $\alpha(t, x) = (\gamma_1(t, x), \beta(t, x))$ for all $t \in J_0$ and $x \in U_0$.

It then follows that for each point $x = (y, v) \in U_0 \subseteq B \times Y$ we have

$$(y, v) = x = \alpha(0, x) = (\gamma_1(0, x), \beta(0, x))$$

and for all $t \in J_0$

$$(\dot{\gamma}_1(t, x), \dot{\beta}(t, x)) = (g(\gamma_1(t, x), \beta(t, x)), 0).$$

The relation $\dot{\beta}(t, x) = 0$ implies that the function $t \mapsto \beta(t, x)$ is constant, hence for all $t \in J_0$ we have $\beta(t, x) = \beta(0, x) = v$. Thus for all $t \in J_0$, $y, v \in B_Y(0, r)$, setting $x = (y, v)$ in the above relations we get

$$\dot{\gamma}_1(t, y, v) = g(\gamma_1(t, y, v), v) \text{ and } \gamma_1(0, y, v) = y.$$

Consequently, the function $\gamma(\cdot, \cdot) := \gamma_1(\cdot, 0, \cdot)\colon J_0 \times B_Y(0, r) \to B$ has the desired properties. Its uniqueness follows by Theorem B.1. ∎

We now use the preceding theorems in order to get the local version of the exponential map of Banach-Lie groups; this will be achieved in Corollary B.7.

PROPOSITION B.3 *Let $r_1 > 0$ and*

$$\Psi\colon B_Y(0, r_1) \to \mathcal{B}(Y)$$

smooth. Then there exist $r \in (0, r_1)$ and $\varepsilon \in (0, 1)$ such that there exists a unique smooth mapping $\gamma\colon (-\varepsilon, \varepsilon) \times B_Y(0, r) \to B_Y(0, r_1)$ with the following property: If $v \in B_Y(0, r)$ and we define $\gamma_v := \gamma(\cdot, v)\colon (-\varepsilon, \varepsilon) \to B_Y(0, r_1)$, then

$$\gamma_v(0) = 0$$

and

$$\dot{\gamma}_v(t) = \Psi(\gamma_v(t))v \text{ whenever } t \in (-\varepsilon, \varepsilon).$$

PROOF Just use Theorem B.2 for $B = B_Y(0, r_1)$ and $g\colon B \times Y \to Y$, $g(y, v) := \Psi(y)v$. ∎

REMARK B.4 In the setting of Proposition B.3 it is not hard to see that we have

$$\gamma(ts, v) = \gamma(t, sv)$$

whenever $t, ts \in (-\varepsilon, \varepsilon)$ and $v, sv \in B_Y(0, r)$.

PROPOSITION B.5 *Let $r_1 > 0$ and*

$$\Psi\colon B_Y(0, r_1) \to \mathcal{B}(Y)$$

smooth. Then there exists $r_2 \in (0, r_1)$ such that there exists a unique smooth mapping

$$\chi\colon B_Y(0, r_2) \to B_Y(0, r_1)$$

with the following properties:

(i) *We have $\chi(0) = 0$ and $\chi'_0 = \Psi(0)$.*

(ii) *For all $v \in B_Y(0, r_2) \setminus \{0\}$, the function*

$$\chi_v\colon (-r_2/\|v\|, r_2/\|v\|) \to B_Y(0, r_1), \quad t \mapsto \chi(tv),$$

satisfies

$$\dot{\chi}_v(t) = \Psi(\chi_v(t))v \text{ whenever } t \in (-r_2/\|v\|, r_2/\|v\|).$$

PROOF Let $r \in (0, r_1)$, $\varepsilon \in (0, 1)$, and $\gamma\colon (-\varepsilon, \varepsilon) \times B_Y(0, r) \to B_Y(0, r_1)$ given by Proposition B.3, and define $r_2 := r\varepsilon/2 (< r)$. Next define

$$\chi\colon B_Y(0, r_2) \to B_Y(0, r_1), \quad \chi(v) := \gamma(\varepsilon/2, (2/\varepsilon)v).$$

Then χ is smooth and $\chi(0) = 0$ according to Proposition B.3.

Now fix $v \in B_Y(0, r_2)$ and let $\chi_v\colon (-r_2/\|v\|, r_2/\|v\|) \to B_Y(0, r_1)$, $\chi_v(t) = \chi(tv)$, as in the statement. We then have by Remark B.4 that

$$\chi_v(t) = \gamma(\varepsilon/2, (2/\varepsilon)tv) = \gamma((\varepsilon/2)t, (2/\varepsilon)v) = \gamma_{(2/\varepsilon)v}((\varepsilon/2)t)$$

whenever $|t| < r_2/\|v\|$, whence

$$\begin{aligned}
\dot{\chi}_v(t) &= \dot{\gamma}_{(2/\varepsilon)v}((\varepsilon/2)t) \\
&= (\varepsilon/2)\Psi(\gamma_{(2/\varepsilon)v}(t))(2/\varepsilon)v \qquad \text{(by Proposition B.3)} \\
&= \Psi(\gamma_{(2/\varepsilon)v}(t))v \\
&= \Psi(\chi_v(t))v,
\end{aligned}$$

as desired. For $t = 0$ we get $\chi_0' v = \Psi(0)v$ whenever $0 \neq v \in B_Y(0, r_2)$, hence the linear operators $\chi_0', \Psi(0) \in \mathcal{B}(Y)$ coincide.

To conclude the proof, let us note that the uniqueness of a smooth function $\chi\colon B_Y(0, r_2) \to B_Y(0, r_1)$ satisfying the conditions (i) and (ii) follows by Theorem B.1. ∎

PROPOSITION B.6 *Let $0 < r_2 < r_1$ and*

$$\mu\colon B_Y(0, r_1) \times B_Y(0, r_1) \to Y$$

a smooth mapping such that

$$\mu(y, 0) = y \text{ for all } y \in B_Y(0, r_1),$$

$\mu(B_Y(0, r_2) \times B_Y(0, r_2)) \subseteq B_Y(0, r_1)$, *and*

$$\mu(\mu(y, z), v) = \mu(y, \mu(z, v))$$

whenever $y, z, v \in B_Y(0, r_2)$. Furthermore, let $\varepsilon > 0$, $y_0 \in Y$ and a smooth mapping $\gamma\colon (-\varepsilon, \varepsilon) \to B_Y(0, r_2)$ such that $\gamma(0) = 0$ and

$$\dot{\gamma}(t) = \partial_2 \mu(\gamma(t), 0) y_0 \text{ whenever } t \in (-\varepsilon, \varepsilon).$$

Then for all $t, s \in (-\varepsilon/2, \varepsilon/2)$ we have

$$\mu(\gamma(t), \gamma(s)) = \gamma(t + s).$$

PROOF First remark that, for all $y, z, v \in B_Y(0, r_2)$ it follows by hypothesis that

$$\partial_2 \mu(\mu(y, z), v) = \partial_2 \mu(y, \mu(z, v)) \partial_2 \mu(z, v).$$

Now fix $s \in (-\varepsilon/2, \varepsilon/2)$ and define

$$\alpha, \beta\colon (-\varepsilon/2, \varepsilon/2) \to Y, \quad \alpha(t) := \mu(\gamma(s), \gamma(t)),\ \beta(t) := \gamma(s + t).$$

Then $\alpha(0) = \mu(\gamma(s), \gamma(0)) = \mu(\gamma(s), 0) = \gamma(s) = \beta(0)$. Also, for all $t \in (-\varepsilon/2, \varepsilon/2)$ we have

$$\dot{\beta}(t) = \dot{\gamma}(s + t) = \partial_2 \mu(\gamma(s + t), 0) y_0 = \partial_2 \mu(\beta(t), 0) y_0,$$

and

$$\begin{aligned}\dot{\alpha}(t) &= \partial_2 \mu(\gamma(s), \gamma(t)) \dot{\gamma}(t) = \partial_2 \mu(\gamma(s), \gamma(t)) \partial_2 \mu(\gamma(t), 0) y_0 \\ &= \partial_2 \mu(\mu(\gamma(s), \gamma(t)), 0) y_0 = \partial_2 \mu(\alpha(t), 0) y_0,\end{aligned}$$

where the second equality follows by the beginning remark. Thus Theorem B.1 (with $B = B_Y(0, r_2)$, $J_1 = J_2 = (-\varepsilon/2, \varepsilon/2)$, $g(t, y) = \partial_2(0, y) y_0$, $\gamma_1 = \alpha$ and $\gamma_2 = \beta$) shows that $\alpha = \beta$ on $(-\varepsilon/2, \varepsilon/2)$. This is just the desired equality, and the proof ends. ∎

The following corollary plays a key role in the construction of the exponential map of Banach-Lie groups (see the proof of Lemma 2.40).

COROLLARY B.7 *Let $0 < r_2 < r_1$ and*

$$\mu\colon B_Y(0,r_1) \times B_Y(0,r_1) \to Y$$

a smooth mapping such that

$$\mu(y,0) = y \text{ for all } y \in B_Y(0,r_1),$$

$\mu\big(B_Y(0,r_2) \times B_Y(0,r_2)\big) \subseteq B_Y(0,r_1)$ *and*

$$\mu(\mu(y,z),v) = \mu(y,\mu(z,v))$$

whenever $y, z, v \in B_Y(0,r_2)$. Then there exists $r_3 \in (0,r_2)$ such that there exists a unique smooth mapping

$$\chi\colon B_Y(0,r_3) \to B_Y(0,r_2)$$

with the following properties:

(i) *We have $\chi(0) = 0$ and $\chi'_0 = \partial_2\mu(0,0)$.*

(ii) *If $v \in B_Y(0,r_3)$ and $\max\{|t|,|s|\} < r_3/(2\|v\|)$, then*

$$\chi\big((t+s)v\big) = \mu\big(\chi(tv),\chi(sv)\big).$$

PROOF Construct r_3 and χ by using Proposition B.5 for

$$\Psi\colon B_Y(0,r_2) \to \mathcal{B}(Y), \quad \Psi(y) = \partial_2\mu(y,0),$$

and then use Proposition B.6 to get the desired property (ii). ∎

We now focus on a number of results eventually leading to the fact that local Lie groups modeled on Banach spaces are real analytic with respect to the local chart provided by the exponential map (see Theorem B.12 below).

PROPOSITION B.8 *Let $0 < r_2 < r_1$ and*

$$\mu\colon B_Y(0,r_1) \times B_Y(0,r_1) \to Y$$

a smooth mapping such that

$$\mu(y,0) = \mu(0,y) = y \text{ for all } y \in B_Y(0,r_1),$$

$\mu\big(B_Y(0,r_2) \times B_Y(0,r_2)\big) \subseteq B_Y(0,r_1)$ *and*

$$\mu(\mu(x,y),z) = \mu(x,\mu(y,z))$$

whenever $x, y, z \in B_Y(0, r_2)$. *For all* $x \in B_Y(0, r_1)$ *denote*

$$a(x) := \partial_2\mu(x, 0) \in \mathcal{B}(Y).$$

Then there exists $r_3 \in (0, r_2)$ *such that* $\mu\big(B_Y(0, r_3) \times B_Y(0, r_3)\big) \subseteq B_Y(0, r_2)$ *and the following conditions are fulfilled:*

(i) *For all* $x \in B_Y(0, r_3)$ *we have* $a(x) \in \mathrm{GL}(Y)$.

(ii) *For all* $x \in B_Y(0, r_3)$ *and* $h, k \in Y$ *we have*

$$(a^{-1})'_0(h, k) - (a^{-1})'_0(k, h) = (a^{-1})'_x(a(x)h, a(x)k) - (a^{-1})'_x(a(x)k, a(x)h).$$

PROOF For all $x, y \in B_Y(0, r_1)$ denote

$$a(x, y) := \partial_2\mu(x, y) \in \mathcal{B}(Y),$$

so that $a(x) = a(x, 0)$.

Since $\mu(0, y) = y$ for all $x \in B_Y(0, r_1)$, it follows that

$$a(0, 0) = \partial_2\mu(0, 0) = \mathrm{id}_Y \in \mathrm{GL}(Y).$$

Since the mapping $a(\cdot, \cdot)\colon B_Y(0, r_1) \times B_Y(0, r_1) \to \mathcal{B}(Y)$ is continuous and $\mathrm{GL}(Y)$ is open in $\mathcal{B}(Y)$, we can find $r_3 \in (0, r_2)$ such that $a\big(B_Y(0, r_3) \times B_Y(0, r_3)\big) \subseteq \mathrm{GL}(Y)$. In particular condition (i) is satisfied. Since $\mu(0, 0) = 0 \in B_Y(0, r_2)$ and μ is also continuous, it follows that, maybe by shrinking r_3, we may assume that $\mu\big(B_Y(0, r_3) \times B_Y(0, r_3)\big) \subseteq B_Y(0, r_2)$ as well.

Now, in order to check condition (ii), differentiate the formula $\mu(\mu(x, y), z) = \mu(x, \mu(y, z))$ with respect to z to get $\partial_2\mu(x, \mu(y, z))\partial_2\mu(y, z) = \partial_2\mu(\mu(x, y), z)$, that is,

$$a(x, \mu(y, z))a(y, z) = a(\mu(x, y), z)$$

for all $x, y, z \in B_Y(0, r_3)$. For $z = 0$ we get

$$a(x, y)a(y) = a(\mu(x, y)), \tag{B.1}$$

whence for $x, y \in B_Y(0, r_3)$ we have

$$a(\mu(x, y))^{-1}a(x, y) = a(y)^{-1}$$

and thus $a(\mu(x, y))^{-1}a(x, y)h = a(y)^{-1}h$ for all $h \in Y$. Differentiating the latter equation with respect to y, we get

$$(a^{-1})'_{\mu(x,y)}(a(x, y)k, a(x, y)h) + a^{-1}(\mu(x, y))\big(\partial_2 a(x, y)(k, h)\big) = (a^{-1})'_y(k, h)$$

whenever $h, k \in Y$. Setting $y = 0$ we get

$$(\forall h, k \in Y) \quad (a^{-1})'_x(a(x)k, a(x)h) + (a^{-1})(x)\big(\partial_2^2\mu(x, 0)(k, h)\big) = (a^{-1})'_0(k, h).$$

Similarly,

$$(\forall h,k\in Y)\quad (a^{-1})'_x(a(x)h,a(x)k)+(a^{-1})(x)\big(\partial_2^2\mu(x,0)(h,k)\big)=(a^{-1})'_0(h,k).$$

Now, by subtracting the latter two equations and taking into account that the continuous bilinear mapping

$$\partial_2^2\mu(x,0)\colon Y\times Y\to Y$$

is symmetric, we get the desired formula in condition (ii). ∎

THEOREM B.9 *Let $0<r_2<r_1$ and*

$$\mu\colon B_Y(0,r_1)\times B_Y(0,r_1)\to Y$$

a smooth mapping such that

$$\mu(y,0)=\mu(0,y)=y \text{ for all } y\in B_Y(0,r_1),$$

$\mu\big(B_Y(0,r_2)\times B_Y(0,r_2)\big)\subseteq B_Y(0,r_1)$ and

$$\mu(\mu(x,y),z)=\mu(x,\mu(y,z))$$

whenever $x,y,z\in B_Y(0,r_2)$. For all $x\in B_Y(0,r_1)$ denote

$$a(x):=\partial_2\mu(x,0)\in\mathcal{B}(Y),\quad b(x):=\partial_1\mu(0,x)\in\mathcal{B}(Y).$$

Then there exists $r_3\in(0,r_2)$ such that $\mu\big(B_Y(0,r_3)\times B_Y(0,r_3)\big)\subseteq B_Y(0,r_2)$ and the following conditions are fulfilled:

(i) *For all $x\in B_Y(0,r_3)$ we have $a(x),b(x)\in\mathrm{GL}(Y)$.*

(ii) *For all $x\in B_Y(0,r_3)$ and $h,k\in Y$ we have*

$$a(x)\big((a^{-1})'_x(h,k)-(a^{-1})'_x(k,h)\big)=-b(x)\big((b^{-1})'_x(h,k)-(b^{-1})'_x(k,h)\big).$$

PROOF As in the first part of the proof of Proposition B.8, we can find $r_3\in(0,r_2)$ such that condition (i) is satisfied. We just have to take into account also the continuity of the mapping

$$b(x,y)=\partial_1\mu(x,y)$$

at $(0,0)\in B_Y(0,r_1)\times B_Y(0,r_1)$, along with the remark that $b(0,0)=\mathrm{id}_Y\in\mathrm{GL}(Y)$.

To check the equation in condition (ii), define

$$(\forall x\in B_Y(0,r_3))\qquad \Gamma(x)\colon Y\times Y\to Y,\ \Gamma(x)(h,k)=-a'_x(h,a^{-1}(x)k),$$

and similarly

$$(\forall x \in B_Y(0, r_3)) \qquad \widetilde{\Gamma}(x)\colon Y \times Y \to Y,\ \widetilde{\Gamma}(x)(h,k) = -b'_x(h, b^{-1}(x)k).$$

By differentiating the equation $a(x)a^{-1}(x)k = k$ with respect to x, we get for all $h, k \in Y$

$$a'_x(h, a^{-1}(x)k) + a(x)(a^{-1})'_x(h,k) = 0$$

and similarly

$$a'_x(k, a^{-1}(x)h) + a(x)(a^{-1})'_x(k,h) = 0.$$

By subtracting the latter two equations we get for all $x \in B_Y(0, r_3)$, $h, k \in Y$,

$$a(x)\big((a^{-1})'_x(h,k) - (a^{-1})'_x(k,h)\big) = \Gamma(x)(h,k) - \Gamma(x)(k,h),$$

and similarly

$$b(x)\big((b^{-1})'_x(h,k) - (b^{-1})'_x(k,h)\big) = \widetilde{\Gamma}(x)(h,k) - \widetilde{\Gamma}(x)(k,h).$$

Hence it suffices to prove that

$$(\forall x \in B_Y(0, r_3))\,(\forall h, k \in Y) \qquad \Gamma(x)(h,k) = \widetilde{\Gamma}(x)(k,h). \tag{B.2}$$

To this end, recall from formula (B.1) in the proof of Proposition B.8 that for all $x, y \in B_Y(0, r_3)$ and $k \in Y$ we have $a(\mu(x,y))k = a(x,y)a(y)k$. By differentiating the latter equation with respect to x, we get for all $h, k \in Y$,

$$a'_{\mu(x,y)}(b(x,y)h, k) = \partial_1 a(x,y)(h, a(y)k).$$

Since $\partial_1 a(x,y) = \partial_1\partial_2\mu(x,y) = \partial_2\partial_1\mu(x,y) = \partial_2 b(x,y)$, we further deduce that

$$a'_{\mu(x,y)}(b(x,y)h, k) = \partial_2 b(x,y)(h, a(y)k).$$

Setting $x = 0$ and taking into account that $\partial_2 b(0,y) = b'_y$, we get the equality $a'_y(b(y)h, k) = b'_y(h, a(y)k)$, that is,

$$(\forall x \in B_Y(0, r_3))\,(\forall h, k \in Y) \quad a'_x(b(x)h, k) = b'_x(h, a(x)k).$$

Since $a(x), b(x) \in \mathcal{B}(Y)$ are invertible operators whenever $x \in B_Y(0, r_3)$, we can use the substitutions $h_0 = b(x)h$ and $k_0 = a(x)k$ to deduce from the above equation that

$$\underbrace{a'_x(h_0, a^{-1}(x)k_0)}_{=\Gamma(x)(h_0,k_0)} \quad = b'_x(b^{-1}(x)h_0, k_0) = \quad \underbrace{b'_x(k_0, b^{-1}(x)h_0)}_{=\widetilde{\Gamma}(x)(k_0,h_0)}$$

where the latter equality follows by the fact that the bilinear mapping

$$b'_x = \partial_1\partial_2\mu(x,0)\colon Y \times Y \to Y$$

is symmetric. Consequently $\Gamma(x)(h_0, k_0) = \widetilde{\Gamma}(x)(k_0, h_0)$, and (B.2) is proved. The proof ends. $\square$

We shall need the following existence theorem concerning solutions of complex ordinary differential equations.

PROPOSITION B.10 *Let E be a complex Banach space, U an open subset of E, c a positive real number, $t_0 \in \mathbb{C}$, and*

$$f\colon B_{\mathbb{C}}(t_0, c) \times U \to E$$

a holomorphic function. Then for all $x_0 \in U$ there exist $b \in (0, c)$ and a unique holomorphic function

$$\alpha\colon B_{\mathbb{C}}(t_0, b) \to U$$

such that

$$(\forall t \in B_{\mathbb{C}}(t_0, b)) \quad \dot{\alpha}(t) = f(t, \alpha(t))$$

and $\alpha(t_0) = x_0$.

PROOF First pick $R \in (0, c)$ and $a \in (0, 1)$ such that $B_E(x_0, a) \subseteq U$ and there exist $L, K \geq 1$ with

$$\sup\{\|f(t, x)\| \mid t \in B_{\mathbb{C}}(t_0, R), x \in B_E(x_0, a)\} \leq K$$

and

$$\sup\{\|\partial_2 f(t, x)\| \mid t \in B_{\mathbb{C}}(t_0, R), x \in B_E(x_0, a)\} \leq L.$$

(The existence of R and a follows since both mappings $f\colon B_{\mathbb{C}}(0, b) \times U \to E$ and $\partial_2 f\colon B_{\mathbb{C}}(0, b) \times U \to \mathcal{B}(E)$ are continuous, hence bounded on some neighborhood of (t_0, x_0).) In particular, the condition on L implies that

$$\big(\forall t \in B_{\mathbb{C}}(t_0, R)\big)\, \big(x_1, x_2 \in B_E(x_0, a)\big) \quad \|f(t, x_1) - f(t, x_2)\| \leq L\|x_1 - x_2\|, \tag{B.3}$$

according to the mean value theorem.

Now pick a real number b such that $0 < b < \min\{R, \frac{a}{LK}\}$, and define

$$M = \{\beta\colon B_{\mathbb{C}}(t_0, b) \to E \mid \beta \text{ holomorphic and } \sup_{t \in B_{\mathbb{C}}(t_0, b)} \|\beta(t) - x_0\| \leq 2a\}$$

and

$$(\forall \beta_1, \beta_2 \in M) \quad \operatorname{dist}(\beta_1, \beta_2) := \sup_{t \in B_{\mathbb{C}}(t_0, b)} \|\beta_1(t) - \beta_2(t)\|,$$

thus making M into a complete metric space (see Theorem A.42).

On the other hand, for each $\beta \in M$, define

$$S\beta\colon B_{\mathbb{C}}(t_0,b) \to E, \quad (S\beta)(t) = x_0 + \int_{t_0}^{t} f(s,\beta(s))\mathrm{d}s.$$

Then Proposition A.43 shows that $S\beta$ is a holomorphic function. Moreover, for all $t \in B_{\mathbb{C}}(t_0,b)$ we have

$$\|(S\beta)(t) - x_0\| \le \int_{t_0}^{t} \|f(s,\beta(s))\|\mathrm{d}s \le K|t-t_0| \le Kb \le 2a,$$

hence $S\beta \in M$. Furthermore, for all $\beta_1, \beta_2 \in M$ and all $t \in B_{\mathbb{C}}(t_0,b)$ we have

$$\begin{aligned}\|(S\beta_1)(t) - (S\beta_2)(t)\| &\le \int_{t_0}^{t} \|f(s,\beta_1(s)) - f(s,\beta_2(s))\|\mathrm{d}s \\ &\le |t-t_0|L \cdot \mathrm{dist}(\beta_1,\beta_2) \le bL \cdot \mathrm{dist}(\beta_1,\beta_2) \\ &\le a \cdot \mathrm{dist}(\beta_1,\beta_2).\end{aligned}$$

Hence, on the complete metric space M, we have a mapping $S\colon M \to M$ such that there exists $a \in (0,1)$ with $\mathrm{dist}(S\beta_1, S\beta_2) \le a \cdot \mathrm{dist}(\beta_1,\beta_2)$ whenever $\beta_1, \beta_2 \in M$. It then follows that S has a unique fixed point $\alpha \in M$. The equation $S\alpha = \alpha$ is clearly equivalent to the required properties of α. ∎

COROLLARY B.11 *Let V be an open subset of the real Banach space X, J an open interval in $\mathbb{R}$ and*

$$g\colon J \times U \to X$$

a real analytic mapping. If $\gamma\colon J \to U$ is a smooth function such that

$$\dot{\gamma}(t) = g(t,\gamma(t)) \text{ whenever } t \in J,$$

then γ is real analytic.

PROOF It easily follows by Theorem A.37 that in order to prove that γ is real analytic, it suffices to show that γ is real analytic on some neighborhood of an arbitrary point $t_0 \in J$. Denote $x_0 = \gamma(t_0) \in U$.

It is clear that the complexification of the real Banach space $\mathbb{R}\times X$ is $\mathbb{C}\times X_{\mathbb{C}}$. Since g is real analytic, it then follows that there exists an open subset W of $\mathbb{C} \times X_{\mathbb{C}}$ and a holomorphic mapping $f\colon W \to X_{\mathbb{C}}$ such that $J \times V \subseteq W$ and $f|_{J\times U} = g$. Since $(t_0, x_0) \in W$ it follows that, by shrinking J and W, we may assume that there exist $c > 0$ and an open subset U of $X_{\mathbb{C}}$ such that

$V \subseteq U$ and $W = B_{\mathbb{C}}(t_0, c) \times U$. It follows by Proposition B.10 that there exist $b \in (0, c)$ and a holomorphic mapping $\alpha: B_{\mathbb{C}}(t_0, b) \to U$ such that $\alpha(t_0) = x_0$ and

$$\alpha'(t) = f(t, \alpha(t)) \text{ whenever } t \in B_{\mathbb{C}}(t_0, b).$$

Then using Theorem B.1 for the functions

$$\alpha|_{(t_0-c,t_0+c)}: (t_0 - c, t_0 + c) \to X_{\mathbb{C}} \text{ and } \gamma: (t_0 - c, t_0 + c) \to X \hookrightarrow X_{\mathbb{C}},$$

we get $\gamma = \alpha|_{(t_0-c,t_0+c)}$, hence γ is real analytic. ∎

Now we are ready to prove the main theorem of this appendix, which concerns analyticity of local Lie groups modeled on Banach spaces. This result is critical for the proof of Theorem 2.42.

THEOREM B.12 *Let* $0 < r_2 < r_1$ *and*

$$\mu: B_Y(0, r_1) \times B_Y(0, r_1) \to Y$$

a smooth mapping such that

$$\mu(y, 0) = \mu(0, y) = y \text{ for all } y \in B_Y(0, r_1),$$

$\mu\big(B_Y(0, r_2) \times B_Y(0, r_2)\big) \subseteq B_Y(0, r_1)$ *and*

$$\mu(\mu(x, y), z) = \mu(x, \mu(y, z))$$

whenever $x, y, z \in B_Y(0, r_2)$. *Moreover assume that*

$$\mu(tx, sx) = (t + s)x \tag{B.4}$$

whenever $0 \neq x \in B_Y(0, r_1)$, $t, s \in \mathbb{R}$, *and* $\max\{|t|, |s|\} < \frac{r_1}{2\|x\|}$. *Then the mapping* μ *is real analytic on some neighborhood of* $(0, 0) \in Y \times Y$.

PROOF The proof has several stages.

1° For all $x, y \in B_Y(0, r_1)$ denote as usual

$$\begin{aligned} a(x, y) &= \partial_2\mu(x, y), \quad & a(x) &= a(x, 0) \\ b(x, y) &= \partial_1\mu(x, y), \quad & b(y) &= b(0, y), \end{aligned}$$

so that $a(x), a(x, y), b(x), b(x, y) \in \mathcal{B}(Y)$.

It then follows by Proposition B.8 and Theorem B.9 that there exists $\tilde{r}_3 \in (0, r_2)$ such that $\mu\big(B_Y(0, \tilde{r}_3), B_Y(0, \tilde{r}_3)\big) \subseteq B_Y(0, r_2)$ and the following assertions hold.

(i) For all $x \in B_Y(0, \tilde{r}_3)$ we have $a(x), b(x) \in \mathrm{GL}(Y)$.

(ii) For all $x \in B_Y(0,\tilde{r}_3)$ and $h,k \in Y$ we have

$$\begin{aligned} S(h,k) :=& (a^{-1})'_0(h,k) - (a^{-1})'_0(k,h) \\ =& (a^{-1})'_x(a(x)h, a(x)k) - (a^{-1})'_x(a(x)k, a(x)h), \end{aligned}$$

and we thus get a skew-symmetric bounded bilinear mapping

$$S\colon Y \times Y \to Y.$$

(iii) For all $x \in B_Y(0,\tilde{r}_3)$ and $h,k \in Y$ we have

$$a(x)\big((a^{-1})'_x(h,k) - (a^{-1})'_x(k,h)\big) = -b(x)\big((b^{-1})'_x(h,k) - (b^{-1})'_x(k,h)\big).$$

Moreover, note that hypothesis (B.4) implies that

$$a(tx)x = b(tx)x = x \text{ whenever } 0 \neq x \in B_Y(0,r_1), |t| < \frac{r_1}{2\|x\|}. \tag{B.5}$$

2° We prove at this stage that the mappings

$$a(\cdot), b(\cdot)\colon B_Y(0,\tilde{r}_3) \to \mathcal{B}(Y)$$

are real analytic. Actually, we are going to consider only the case of $a(\cdot)$, since the case of $b(\cdot)$ can be treated similarly.

To prove that $a(\cdot)$ is real analytic, we denote

$$(\forall x \in Y) \quad S_x := S(x,\cdot) \in \mathcal{B}(Y)$$

and we will prove that

$$(\forall x \in B_Y(0,\tilde{r}_3)) \quad a^{-1}(x) = \sum_{n=1}^{\infty} \frac{1}{n!} S_x^{n-1} = \Theta(S_x), \tag{B.6}$$

where $\Theta\colon \mathbb{C} \to \mathbb{C}$ is the entire function defined by $\Theta(z) = \sum\limits_{n=1}^{\infty} \frac{1}{n!} z^{n-1} = (\mathrm{e}^z - 1)/z$. Since the mapping $Y \to \mathcal{B}(Y)$, $x \mapsto S_x$, is real analytic (being linear), while the mapping $\mathcal{B}(Y) \to \mathcal{B}(Y)$, $T \mapsto \Theta(T)$, is real analytic by Remark A.47(a), it will follow by Proposition A.38 that their composition is real analytic. But (B.6) shows that the corresponding composition is just $a^{-1}(\cdot)$. On the other hand, the inversion mapping $\eta\colon \mathrm{GL}(Y) \to \mathrm{GL}(Y)$, $T \mapsto T^{-1}$, is real analytic by Remark A.20 along with Proposition A.46. Hence, by Proposition A.38 again, $a(\cdot) = \eta \circ (a^{-1})(\cdot)$ is real analytic, as desired.

Now, to prove (B.6), fix $x,y \in B_Y(0,\tilde{r}_3)$ and an open interval $I \subseteq \mathbb{R}$ such that $0,1 \in I$ and $tx \in B_Y(0,\tilde{r}_3)$ whenever $t \in I$. Then define

$$\varphi\colon I \to \mathcal{B}(Y), \quad \varphi(t) = t a^{-1}(tx),$$

and

$$\psi\colon I \to Y, \quad \psi(t) = \varphi(t)y = ta^{-1}(tx)y.$$

Then for all $t \in I$ we have

$$\dot{\psi}(t) = a^{-1}(tx)y + t(a^{-1})'_{tx}(x, y).$$

On the other hand, by (B.5) along with assertion (ii) in stage 1° of the present proof, we have

$$\begin{aligned}(a^{-1})'_{tx}(x, y) - (a^{-1})'_{tx}(y, x) =&(a^{-1})'_{tx}\big(a(tx)x, a(tx)(a^{-1}(tx)y)\big)\\ &- (a^{-1})'_{tx}\big(a(tx)(a^{-1}(tx)y), a(tx)x\big)\\ =&(a^{-1})'_0(x, a^{-1}(tx)y) - (a^{-1})'_0(a^{-1}(tx)y, x)\\ =&S(x, a^{-1}(tx)y)\\ =&S_x(a^{-1}(tx)y),\end{aligned}$$

so that

$$\dot{\psi}(t) = a^{-1}(tx)y + tS_x a^{-1}(tx)y = a^{-1}(tx)y + t(a^{-1})'_{tx}(y, x) + S_x\varphi(t)y.$$

If we write (B.5) under the form $a^{-1}(tx)x = x$ and then differentiate this equation with respect to x, we get $(a^{-1})'_{tx}(ty, x) + a^{-1}(tx)y = y$ for all $y \in Y$, hence

$$(\forall y \in Y) \quad \dot{\psi}(t) = y + S_x\varphi(t)y,$$

whence

$$(\forall t \in I) \quad \dot{\varphi}(t) = \mathrm{id}_Y + S_x\varphi(t).$$

Since $\varphi(0) = 0$, we get by Theorem B.1

$$(\forall t \in I) \quad \varphi(t) = \sum_{n=1}^{\infty} \frac{t^n}{n!} S_x^{n-1}.$$

We recall that $\varphi(t) = ta^{-1}(tx)$ and $1 \in I$, hence the above equality for $t = 1$ shows that (B.6) holds.

3° At this stage we prove that μ is real analytic on $B_Y(0, r_3) \times B_Y(0, r_3)$, where $r_3 \in (0, \tilde{r_3})$ is chosen so that $\mu(B_Y(0, r_3) \times B_Y(0, r_3)) \subseteq B_Y(0, \tilde{r}_3)$. According to Theorem A.41 it suffices to prove that, for arbitrary $u, v, w, z \in B_Y(0, r_3)$, the function

$$\gamma\colon D_{u,v,w,z} \to Y, \quad \gamma(t) = \mu(u + tv, w + tz),$$

is real analytic, where

$$D_{u,v,w,z} := \{t \in \mathbb{R} \mid u + tv, w + tz \in B_Y(0, r_3)\}.$$

To this end, we will use Corollary B.11.

We have

$$\dot{\gamma}(t) = b(u + tv, w + tz)v + a(u + tv, w + tz)z.$$

On the other hand, by formula (B.1) (see the proof of Proposition B.8) we have $a(u+tv, w+tz) = a(\mu(u+tv, w+tz))a^{-1}(w+tz)$. Similarly to (B.1) we have $b(x, y)b(x) = b(\mu(x, y))$ whenever $x, y \in B_Y(0, \tilde{r}_3)$, whence $b(u+tv, w+tz) = b(\mu(u + tv, w + tz))b^{-1}(u + tv)$, so that

$$\begin{aligned}\dot{\gamma}(t) &= b(\mu(u + tv, w + tz))b^{-1}(u + tv)v + a(\mu(u + tv, w + tz))a^{-1}(w + tz)z \\ &= b(\gamma(t))b^{-1}(u + tv)v + a(\gamma(t))a^{-1}(w + tz)z.\end{aligned}$$

We have seen at the beginning of stage 2° that all of the mappings $a(\cdot)$, $a^{-1}(\cdot)$, $b(\cdot)$ and $b^{-1}(\cdot)$ are real analytic, it follows that the mapping

$$f\colon D_{u,v,w,z} \times B_Y(0, \tilde{r}_3) \to Y, f(t, x) = b(x)b^{-1}(u + tv)v + a(x)a^{-1}(w + tz)z,$$

is real analytic as well. Since

$$(\forall t \in D_{u,v,w,z}) \quad \dot{\gamma}(t) = f(t, \gamma(t)),$$

it then follows by Corollary B.11 that $\gamma\colon D_{u,v,w,z} \to Y$ is real analytic, and the proof ends. ∎

We now turn to some facts that hold in the more general context of locally convex spaces and are needed in Chapter 2.

REMARK B.13 Let X be a real locally convex space, V an open subset of X, and $x_0, y_0 \in V$. Also let $\mu\colon V \times V \to X$ be a smooth mapping. Then the following assertions hold:

(a) The *linear* mapping $\mu'_{(x_0,y_0)}\colon X \times X \to X$ has the property

$$(\forall u, v \in X) \quad \mu'_{(x_0,y_0)}(u, v) = \partial_1\mu(x_0, y_0)u + \partial_2\mu(x_0, y_0)v.$$

(b) The *bilinear* mapping $\mu''_{(x_0,y_0)}\colon (X \times X) \times (X \times X) \to X$ has the property

$$\begin{aligned}\mu''_{(x_0,y_0)}((u, v), (u, v)) =&\partial_1^2\mu(x_0, y_0)(u, u) + 2\partial_1\partial_2\mu(x_0, y_0)(u, v) \\ &+ \partial_2^2\mu(x_0, y_0)(v, v),\end{aligned}$$

whenever $u, v \in X$.

(c) If moreover V is convex, then the mapping $R\colon V \times V \to X$ defined by the equation

$$(\forall z \in V \times V) \quad \mu(z) = \mu(z_0) + \mu'_{z_0}(z - z_0) + \frac{1}{2!}\mu''_{z_0}(z - z_0, z - z_0) + R(z),$$

has the properties

$$R(z_0) = 0, \; R'_{z_0} = 0, \; R''_{z_0} = 0,$$

where $z_0 := (x_0, y_0) \in V \times V$.

PROPOSITION B.14 *Let X be a real locally convex space, V_1 an open neighborhood of $0 \in X$, $\mu\colon V_1 \times V_1 \to X$ a smooth mapping such that*

$$(\forall x \in V_1) \quad \mu(x,0) = \mu(0,x) = x,$$

and $\eta\colon V_1 \to V_1$ a smooth mapping with the property that $\eta(0) = 0$ and

$$(\forall x \in V_1) \quad \mu\big(x, \eta(x)\big) = 0.$$

Moreover consider an open neighborhood V_2 of $0 \in X$ such that $V_2 \subseteq V_1$ and $\mu(V_2 \times V_2) \subseteq V_1$, and define

$$\psi\colon V_2 \times V_2 \to X, \quad \psi(x,y) = \mu(\mu(x,y), \eta(x)).$$

For $u, v \in V_2$, define

$$\tilde{u}, \tilde{v}\colon V_2 \to X, \quad \tilde{u}(x) = \partial_2\mu(x,0)u, \ \tilde{v}(x) = \partial_2\mu(x,0)v.$$

Then

$$\partial_1\partial_2\psi(0,0)(v,u) = \partial_1\partial_2\mu(0,0)(v,u) - \partial_1\partial_2\mu(0,0)(u,v) = (\tilde{v})'_0 u - (\tilde{u})'_0 v.$$

PROOF The second of the desired equalities clearly follows by the very definition of $\tilde{u}$ and $\tilde{v}$. Next, we are going to prove that

$$\partial_1\psi(0,v) = \partial_1\mu(0,v) - \partial_2\mu(v,0),$$

which implies the first of the asserted equalities. To prove the above equality, first differentiate the equations $\mu(x,0) = \mu(0,x) = x$, to get

$$\partial_1\mu(x,0) = \partial_2\mu(0,x) = \mathrm{id}_X.$$

Then differentiate the equation $\mu(x,\eta(x)) = 0$ to obtain that $\partial_1\mu(x,\eta(x)) + \partial_2\mu(x,\eta(x))\eta'_x = 0$. For $x = 0$, we get $\mathrm{id}_X + \mathrm{id}_X\eta'_0 = 0$, whence $\eta'_0 = -\mathrm{id}_X$.

Now the definition of ψ implies that for $v \in V_2$ we have

$$\begin{aligned}\partial_1\psi(0,v) &= \partial_1\mu(\mu(0,v),\eta(0))\partial_1\mu(0,v) + \partial_2\mu(\mu(0,v),\eta(0))\eta'_0 \\ &= \partial_1\mu(0,v) - \partial_2\mu(v,0),\end{aligned}$$

whence the first of the desired equalities clearly follows by computing the derivative with respect to v at $v = 0$. ∎

Notes

Several results contained in this appendix are updated versions of some of the basic facts underlying the paper [Ma62], where the basic theory of Banach-Lie groups is developed following the pattern of finite-dimensional Lie theory.

Our Proposition B.8 is inspired by Satz 4.2, while Theorem B.9 is essentially Satz 4.1 at page 241 in [Ma62]. Moreover, Theorem B.12 is the essential result contained in Satz 7.1 in [Ma62]. It says that every local Banach-Lie group is analytic.

Proposition B.14 contains some calculations carried out in section 5 of [Mi84].

For a good exposition of the needed elements of the theory of ordinary differential equations in Banach spaces, we refer to [La01]. See also Chapter 5 in [Up85] for an exposition of that theory in the context of analytic functions.

Appendix C

Topological Groups

Abstract. In the present appendix, we develop the basic facts concerning topological groups that are needed in the main body of the book. The most important results contained here are the fact that the group topology is uniquely determined by its restriction to a neighborhood of $\mathbf{1}$ (Theorem C.11), and the theorem concerning the construction of homomorphisms from simply connected topological groups (Theorem C.19).

DEFINITION C.1 Let G be a group. A *group topology* on G is a topology τ on G making the map

$$G \times G \to G, \qquad (a,b) \mapsto ab^{-1}$$

into a continuous map.

A *topological group* is a group equipped with a group topology.

REMARK C.2

(a) In the framework of Definition C.1, the condition that τ is a group topology is equivalent to the requirement that both the multiplication map

$$m\colon G \times G \to G, \qquad (a,b) \mapsto ab,$$

and the inversion map

$$\eta\colon G \to G, \qquad a \mapsto a^{-1},$$

are continuous.

(b) The discrete topology of any group is always a group topology. Thus, every group admits at least one group topology. On the other hand, there can exist several group topologies on a given group. For instance, the additive group $(\mathbb{R}, +)$ has at least two group topologies: the discrete topology and the usual one.

LEMMA C.3 *Assume that E is a set and for every $x \in E$ we have singled out a set $\mathcal{V}(x)$ of subsets of E such that the following conditions are fulfilled.*

(V1) *If* $x \in E$, $V \in \mathcal{V}(x)$, *and* $V \subseteq U \subseteq E$, *then* $U \in \mathcal{V}(x)$.

(V2) *If* $x \in E$ *and* $V_1, V_2 \in \mathcal{V}(x)$, *then* $V_1 \cap V_2 \in \mathcal{V}(x)$.

(V3) *If* $x \in E$ *and* $V \in \mathcal{V}(x)$, *then* $x \in V$.

(V4) *If* $x \in E$ *and* $V \in \mathcal{V}(x)$, *then there exists* $W \in \mathcal{V}(x)$ *such that for all* $y \in W$ *we have* $V \in \mathcal{V}(y)$.

Next denote

$$\tau = \{D \mid D \subseteq E; (\forall x \in D)(\exists V \in \mathcal{V}(x))\ V \subseteq D\}.$$

Then τ *is the unique topology on* E *such that, for all* $x \in E$, $\mathcal{V}(x)$ *is the set of neighborhoods of* x *with respect to* τ.

PROOF See Proposition 2 in §1, no. 2, in Chapitre I in [Bo71a]. ∎

NOTATION C.4 If G is a group, $x \in G$, and $A, B \subseteq G$, we denote $AB := \{ab \mid a \in A, b \in B\}$, $xA := \{x\}A$, $Ax := A\{x\}$, and $A^{-1} := \{a^{-1} \mid a \in A\}$. Whenever it is not otherwise stated, we denote by $\mathbf{1}$ the unit element of any group.

The following statement is a first step towards the method to construct group topologies on a given group, starting from local structures around $\mathbf{1}$. (See Theorem C.11.)

PROPOSITION C.5 *Assume that* G *is a group and* $\mathcal{V}$ *is a set of subsets of* G *such that the following conditions are satisfied.*

(GV0) *If* $U_1, U_2 \in \mathcal{V}$, *then* $U_1 \cap U_2 \in \mathcal{V}$. *If* $V \in \mathcal{V}$ *and* $V \subseteq U \subseteq G$, *then* $U \in \mathcal{V}$.

(GV1) *If* $U \in \mathcal{V}$, *then there exists* $V \in \mathcal{V}$ *with* $VV \subseteq U$.

(GV2) *For all* $U \in \mathcal{V}$ *we have* $U^{-1} \in \mathcal{V}$.

(GV3) *For all* $U \in \mathcal{V}$ *we have* $\mathbf{1} \in U$.

(GV4) *For all* $U \in \mathcal{V}$ *and* $a \in G$ *we have* $aUa^{-1} \in \mathcal{V}$.

Then there exists a unique group topology τ *on* G *such that* $\mathcal{V}$ *is the set of all neighborhoods of* $\mathbf{1} \in G$ *with respect to* τ. *Moreover, for each* $a \in G$, *we have*

$$\mathcal{V}(a) := \{aV \mid V \in \mathcal{V}\} = \{Va \mid V \in \mathcal{V}\}$$

and this is the set of all neighborhoods of a *with respect to* τ.

PROOF The proof has several stages.

1° To prove the existence and uniqueness of the topology τ, we are going to use Lemma C.3 for $\mathcal{V}(x)$, $x \in G$, as in the statement. So we have to check that conditions (V1)–(V4) in Lemma C.3 are satisfied.

To this end, note that both (V1) and (V2) follow from hypothesis (GV0), while (V3) follows from (GV3). To prove (V4), first note that, for arbitrary $x \in G$ and $V \subseteq G$, we have $V \in \mathcal{V}(x)$ if and only if $x^{-1}V \in \mathcal{V}$. This remark shows that condition (V4) in Lemma C.3 is equivalent to the following: for all $x \in G$ and $V_0 \in \mathcal{V}$ there exists $W_0 \in \mathcal{V}$ such that for all $y \in xW_0$ we have $y^{-1}xV_0 \in \mathcal{V}$. To prove this, note that hypothesis (GV1) implies that there exists $W_0 \in \mathcal{V}$ with $W_0W_0 \subseteq V_0$. Then for each $y \in xW_0$ we have $x^{-1}y \in W_0$, hence $x^{-1}y \in W_0$, hence $x^{-1}yW_0 \subseteq W_0W_0 \subseteq V_0$. Thus $W_0 \subseteq y^{-1}xV_0$, which implies by hypothesis (GV2) that $y^{-1}xV_0 \in \mathcal{V}$, as desired.

2° We now show that the topology τ constructed at stage 1° by means of Lemma C.3 is a group topology. To this end, first recall from Lemma C.3 that $\mathcal{V}(x)$ ($= \{xV \mid V \in \mathcal{V}\}$ in the present situation) is the set of all neighborhoods of x, for all $x \in G$. Thus, in order to prove that the map

$$G \times G \to G, \qquad (a, b) \mapsto ab^{-1}$$

is continuous, it suffices to check that the following statement holds: for arbitrary $a, b \in G$ and $U \in \mathcal{V}$, there exists $W \in \mathcal{V}$ such that

$$(aW)(bW)^{-1} \subseteq (ab^{-1})U.$$

Note that the above inclusion is equivalent to $aWW^{-1}b^{-1} \subseteq ab^{-1}U$, and further to

$$WW^{-1} \subseteq b^{-1}Ub.$$

On the other hand, we have by hypothesis (GV4) that $b^{-1}Ub \in \mathcal{V}$, hence, according to hypothesis (GV1), there exists $W_1 \in \mathcal{V}$ such that

$$W_1W_1 \subseteq b^{-1}Ub.$$

Now, for $W := W_1 \cap (W_1)^{-1} \in \mathcal{V}$ (use both hypotheses (GV0) and (GV2)), we get $WW^{-1} \subseteq W_1W_1 \subseteq b^{-1}Ub$, as desired.

3° To conclude the proof, note that the second of the equalities in

$$\mathcal{V}(a) = \{aV \mid V \in \mathcal{V}\} = \{Wa \mid W \in \mathcal{V}\}$$

follows by hypothesis (GV4) for each $a \in G$. ∎

Conditions (GBV0)–(GBV4) in the following auxiliary result are usually easier to check than conditions (GV0)–(GV4) in Proposition C.5.

LEMMA C.6 *Assume that G is a group and $\mathcal{B}$ is a set of subsets of G satisfying the following conditions:*

(GBV0) *For all* $U_1, U_2 \in \mathcal{B}$ *there exists* $U_0 \in \mathcal{B}$ *with* $U_0 \subseteq U_1 \cap U_2$.

(GBV1) *For each* $U \in \mathcal{B}$ *there exists* $V \in \mathcal{B}$ *such that* $VV \subseteq U$.

(GBV2) *For each* $U \in \mathcal{B}$ *there exists* $V \in \mathcal{B}$ *with* $V^{-1} \subseteq U$.

(GBV3) *For all* $U \in \mathcal{B}$ *we have* $\mathbf{1} \in U$.

(GBV4) *If* $U \in \mathcal{B}$ *and* $a \in G$, *then there exists* $V \in \mathcal{B}$ *with* $aVa^{-1} \subseteq U$.

Then

$$\tau = \{D \mid D \subseteq G;\ (\forall a \in D)(\exists V \in \mathcal{B})\, aV \subseteq D\}$$

is the unique group topology on G *such that* $\mathcal{B}$ *is a basis of neighborhoods of* $\mathbf{1} \in G$ *with respect to* τ.

PROOF It is easy to check that $\mathcal{V} := \{U \mid U \subseteq G;\ (\exists V \in \mathcal{B})\, V \subseteq U\}$ satisfies conditions (GV0)–(GV4) in Proposition C.5. ∎

NOTATION C.7 If G is a group and $A \subseteq G$, then we denote by $\langle A \rangle$ the subgroup of G generated by A, i.e., the smallest subgroup of G that contains A.

REMARK C.8 If G is a group and A is a subset of G, then we have $\langle A \rangle = \{\mathbf{1}\} \cup \bigcup_{n=1}^{\infty} \{a_1 \cdots a_n \mid a_1, \dots a_n \in A \cup A^{-1}\}$.

REMARK C.9

(a) If G is a connected topological group and U is a neighborhood of $\mathbf{1} \in G$, then $\langle U \rangle = G$.

(b) If G is a topological group and there exists a connected neighborhood U of $\mathbf{1} \in G$ such that $\langle U \rangle = G$, then G is connected.

DEFINITION C.10 Let T_1 and T_2 be topological spaces. An *embedding* (of topological spaces) of T_1 into T_2 is a mapping $f\colon T_1 \to T_2$ that induces a homeomorphism $T_1 \to f(T_1)$, provided we view $f(T_1)$ as a topological subspace of T_2.

The following theorem provides a useful way to endow a group with a group topology. The results of this type are particularly useful in Lie theory, inasmuch as they allow one to extend local structures to global ones.

THEOREM C.11 *Let G be a group with the multiplication map*

$$m\colon G \times G \to G, \qquad (x, y) \mapsto xy,$$

and $K \subseteq G$ such that

$$\mathbf{1} \in K = K^{-1} \text{ and } \langle K \rangle = G.$$

Assume that the subset K of G is equipped with a Hausdorff topology such that the inversion map

$$K \to K, \qquad x \mapsto x^{-1},$$

is continuous and there exists an open set $V_0 \subseteq K \times K$ satisfying the following conditions:

(a) $m(V_0) \subseteq K$,

(b) $m|_{V_0}\colon V_0 \to K$ *is continuous, and*

(c) *for all $x \in K$ we have $(x, x^{-1}), (x, \mathbf{1}), (\mathbf{1}, x) \in V_0$.*

Then there exists a unique group topology on G making the inclusion map

$$K \hookrightarrow G$$

into an embedding of topological spaces such that K is an open subset of G.

PROOF The proof has several stages.

1° To construct the group topology of G, we will make use of Lemma C.6. To this end, we check conditions (GBV0)–(GBV4) in Lemma C.6 for the set of subsets of G defined by

$$\mathcal{B} := \{W \mid W \subseteq K;\ W \text{ is a neighborhood of } \mathbf{1} \in K\}.$$

Conditions (GBV0) and (GBV3) are obvious.

For (GBV1), note that V_0 is a neighborhood of $(\mathbf{1}, \mathbf{1}) \in K \times K$. Since the mapping $m|_{V_0}\colon V_0 \to K$ is continuous and $m(\mathbf{1}, \mathbf{1}) = \mathbf{1}$, it then easily follows that for each $W \in \mathcal{B}$ there exists $W_1 \in \mathcal{B}$ with $W_1 W_1 \subseteq W$.

To see that (GBV2) holds, we use the fact that the inversion mapping $\eta\colon K \to K$, $x \mapsto x^{-1}$, is continuous. Since $\eta^2 = \mathrm{id}_K$, it follows that η is actually a homeomorphism of K onto itself, and thus for each $W \in \mathcal{B}$ we have $W^{-1}(= \eta(W)) \in \mathcal{B}$ as well.

In order to check condition (GBV3) in Lemma C.6, we first note that since $K = K^{-1}$ and $\langle K \rangle = G$, it follows by Remark C.8 that

$$G = \bigcup_{n=1}^{\infty} \underbrace{K \cdots K}_{n \text{ times}}.$$

One then easily shows by induction that it suffices to check (GBV3) only for $a \in K$. To do this, let $a \in K$ and $W \in \mathcal{B}$ arbitrary. By hypothesis (c) we have $(a, a^{-1}) \in V_0$. Since V_0 is open in $K \times K$ and $m|_{V_0}$ is continuous, it then follows that for some neighborhood U_0 of $a \in K$ we have both

$$U_1 \times \{a^{-1}\} \subseteq V_0 \text{ and } U_1 a^{-1} \subseteq W.$$

On the other hand, also by hypothesis (c) we have $(a, \mathbf{1}) \in V_0$. Again by the continuity of $m|_{V_0}$, there exists $W_1 \in \mathcal{B}$ such that

$$\{a\} \times W_1 \subseteq V_0 \text{ and } aW_1 \subseteq U_1.$$

Then $aW_1a^{-1} \subseteq U_1a^{-1} \subseteq W$, as desired in condition (GBV3) in Lemma C.6.

Consequently, we can use Lemma C.6 to make G into a topological group with the group topology τ defined by

$$\tau = \{D \mid D \subseteq G;\ (\forall a \in D)(\exists W \in \mathcal{B})\, aW \subseteq D\}.$$

Note that, in particular, we have $K \in \tau$. In fact, let $a \in K$. We have $(a, \mathbf{1}) \in V_0$ by hypothesis (c), hence, using as above the continuity of $m|_{V_0}$, we can find $W \in \mathcal{B}$ such that ($\{a\} \times W \subseteq V_0$ and) $aW \subseteq K$.

2° We now prove that the inclusion mapping

$$\iota \colon K \hookrightarrow G$$

is an embedding of topological spaces. To this end, we have to prove that it is both an open mapping (i.e., it maps every open subset of K onto an open subset of G) and a continuous mapping.

To see that ι is an open mapping, it clearly suffices to show that, for every neighborhood U of an arbitrary $k \in K$, the set $\iota(U)$ is a neighborhood of $\iota(k)(= k)\in G$. To this end, we repeat the proof of the fact that $K \in \tau$: we have $(k, \mathbf{1}) \in V_0$, hence the continuity of $m|_{V_0}$ shows that for some $W \in \mathcal{B}$ we have $kW \subseteq U = \iota(U)$. Thus $\iota(U)$ is a neighborhood of $\iota(k)$, according to the above definition of the topology τ.

Now, to prove that ι is continuous, let $D \in \tau$ arbitrary. We have $\iota^{-1}(D) = K \cap D$, hence we have to prove that $K \cap D \in \tau$. To this end, let $k \in K \cap D$ arbitrary. The fact that $k \in D$ shows that for some $W_1 \in \mathcal{B}$ we have

$$kW_1 \in \mathcal{B}.$$

On the other hand, the fact that $k \in K$ implies as above (using that $(k, \mathbf{1}) \in V_0$ along with the continuity of the multiplication map $m|_{V_0}$) that for some $W_2 \in \mathcal{B}$ we have both $\{k\} \times W_2 \subseteq V_0$ and

$$kW_2 \subseteq K.$$

Then

$$k \cdot (W_1 \cap W_2) = kW_1 \cap kW_2 \subseteq D \cap K.$$

Since $W_1, W_2 \in \mathcal{B}$, we have $W \subseteq W_1 \cap W_2$ for some $W \in \mathcal{B}$ (see condition (GBV0) in Lemma C.6), hence $kW \subseteq D \cap K$, and this is just what is needed in order to have $D \cap K \in \tau$.

3° The uniqueness assertion is an easy consequence of the corresponding assertion in Lemma C.6. □

Our next aim is to describe one of the basic methods to construct homomorphisms from simply connected groups into arbitrary groups (see Theorem C.19 below). The notion of simply connected space (Definition C.15(c)) needs the idea of covering, in the sense of the following definition.

DEFINITION C.12 Let T and S be topological spaces and $f\colon T \to S$ a continuous mapping. We say that f is a *covering mapping* if for every $s \in S$ there exists an open neighborhood W of s such that, for some family $\{A_i\}_{i\in I}$ of pairwise disjoint open subsets of T, the following conditions are satisfied:

1° We have $f^{-1}(W) = \bigcup_{i\in I} A_i$.

2° For every $i \in I$ the mapping $f|_{A_i}\colon A_i \to W$ is a homeomorphism provided A_i and W are equipped with the topologies inherited from T and S, respectively.

REMARK C.13 Let G and H be topological groups and $\varphi\colon G \to H$ a group homomorphism.

(a) If there exists an open neighborhood U of $\mathbf{1} \in G$ such that $\varphi|_U$ is continuous, then φ is continuous.

(b) If there exists an open neighborhood W of $\mathbf{1} \in G$ such that $\varphi(W)$ is an open subset of $\mathbf{1} \in H$ and $\varphi|_W\colon W \to \varphi(W)$ is a homeomorphism when $\varphi(W)$ is equipped with the topology inherited from H, then φ is a covering map.

DEFINITION C.14 Let T be topological space. If X is another topological space and $f_0, f_1\colon X \to T$ are two continuous mappings, we say that f_1 and f_2 are *homotopic* if there exists a continuous mapping $H\colon [0,1] \times X \to T$ such that for all $x \in X$ we have $H(0,x) = f_0(x)$ and $H(1,x) = f_1(x)$. In this case, H is said to be a *homotopy connecting* f_0 *and* f_1.

DEFINITION C.15 Let T be topological space.

(a) We say that T is *connected* if $\emptyset$ and T are the only subsets of T which are simultaneously closed and open.

(b) We say that T is *locally connected* if every point of T has a basis of connected neighborhoods.

(c) We say that T is *simply connected* if it is connected and locally connected, and, whenever $h\colon P \to S$ is a covering mapping, $f\colon T \to S$ is continuous, $t_0 \in T$, $p_0 \in P$, $h(p_0) = f(t_0)$, it follows that there exists a unique continuous mapping $\widetilde{f}\colon T \to P$ such that the diagram

$$\begin{array}{ccc} P & \xleftarrow{\widetilde{f}} & T \\ {\scriptstyle h}\big\downarrow & \swarrow_{f} & \\ S & & \end{array}$$

is commutative and $\widetilde{f}(t_0) = p_0$.

(d) We say that T is *pathwise connected* if for all $t_0, t_1 \in T$ there exists a continuous mapping (that is a *path*) $\gamma\colon [0,1] \to T$ such that $\gamma(0) = t_0$ and $\gamma(1) = t_1$.

(e) We say that T is *locally pathwise connected* if every point of T has a pathwise connected neighborhood.

(f) We say that T is *pathwise simply connected* if it is pathwise connected and locally pathwise connected, and every continuous path $\gamma\colon [0,1] \to T$ with $\gamma(0) = \gamma(1)$ is homotopic to a constant map $[0,1] \to T$.

REMARK C.16 Let G be a topological group. If $\mathbf{1} \in G$ has a basis of connected neighborhoods (respectively, a pathwise connected neighborhood), then G is locally connected (respectively, locally pathwise connected).

The following theorem provides the main tool used to check that a certain space is simply connected.

THEOREM C.17 *Every pathwise simply connected space is simply connected.*

PROOF See Theorem 2.1 in Chapter IV in [Ho65]. ∎

The next theorem describes a very important property of simply connected spaces, and will play a key role in the proof of Theorem C.19 below.

THEOREM C.18 *Let P and S be topological spaces and $h\colon P \to S$ a covering mapping. If*

(i) *P is connected and locally connected, and*

(ii) *S is simply connected,*

then f is a homeomorphism.

PROOF See Theorem 1.4 in Chapter IV in [Ho65]. The idea is to use condition (c) in Definition C.15 for $T = S$ and $f = \mathrm{id}_S$, in order to construct a continuous inverse of h. ∎

We are now ready to describe the main method to construct group homomorphisms defined on simply connected groups.

THEOREM C.19 *Let G be a simply connected topological group and H an arbitrary group. Suppose that W is a connected open neighborhood of $\mathbf{1} \in G$ such that $W = W^{-1}$ and $f\colon W \to H$ is a mapping such that*

$$f(xy) = f(x)f(y) \text{ whenever } x, y, xy \in W.$$

Then there exists a unique group homomorphism $\varphi\colon G \to H$ such that $\varphi|_W = f$.

PROOF Denote

$$K := \{(g, f(g)) \mid g \in W\} \subseteq G \times H$$

and endow K with the unique topology making the bijection

$$\beta\colon W \to K, \quad g \mapsto (g, f(g))$$

into a homeomorphism. Then denote by E the subgroup of $G \times H$ generated by K.

Using Theorem C.11, we are going to make E into a connected topological group such that K is an open neighborhood of $\mathbf{1} \in E$. To this end, denote by $m\colon G \times G \to G$ be the multiplication in G. Then $W_0 := m^{-1}(W) \cap (W \times W)$ is an open subset of $W \times W$ such that conditions (a)–(c) in Theorem C.11 are satisfied (with K replaced by W and V_0 replaced by W_0). Since β is a homeomorphism, it then follows that $V_0 := \{(\beta(g_1), \beta(g_2)) \mid (g_1, g_2) \in W_0\}$ and K also satisfy conditions (a)–(c) in Theorem C.11, hence the group $E = \langle K \rangle$ ($\subseteq G \times H$) has a unique group topology such that K is an open neighborhood of $\mathbf{1} \in E$. Since W is connected and K is homeomorphic to W, it follows that K is connected, and it then follows by Remark C.9(b) that the topological group E is connected.

Now consider the mapping

$$\pi\colon E \to G, (g, h) \mapsto g$$

which is the restriction to E of the natural projection $\mathrm{pr}_1 \colon G \times H \to G$. Since the latter projection is a group homomorphism, it follows that π is a group homomorphism as well. Moreover, note that

$$\pi|_K = \beta^{-1} \colon K \to W$$

Since β is a homeomorphism and K is an open neighborhood of $\mathbf{1} \in E$, it then follows by Remark C.13(b) that π is a covering map of E onto G. But E is connected and locally connected by Remark C.16, while G is simply connected, hence Theorem C.18 shows that π is a homeomorphism.

In particular, π is bijective, and then $\pi^{-1} \colon G \to E$ is a group isomorphism. For every $g \in W$ we have

$$\pi^{-1}(g) = \beta(g) = (g, f(g)),$$

hence for the group homomorphism

$$\varphi := \mathrm{pr}_2 \circ \pi^{-1} \colon G \to H$$

we have $\varphi|_W = f$. (Here $\mathrm{pr}_2 \colon G \times H \to H$ stands for the natural projection, which is a group homomorphism.)

The uniqueness of the group homomorphism φ follows since we have by Remark C.9(a) that $\langle W \rangle = G$. ∎

Notes

Theorem C.11 appears explicitly as Lemma II.2 in the paper [Ne02d]. See also pages 263–265 in [Hof68]. The basic idea underlying this result is that of local (topological) group. See e.g., page 209 in [Sw65].

Theorem C.19 is sometimes called the "monodromy theorem." It appears e.g., as Proposition 5.60 in the notes by K. H. Hoffman [Hof68], or as Theorem 3.1 in [Ho65]. See Theorem 1.7 in Chapter IV in [Ho65] for a more general result of this type.

Among the basic references for the topic of topological groups, we mention the books [Bo71a] and [Ho65].

References

[ACLM84] M.C. Abbati, R. Cirelli, P. Lanzavecchia, A. Manià, Pure states of general quantum-mechanical systems as Kähler bundles, *Nuovo Cimento B (11)* **83**(1984), no. 1, 43–60.

[1] R. Abraham, J.E. Marsden, T. Ratiu, *Manifolds, Tensor Analysis, and Applications* (second edition), Applied Mathematical Sciences 75, Springer-Verlag, New York, 1988.

[Al74] E. Albrecht, Funktionalkalküle in mehreren Veränderlichen für stetige lineare Operatoren auf Banachräumen, *manuscr. math.* **14**(1974), 1–40.

[Al78] E. Albrecht, On some classes of generalized spectral operators, *Arch. Math. (Basel)* **30**(1978), 297–303.

[AS74] R.K. Amayo, I. Stewart, *Infinite-dimensional Lie Algebras*, Noordhoff International Publishing, Leyden, 1974.

[An69] R.F.V. Anderson, The Weyl functional calculus, *J. Functional Analysis* **4**(1969), 240–267.

[An70] R.F.V. Anderson, On the Weyl functional calculus, *J. Functional Analysis* **6**(1970), 110-115.

[And03] E. Andruchow, A nonsmooth exponential, *Studia Math.* **155** (2003), no. 3, 265–271.

[AC04] E. Andruchow, G. Corach, Differential geometry of partial isometries and partial unitaries, *Illinois J. Math.* **48**(2004), no. 1, 97–120.

[ACMS97] E. Andruchow, G. Corach, M. Milman, D. Stojanoff, Geodesics and interpolation, *Rev. Un. Mat. Argentina* **40**(1997), no. 3-4, 83–91.

[ACS95a] E. Andruchow, G. Corach, D. Stojanoff, A geometric characterization of nuclearity and injectivity, *J. Funct. Anal.* **133**(1995), no. 2, 474–494.

[ACS95b] E. Andruchow, G. Corach, D. Stojanoff, The homogeneous space of representations of a nuclear C^*-algebra, in: *Harmonic Analysis and Operator Theory (Caracas, 1994)*, Contemp. Math., 189, Amer. Math. Soc., Providence, RI, 1995, pp. 37–53.

[ACS99] E. Andruchow, G. Corach, D. Stojanoff, Geometry of the sphere of a Hilbert module, *Math. Proc. Cambridge Philos. Soc.* **127**(1999), no. 2, 295–315.

[ACS00] E. Andruchow, G. Corach, D. Stojanoff, Projective spaces of a C^*-algebra, *Integral Equations Operator Theory* **37**(2000), no. 2, 143–168.

[ACS01] E. Andruchow, G. Corach, D. Stojanoff, Projective space of a C^*-module, *Infin. Dimens. Anal. Quantum Probab. Relat. Top.* **4**(2001), no. 3, 289–307.

[AFHPS90] E. Andruchow, L.A. Fialkow, D.A. Herrero, M.B. Pecuch Herrero, D. Stojanoff, Joint similarity orbits with local cross sections, *Integral Equations Operator Theory* **13**(1990), no. 1, 1–48.

[ALRS97] E. Andruchow, A. Larotonda, L. Recht, D. Stojanoff, Infinite-dimensional homogeneous reductive spaces and finite index conditional expectations, *Illinois J. Math.* **41**(1997), no. 1, 54–76.

[ARS93] E. Andruchow, L. Recht, D. Stojanoff, The space of spectral measures is a homogeneous reductive space, *Integral Equations Operator Theory* **16**(1993), no. 1, 1–14.

[AS89] E. Andruchow, D. Stojanoff, Differentiable structure of similarity orbits, *J. Operator Theory* **21**(2) (1989), 349–366.

[AS91] E. Andruchow, D. Stojanoff, Geometry of unitary orbits, *J. Operator Theory* **26**(1991), no. 1, 25–41.

[AS94] E. Andruchow, D. Stojanoff, Geometry of conditional expectations and finite index, *Internat. J. Math.* **5**(1994), no. 2, 169–178.

[AS99] E. Andruchow, D. Stojanoff, Geometry of oblique projections, *Studia Math.* **137**(1999), no. 1, 61–79.

[AS03] E. Andruchow, D. Stojanoff, Nilpotents in finite algebras, *Integral Equations Operator Theory* **45**(2003), no. 3, 251–267.

[AV96] E. Andruchow, A. Varela, Geometry and the Jones projection of a state, *Integral Equations Operator Theory* **25**(1996), no. 2, 129–146.

[AV99] E. Andruchow, A. Varela, Weight centralizer expectations with finite index, *Math. Scand.* **84**(1999), no. 2, 243–260.

[AV01] E. Andruchow, A. Varela, Fibre bundles over orbits of states, in: *Margarita mathematica*, Univ. La Rioja, Logroño, 2001, pp. 635–659.

[AV02] E. Andruchow, A. Varela, Homotopy of state orbits, *J. Operator Theory* **48**(2002), 419–430.

[AFHV84] C. Apostol, L.A. Fialkow, D.A. Herrero, D. Voiculescu, *Approximation of Hilbert Space Operators*, vol II, Research Notes in Mathematics, 102. Pitman (Advanced Publishing Program), Boston, MA, 1984.

[AK99] J. Arazy, W. Kaup, On the holomorphic rigidity of linear operators on complex Banach spaces, *Quart. J. Math. Oxford Ser. (2)* **50**(1999), no. 199, 249–277.

[ArS01] M. Argerami, D. Stojanoff, Orbits of conditional expectations, *Illinois J. Math.* **45**(2001), no. 1, 243–263.

[Ba72] V.K. Balachandran, Real L^*-algebras, *Indian J. Pure Appl. Math.* **3**(1972), 1224–1246.

[Bam99] D. Bambusi, On the Darboux theorem for weak symplectic manifolds, *Proc. Amer. Math. Soc.* **127**(1999), no. 11, 3383–3391.

[Bea83] A. Beauville, Variétés Kähleriennes dont la première classe de Chern est nulle, *J. Differential Geom.* **18**(1983), no. 4, 755–782

[BlDa90] Ya.I. Belopol'skaya, Yu.L. Dalecky, *Stochastic Equations and Differential Geometry*, Mathematics and its Applications (Soviet Series), vol. 30. Kluwer Academic Publishers Group, Dordrecht, 1990.

[Be02a] D. Beltiţă, Spectral theory within the framework of locally solvable Lie algebras, in: A. Strasburger, J. Hilgert, K.-H. Neeb, W. Wojyyński (eds.), *Geometry and Analysis on Finite and Infinite-dimensional Lie Groups (Będlewo, 2000)*, Banach Center Publ. vol. 55, Polish Acad. Sci., Warsaw, 2002, pp. 13–25.

[Be02b] D. Beltiţă, Spectra for solvable Lie algebras of bundle endomorphisms, *Math. Ann.* **324**(2002), no. 2, 405–429.

[Be03] D. Beltiţă, Complex homogeneous spaces of pseudo-restricted groups, *Math. Research Letters* **10**(2003), no. 4, 459–467.

[Be04] D. Beltiţă, Asymptotic products and enlargibility of Banach-Lie algebras, *J. Lie Theory* **14**(2004), no. 1, 215–226.

[Be05a] D. Beltiţă, On Banach-Lie algebras, spectral decompositions and complex polarizations, in: D. Gaşpar, I. Gohberg, D. Timotin, F.-H. Vasilescu, L. Zsido (eds.), *Recent Advances in Operator Theory, Operator Algebras, and Their Applications. XIXth International Conference on Operator Theory, Timisoara (Romania), 2002.* Operator Theory: Advances and Applications, 153. Birkhäuser Verlag Basel, 2005, pp. 13–38.

[Be05b] D. Beltiţă, Integrability of analytic almost complex structures on Banach manifolds, *Ann. Global Anal. Geom.* **28**(2005), no. 1, 59–73.

[BP05] D. Beltiţă, B. Prunaru, Amenability, completely bounded projections, dynamical systems and smooth orbits (see preprint math.OA/0504313).

[BR04] D. Beltiţă, T.S. Ratiu, Symplectic leaves in real Banach Lie-Poisson spaces, *Geom. Funct. Anal.* **15**(2005), no. 4 (to appear) (see preprint math.SG/0403345).

[BR05] D. Beltiţă, T.S. Ratiu, Geometric representation theory for unitary groups of operator algebras (see preprint math.RT/0501057).

[BS01] D. Beltiţă, M. Şabac, *Lie Algebras of Bounded Operators*, Operator Theory: Advances and Applications, 120. Birkhäuser Verlag, Basel, 2001.

[Bi38] G. Birkhoff, Analytical groups, *Trans. Amer. Math. Soc.* **43** (1938), no. 1, 61–101.

[Boa54] R.P. Boas, Jr., *Entire Functions*, Academic Press Inc., New York, 1954.

[BS71a] J. Bochnak, J. Siciak, Polynomials and multilinear mappings in topological vector spaces, *Studia Math.* **39**(1971), 59–76.

[BS71b] J. Bochnak, J. Siciak, Analytic functions in topological vector spaces, *Studia Math.* **39**(1971), 77–112.

[Bog96] F.A. Bogomolov, On Guan's examples of simply connected non-Kähler compact complex manifolds, *Amer. J. Math.* **118**(1996), no. 5, 1037–1046.

[Bon04] P. Bona, Some considerations on topologies of infinite dimensional unitary coadjoint orbits, *J. Geom. Phys.* **51**(2004), no. 2, 256–268.

[BD71] F.F. Bonsall, J. Duncan, *Numerical Ranges of Operators on Normed Spaces and of Elements of Normed Algebras*, London Mathematical Society Lecture Note Series, 2. Cambridge University Press, London-New York, 1971.

[Bo71a] N. Bourbaki, *Topologie Generale* (Chapitres I à IV, Hermann, Paris, 1971.

[Bo71b] N. Bourbaki, *Groupes et Algébres de Lie* (Chapitre I), Hermann, Paris, 1971.

[Bo72] N. Bourbaki, *Groupes et Algébres de Lie* (Chapitres II et III), Hermann, Paris, 1972.

[BW76] J. Brüning, W. Willgerodt, Eine Verallgemeinerung eines Satzes von N. Kuiper, *Math. Ann.* **220**(1976), 47–58.

[CMMR94] M. Cabrera, A. El Marrakchi, J. Martínez, Á. Rodríguez Palacios, An Allison-Kantor-Koecher-Tits construction for Lie H^*-algebras *J. Algebra* **164**(1994), 361–408.

[CM74] P.R. Chernoff, J.E. Marsden, *Properties of Infinite Dimensional Hamiltonian Systems*, Lecture Notes in Mathematics, Vol. 425, Springer-Verlag, Berlin-New York, 1974.

[CI00] C.-H. Chu, J.-M. Isidro, Manifolds of tripotents in JB*-triples, *Math. Z.* **233**(2000), no. 4, 741–754.

[CGM03] R. Cirelli, M. Gatti, A. Manià, The pure state space of quantum mechanics as Hermitian symmetric space, *J. Geom. Phys.* **45**(2003), no. 3-4, 267–284.

[CL84] R. Cirelli, P. Lanzavecchia, Hamiltonian vector fields in quantum mechanics, *Nuovo Cimento B (11)* **79**(1984), no. 2, 271–283.

[CLM83] R. Cirelli, P. Lanzavecchia, A. Manià, Normal pure states of the von Neumann algebra of bounded operators as Kähler manifold, *J. Phys. A* **16**(1983), no. 16, 3829–3835.

[Co88] F. Cobos, Duality and Lorentz-Marcinkiewicz operator spaces, *Math. Scand.* **63**(1988), 261–267.

[CF68] I. Colojoară, C. Foiaş, *Theory of Generalized Spectral Operators*, Mathematics and its Applications, Vol. 9. Gordon and Breach, Science Publishers, New York-London-Paris, 1968.

[Con76] A. Connes, Classification of injective factors. Cases II_1, II_∞, III_λ, $\lambda \neq 1$, *Ann. of Math. (2)* **104**(1976), no. 1, 73–115.

[CG98] G. Corach, J.E. Galé, Averaging with virtual diagonals and geometry of representations, in: *Banach Algebras '97 (Blaubeuren)*, de Gruyter, Berlin, 1998, pp. 87–100.

[CG99] G. Corach, J.E. Galé, On amenability and geometry of spaces of bounded representations, *J. London Math. Soc. (2)* **59**(1999), no. 1, 311–329.

[CM99] G. Corach, A. Maestripieri, Differential and metrical structure of positive operators, *Positivity* **3**(1999), no. 4, 297–315.

[CM00a] G. Corach, A. Maestripieri, Positive operators on Hilbert space: a geometrical view point, *Bol. Acad. Nac. Cienc. (Córdoba)* **65**(2000), 81–94.

[CM00b] G. Corach, A. Maestripieri, Differential geometry on Thompson's components of positive operators, *Rep. Math. Phys.* **45**(2000), no. 1, 23–37.

[CMS04] G. Corach, A. Maestripieri, D. Stojanoff, Orbits of positive operators from a differentiable viewpoint, *Positivity* **8**(2004), no. 1, 31–48.

[CPR88] G. Corach, H. Porta, L. Recht, Multiplicative integrals and geometry of spaces of projections, *Rev. Un. Mat. Argentina* **34**(1988), 132–149.

[CPR90a] G. Corach, H. Porta, L. Recht, Differential geometry of systems of projections in Banach algebras, *Pacific J. Math.* **143**(1990), no. 2, 209–228

[CPR90b] G. Corach, H. Porta, L. Recht, Differential geometry of spaces of relatively regular operators, *Integral Equations Operator Theory* **13**(1990), no. 6, 771–794.

[CPR93a] G. Corach, H. Porta, L. Recht, The geometry of the space of selfadjoint invertible elements in a C^*-algebra, *Integral Equations Operator Theory* **16**(1993), no. 3, 333–359.

[CPR93b] G. Corach, H. Porta, L. Recht, The geometry of spaces of projections in C^*-algebras, *Adv. Math.* **101**(1993), no. 1, 59–77.

[CPR94] G. Corach, H. Porta, L. Recht, Convexity of the geodesic distance on spaces of positive operators, *Illinois J. Math.* **38**(1994), no. 1, 87–94.

[CD78] M.J. Cowen, R.G. Douglas, Complex geometry and operator theory, *Acta Math.* **141**(1978), no. 3-4, 187–261.

[CR85] J.A. Cuenca Mira, Á. Rodríguez Palacios, Isomorphisms of H^*-algebras, *Math. Proc. Cambridge Philos. Soc.* **97**(1985), 93–99.

[CGM90] J.A. Cuenca Mira, A. García Martín, C. Martínez Gonzalez, Structure theory for L^*-algebras, *Math. Proc. Camb. Philos. Soc.* **107**(1990), 361–365.

[CR87] J.A. Cuenca Mira, Á. Rodríguez Palacios, Structure theory for noncommutative Jordan H^*-algebras, *J. Algebra* **106**(1987), 1–14.

[CS94] J.A. Cuenca Mira, A. Sánchez Sánchez, Structure theory for real noncommutative Jordan H^*-algebras, *J. Algebra* **164**(1994), 481–499.

[DF79] D. Deckard, L.A. Fialkow, Characterization of Hilbert space operators with unitary cross sections, *J. Operator Theory* **2** (1979), no. 2, 153–158.

[Din99] S. Dineen, *Complex Analysis on Infinite-dimensional Spaces*, Springer Monographs in Mathematics. Springer-Verlag London, Ltd., London, 1999.

[DM98] S. Dineen, P. Mellon, Holomorphic functions on symmetric Banach manifolds of compact type are constant, *Math. Z.* **229**(1998), no. 4, 753–765.

[Di74] J. Dixmier, *Algèbres Enveloppantes*, Cahiers scientifiques, fasc. 37, Gauthier-Villars Éditeur, Paris-Bruxelles-Montreal, 1974.

[Dj75] D.Z. Djocović, An elementary proof of the Baker-Campbell-Hausdorff-Dynkin formula, *Math. Z.* **143**(1975), 209–211.

[DG92] J. Dorfmeister, Z.-D. Guan, Classification of compact homogeneous pseudo-Kähler manifolds, *Comment. Math. Helvetici* **67**(1992), 499–513.

[DN88] J. Dorfmeister, K. Nakajima, The fundamental conjecture for homogeneous Kähler manifolds, *Acta Math.* **161**(1988), 23–70.

[Do65] A. Douady, Un espace de Banach dont le groupe linéaire n'est pas connexe, *Nederl. Akad. Wetensch. Proc. Ser. A 68 = Indag. Math.* **27**(1965), 787–789.

[DL66] A. Douady, M. Lazard, Espaces fibrés en algèbres de Lie et en groupes, *Invent. math.* **1**(1966), 133–151.

[DEG98] M.J. Dupré, J.-C. Evard, J.F. Glazebrook, Smooth parametrization of subspaces in a Banach space, *Rev. Un. Mat. Argentina* **41**(1998), no. 2, 1–13.

[DGi83] M.J. Dupré, R.M. Gillette, Banach bundles, Banach modules and automorphisms of C^*-algebras, Research Notes in Mathematics, 92. Pitman (Advanced Publishing Program), Boston, MA, 1983.

[DG00] M.J. Dupré, J.F. Glazebrook, Infinite dimensional manifold structures on principal bundles, *J. Lie Theory* **10**(2000), 359–373.

[DG01] M.J. Dupré, J.F. Glazebrook, The Stiefel bundle of a Banach algebra, *Integral Equations Operator Theory* **41**(2001), no. 3, 264–287.

[DMR00] C.E. Durán, L.E. Mata-Lorenzo, L. Recht, Natural variational problems in the Grassmann manifold of a C^*-algebra with trace, *Adv. Math.* **154**(2000), no. 1, 196–228.

[DMR04] C.E. Durán, L.E. Mata-Lorenzo, L. Recht, Metric geometry in homogeneous spaces of the unitary group of a C^*-algebra: Part I—minimal curves, *Adv. Math.* **184**(2004), no. 2, 342–366.

[DFWW01] K. Dykema, T. Figiel, G. Weiss, M. Wodzicki, Commutator structure of operator ideals, *Adv. Math.* **185**(2004), no. 1, 1–79.

[Es87] J. Eschmeier, *Analytische Dualität und Tensorprodukte in der mehrdimensionalen Spektraltheorie*, Schriftenreihe des Mathematischen Institut der Universität Münster, 2. Serie, Heft 42, 1987.

[EK64] W.T. van Est, Th.J. Korthagen, Non-enlargible Lie algebras, *Proc. Kon. Ned. Akad. v. Wet. A67= Indag. Math.* **26**(1964), 15–31.

[ES73] W.T. van Est, S. Świerczkowski, The path functor and faithful representability of Banach Lie algebras, *J. Austral. Math. Soc.* **16**(1973), 54–69.

[FW71] P.A. Fillmore, J.P. Williams, On operator ranges, *Adv. Math.* **7**(1971), 254–281.

[Fr85] D.S. Freed, Flag manifolds and infinite-dimensional Kähler geometry, in: *Infinite-dimensional groups with applications (Berkeley, Calif., 1984)*, Math. Sci. Res. Inst. Publ., 4, Springer, New York, 1985, pp. 83–124.

[GL74] Ch. Gapaillard, P.T. Lai, Remarques sur les propriétés de dualité et d'interpolation des idéaux de R. Schatten, *Studia Math.* **49**(1974), 129–138.

[GG04] A. Gheondea, S. Gudder, Sequential product of quantum effects, *Proc. Amer. Math. Soc.* **132**(2004), no. 2, 503–512.

[Gl02a] H. Glöckner, Infinite-dimensional Lie groups without completeness restrictions, in: A. Strasburger, J. Hilgert, K.-H. Neeb, W. Wojyyński (eds.), *Geometry and Analysis on Finite and Infinite-dimensional Lie Groups (Będlewo, 2000)*, Banach Center Publ. vol. 55, Polish Acad. Sci., Warsaw, 2002, pp. 43–59.

[Gl02b] H. Glöckner, Algebras whose groups of units are Lie groups, *Studia Math.* **153**(2002), 147–177.

[GN03] H. Glöckner, K.-H. Neeb, Banach-Lie quotients, enlargibility, and universal complexifications, *J. Reine Angew. Math.* **560**(2003), 1–28.

[GK69] I.C. Gohberg, M.G. Kreĭn, *Introduction to the Theory of Linear Nonselfadjoint Operators*, Translations of Mathematical Monographs, vol. 18, American Mathematical Society, Providence, R.I., 1969.

[Go56] K. Goldberg, The formal power series for $\log e^x e^y$, *Duke Math. J.* **23**(1956), 13–21.

[Gr84] B. Gramsch, Relative Inversion in der Störungstheorie von Operatoren und Ψ-Algebren, *Math. Ann.* **269**(1984), no. 1, 27–71.

[Gu94] D. Guan, Examples of compact holomorphic symplectic manifolds which admit no Kähler structure, in: *Geometry and Analysis on Complex Manifolds*, World Sci. Publishing, River Edge, NJ, 1994, pp. 63–74.

[Gu95a] D. Guan, Examples of compact holomorphic symplectic manifolds which are not Kählerian II, *Invent. Math.* **121**(1995), no. 1, 135–145.

[Gu95b] D. Guan, Examples of compact holomorphic symplectic manifolds which are not Kählerian III, *Internat. J. Math.* **6**(1995), no. 5, 709–718.

[Gu97a] D. Guan, Classification of compact homogeneous spaces with invariant symplectic structures, *Electron. Res. Announc. Amer. Math. Soc.* **3**(1997), 52–54.

[Gu97b] D. Guan, Classification of compact complex homogeneous spaces with invariant volumes, *Electron. Res. Announc. Amer. Math. Soc.* **3**(1997), 90–92

[Gu02] D. Guan, Classification of compact complex homogeneous spaces with invariant volumes, *Trans. Amer. Math. Soc.* **354**(2002), no. 11, 4493–4504.

[Gu04] D. Guan, Toward a classification of compact complex homogeneous spaces, *J. Algebra* **273**(2004), no. 1, 33–59.

[Ha82] P.R. Halmos, *A Hilbert Space Problem Book*, Graduate texts in mathematics, vol. 19, Springer-Verlag, New York-Heidelberg-Berlin, 1982.

[Ht82] R.S. Hamilton, The inverse function theorem of Nash and Moser, Bull. Amer. Math. Soc. **7**(1982), 65–222.

[dlH72] P. de la Harpe, *Classical Banach-Lie Algebras and Banach-Lie Groups of Operators in Hilbert Space*, Lecture Notes in Mathematics, Vol. 285. Springer-Verlag, Berlin-New York, 1972.

[HK77] L.A. Harris, W. Kaup, Linear algebraic groups in infinite dimensions, *Illinois J. Math.* **21**(1977), 666–674.

[Hel62] S. Helgason, *Differential Geometry and Symmetric Spaces*, Academic Press, New York-London, 1962.

[HH94a] G.F. Helminck, A.G. Helminck, Holomorphic line bundles over Hilbert flag varieties, in: *Algebraic groups and their generalizations: quantum and infinite-dimensional methods (University Park, PA, 1991)*, Proc. Sympos. Pure Math., 56, Part 2, Amer. Math. Soc., Providence, RI, 1994, pp. 349–375.

[HH94b] G.F. Helminck, A.G. Helminck, The structure of Hilbert flag varieties, *Publ. Res. Inst. Math. Sci.* **30**(1994), no. 3, 401–441.

[HH95] G.F. Helminck, A.G. Helminck, Infinite-dimensional flag manifolds in integrable systems. Geometric and algebraic structures in differential equations, *Acta Appl. Math.* **41**(1995), no. 1-3, 99–121.

[HH02] G.F. Helminck, A.G. Helminck, Hilbert flag varieties and their Kähler structure, in: Foundations of quantum theory (Krákow/ Bregenz, 2001), *J. Phys. A* **35**(2002), no. 40, 8531–8550.

[Her82] D.A. Herrero, *Approximation of Hilbert Space Operators.* Vol. I, Research Notes in Mathematics, 72. Pitman (Advanced Publishing Program), Boston, MA, 1982.

[He89] M. Hervé, *Analyticity in Infinite-dimensional Spaces*, de Gruyter Studies in Mathematics, 10. Walter de Gruyter & Co., Berlin, 1989.

[Hl90] M. Hladnik, Spectrality of elementary operators, *J. Austral. Math. Soc. Ser. A* **49**(1990), 327–346.

[Ho65] G. Hochschild, *The Structure of Lie Groups*, Holden-Day, San Francisco, 1965.

[Hof68] K.H. Hofmann, *Introduction to the Theory of Compact Groups. Part I*, Tulane University, 1968.

[Ho90] L. Hörmander, *The Analysis of Linear Partial Differential Operators I* (second edition), Springer Study Edition, Springer-Verlag, Berlin, 1990.

[IM02] J.M. Isidro, M. Mackey, The manifold of finite rank projections in the algebra $\mathfrak{L}(H)$ of bounded linear operators, *Expo. Math.* **20**(2002), no. 2, 97–116.

[Ja62] N. Jacobson, *Lie Algebras*, Wiley-Interscience, New York, 1962.

[Kac90] V.G. Kac, *Infinite-dimensional Lie Algebras* (third edition), Cambridge University Press, Cambridge, 1990.

[Ka75] W. Kaup, Über die Automorphismen Grassmannscher Mannigfaltigkeiten unendlicher Dimension, *Math. Z.* **144**(1975), no. 2, 75–96.

[Ka77] W. Kaup, Algebraic characterization of symmetric complex Banach manifolds, *Math. Ann.* **228**(1977), no. 1, 39–64.

[Ka81] W. Kaup, Über die Klassifikation der symmetrischen hermiteschen Mannigfaltigkeiten unendlicher Dimension I, *Math. Ann.* **257**(1981), no. 4, 463–486.

[Ka83a] W. Kaup, Über die Klassifikation der symmetrischen hermiteschen Mannigfaltigkeiten unendlicher Dimension II, *Math. Ann.* **262**(1983), no. 1, 57–75.

[Ka83b] W. Kaup, A Riemann mapping theorem for bounded symmetric domains in complex Banach spaces, *Math. Z.* **183**(1983), no. 4, 503–529.

[Ka84] W. Kaup, Contractive projections on Jordan C^*-algebras and generalizations, *Math. Scand.* **54**(1984), no. 1, 95–100.

[Ka01] W. Kaup, On Grassmannians associated with JB*-triples, *Math. Z.* **236**(2001), no. 3, 567–584.

[KV90] W. Kaup, J.-P. Vigué, Symmetry and local conjugacy on complex manifolds, *Math. Ann.* **286**(1990), no. 1-3, 329–340.

[Ke74] H.H. Keller, *Differential Calculus in Locally Convex Spaces*, Lecture notes in mathematics, vol. 417, Springer-Verlag, Berlin-Heidelberg-New York, 1974.

[KK02] K.-T. Kim, S.G. Krantz, Characterization of the Hilbert ball by its automorphism group, *Trans. Amer. Math. Soc.* **354** (2002), no. 7, 2797–2818.

[KK04] K.-T. Kim, S.G. Krantz, Normal families of holomorphic functions and mappings on a Banach space, *Expo. Math.* **21** (2003), no. 3, 193–218.

[KMa03] K.-T. Kim, D. Ma, Characterization of the Hilbert ball by its automorphisms, *J. Korean Math. Soc.* **40** (2003), no. 3, 503–516.

[Ki76] A.A. Kirillov, *Elements of the Theory of Representations*, Grundlehren der Mathematischen Wissenschaften, Band 220. Springer-Verlag, Berlin-New York, 1976.

[Ki04] A.A. Kirillov, *Lectures on the Orbit Method*, Graduate Studies in Mathematics, 64. American Mathematical Society, Providence, RI, 2004.

[Kn96] A.W. Knapp, *Lie Groups Beyond an Introduction*, Progress in Mathematics, vol. 140. Birkhäuser Boston, Inc., Boston, MA, 1996.

[Ko70] B. Kostant, Quantization and unitary representations. I. Prequantization, in: *Lectures in Modern Analysis and Applications* III, Lecture Notes in Math., Vol. 170, Springer, Berlin, 1970, pp. 87–208.

[Kr92] S.G. Krantz, *Function Theory of Several Complex Variables*. Second edition, The Wadsworth & Brooks/Cole Mathematics Series.

Wadsworth & Brooks/Cole Advanced Books & Software, Pacific Grove, CA, 1992.

[KM97] A. Kriegl, P.W. Michor, *The Convenient Setting of Global Analysis*, Mathematical Surveys and Monographs, 53, Amer. Math. Soc., Providence, 1997.

[Ku65] N.H. Kuiper, The homotopy type of the unitary group of Hilbert spaces, *Topology* **3**(1965), 19–30.

[Kuo75] H. Kuo, *Gaussian Measures in Banach Spaces*, Lecture Notes in Mathematics, Vol. 463. Springer-Verlag, Berlin, 1975.

[La01] S. Lang, *Fundamentals of Differential Geometry* (corrected second printing), Graduate Texts in Mathematics, 191. Springer-Verlag, New-York, 2001.

[LR95] A. Larotonda, L. Recht, The orbit of a conditional expectation as a reductive homogeneous space, in: *Volume in homage to Dr. Rodolfo A. Ricabarra* (Spanish), Vol. Homenaje, 1, Univ. Nac. del Sur, Bahía Blanca, 1995, pp. 61–73.

[Le98] L. Lempert, The Dolbeault complex in infinite dimensions I, *J. Amer. Math. Soc.* **11**(1998), no. 3, 485–520.

[Le99] L. Lempert, The Dolbeault complex in infinite dimensions II, *J. Amer. Math. Soc.* **12**(1999), no. 3, 775–793.

[Le00] L. Lempert, The Dolbeault complex in infinite dimensions III. Sheaf cohomology in Banach spaces, *Invent. Math.* **142**(2000), no. 3, 579–603.

[Lie1890] S. Lie, *Theorie der Transformationsgruppen. Zweiter Abschnitt. Unter Mitwirkung von Friedrich Engel*, Leipzig. B. G. Teubner, Leipzig, 1890. Reprinted by Chelsea Publishing Company, New York, 1970.

[Lo89a] K. Lorentz, On the structure of the similarity orbits of Jordan operators as analytic homogeneous manifolds, *Integral Equations Operator Theory* **12**(1989), no. 3, 435–443.

[Lo89b] K. Lorentz, On the local structure of the similarity orbits of Jordan elements in operator algebras, *Ann. Univ. Sarav. Ser. Math.* **2**(1989), no. 3, 159–189.

[Lo91] K. Lorentz, On the rational homogeneous manifold structure of the similarity orbits of Jordan elements in operator algebras, in: *Topics in Matrix and Operator Theory (Rotterdam, 1989)*, Oper. Theory Adv. Appl., 50, Birkhäuser, Basel, 1991, pp. 293–306.

[Lo95] K. Lorentz, Characterization of Jordan elements in Ψ^*-algebras, *J. Operator Theory* **33**(1995), no. 1, 117–158.

[Mac00] M. Mackey, The Grassmannian manifold associated to a bounded symmetric domain, in: *Finite or Infinite Dimensional Complex Analysis (Fukuoka, 1999)*, Lecture Notes in Pure and Appl. Math., 214, Dekker, New York, 2000, pp. 317–323.

[MM01] M. Mackey, P. Mellon, Compact-like manifolds associated to JB*-triples, *manuscripta math.* **106**(2001), no. 2, 203–212.

[Ma62] B. Maissen, Lie-Gruppen mit Banachräumen als Parameterräume, *Acta Math.* **108**(1962), 229–270.

[Md72] J.E. Marsden, Darboux's theorem fails for weak symplectic forms, *Proc. Amer. Math. Soc.* **32**(1972), 590–592.

[Mr85] M. Martin, Hermitian geometry and involutive algebras, *Math. Z.* **188**(1985), no. 3, 359–382.

[Mr87] M. Martin, An operator theoretic approach to analytic functions into a Grassmann manifold, *Math. Balkanica (N.S.)* **1**(1987), no. 1, 45–58.

[Mr90] M. Martin, Projective representations of compact groups in C^*-algebras, in: *Linear Operators in Function Spaces (Timişoara, 1988)*, Oper. Theory Adv. Appl., 43, Birkhäuser, Basel, 1990, pp. 237–253.

[MS95] M. Martin, N. Salinas, Differential geometry of generalized Grassmann manifolds in C^*-algebras, in: *Operator Theory and Boundary Eigenvalue Problems (Vienna, 1993)*, Oper. Theory Adv. Appl., 80, Birkhäuser, Basel, 1995, pp. 206–243.

[MS97] M. Martin, N. Salinas, Flag manifolds and the Cowen-Douglas theory, *J. Operator Theory* **38**(1997), 329-365.

[MS98] M. Martin, N. Salinas, The canonical complex structure of flag manifolds in a C^*-algebra. In: Bercovici, Hari (ed.) et al., *Nonselfadjoint Operator Algebras, Operator Theory, and Related Topics. The Carl M. Pearcy Anniversary Volume on the Occasion of His 60th Birthday*, Birkhäuser Verlag, Oper. Theory Adv. Appl., 104, Basel, 1998, pp. 173-187.

[MR92] L.E. Mata-Lorenzo, L. Recht, Infinite-dimensional homogeneous reductive spaces, *Acta Cient. Venezolana* **43**(1992), no. 2, 76–90.

[Ma78] K. Mattila, Normal operators and proper boundary points of the spectra of operators on a Banach space, *Ann. Acad. Sci. Fenn. Ser. A I Math. Dissertationes* **19**(1978), 48 pp.

[Me93a] P. Mellon, Holomorphic curvature of infinite-dimensional symmetric complex Banach manifolds of compact type, *Ann. Acad. Sci. Fenn. Ser. A I Math.* **18**(1993), no. 2, 299–306.

[Me93b] P. Mellon, Dual manifolds of JB*-triples of the form $C(X, U)$, *Proc. Roy. Irish Acad. Sect. A* **93**(1993), no. 1, 27–42.

[Me96] P. Mellon, Symmetric manifolds of compact type associated to the JB*-triples $C_0(X, Z)$, *Math. Scand.* **78**(1996), no. 1, 19–36.

[Mi84] J. Milnor, Remarks on infinite-dimensional Lie groups, in: B.DeWitt, R. Stora (eds.), *Relativity, groups and topology, II* (Les Houches, 1983), North-Holland, Amsterdam, 1984, pp. 1007-1057.

[MP95] R.V. Moody, A. Pianzola, *Lie Algebras with Triangular Decompositions*, Canadian Mathematical Society Series of Monographs and Advanced Texts. A Wiley-Interscience Publication. John Wiley & Sons, Inc., New York, 1995.

[Mu86] J. Mujica, *Complex Analysis in Banach Spaces*, North-Holland Mathematics Studies, 120. Notas de Matemática, 107. North-Holland Publishing Co., Amsterdam, 1986.

[NRW01] L. Natarajan, E. Rodríguez-Carrington, J.A. Wolf, The Bott-Borel-Weil theorem for direct limit groups, *Trans. Amer. Math. Soc.* **353** (2001), no. 11, 4583–4622.

[Ne00] K.-H. Neeb, *Holomorphy and Convexity in Lie Theory*, de Gruyter Expositions in Mathematics 28, Walter de Gruyter & Co., Berlin, 2000.

[Ne01a] K.-H. Neeb, Infinite-dimensional groups and their representations, in: A. Huckleberry, T. Wurzbacher (eds.), *Infinite-dimensional Kähler Manifolds (Oberwolfach, 1995)*, DMV-Seminar, vol. 31, Birkhäuser-Verlag, Basel, 2001, pp. 131-178.

[Ne01b] K.-H. Neeb, Borel-Weil theory for loop groups, in: A. Huckleberry, T. Wurzbacher (eds.), *Infinite-dimensional Kähler Manifolds (Oberwolfach, 1995)*, DMV-Seminar, vol. 31, Birkhäuser-Verlag, Basel, 2001, pp. 179–229.

[Ne02a] K.-H. Neeb, Highest weight representations and infinite-dimensional Kähler manifolds, in: *Recent Advances in Lie Theory (Vigo, 2000)*, Res. Exp. Math. 25, Heldermann, Lemgo, 2002, pp. 367–392.

[Ne02b] K.-H. Neeb, Classical Hilbert-Lie groups, their extensions and their homotopy groups, in: A. Strasburger, J. Hilgert, K.-H. Neeb, W. Wojyyński (eds.), *Geometry and Analysis on Finite and Infinite-dimensional Lie Groups (Będlewo, 2000)*, Banach Center Publ., vol. 55, Polish Acad. Sci. Warsaw, 2002, pp. 87–151.

[Ne02c] K.-H. Neeb, A Cartan-Hadamard theorem for Banach-Finsler manifolds, *Geom. Dedicata* **95**(2002), 115–156.

[Ne02d] K.-H. Neeb, Central extensions of infinite dimensional Lie groups, *Annales de l'Inst. Fourier* **52**(2002), 1365–1342.

[Ne04] K.-H. Neeb, Infinite-dimensional groups and their representations, in: *Lie Theory*, Progr. Math. 228, Birkhäuser, Boston, MA, 2004, pp. 213–328.

[NO98] K.-H. Neeb, B. Ørsted, Unitary highest weight representations in Hilbert spaces of holomorphic functions on infinite-dimensional domains, *J. Funct. Anal.* **156**(1998), no. 1, 263–300.

[Nel69] E. Nelson, *Topics in dynamics, I: Flows*, Princeton University Press and the University of Tokyo Press, Princeton, 1969.

[Ni97] L.I Nicolaescu, Generalized symplectic geometries and the index of families of elliptic problems, *Memoirs Amer. Math. Soc.* **128**(1997), no. 609.

[NT82] M. Newman, R.C. Thompson, Numerical values of Goldberg's coefficients in the series for $\log(e^x e^y)$, *Math. Comp.* **48**(1987), 265–271.

[OR03] A. Odzijewicz, T.S. Ratiu, Banach Lie-Poisson spaces and reduction, *Comm. Math. Phys.* **243**(2003), no. 1, 1–54.

[Om74] H. Omori, *Infinite Dimensional Lie Transformation Groups*, Lecture Notes in Mathematics, Vol. 427, Springer-Verlag, Berlin-New York, 1974.

[Om97] H. Omori, *Infinite-dimensional Lie Groups* (translated from the 1979 Japanese original and revised by the author), Translations of Mathematical Monographs, 158, American Mathematical Society, Providence, RI, 1997.

[OR04] J.-P. Ortega, T.S. Ratiu, *Momentum Maps and Hamiltonian Reduction*, Progress in Mathematics, 222. Birkhäuser Boston, Inc., Boston, MA, 2004.

[Or80] B. Ørsted, A model for an interacting quantum field, *J. Funct. Anal.* **36**(1980), no. 1, 53–71.

[Pa65] R.S. Palais, On the homotopy type of certain groups of operators, *Topology* **3**(1965), 271–279.

[Pa66] R.S. Palais, Homotopy theory of infinite-dimensional manifolds, *Topology* **5**(1966), 1–16.

[Pt00] I. Patyi, On the $\overline{\partial}$-equation in a Banach space, *Bull. Soc. Math. France* **128**(2000), no. 3, 391–406.

[Pc84] M. Pecuch Herrero, Global cross sections of unitary and similarity orbits of Hilbert space operators, *J. Operator Theory* **12**(1984), no. 2, 265–283.

[Pe88] V.G. Pestov, Fermeture nonstandard des algèbres et groupes de Lie banachiques, *C. R. Acad. Sci. Paris Sèr. I Math.* **306**(1988), 643–645.

[Pe92] V.G. Pestov, Nonstandard hulls of Banch-Lie groups and algebras, *Nova J. Algebra Geom.* **1**(1992), 371–381.

[Pit78] A. Pietsch, *Operator Ideals*, VEB Deutscher Verlag der Wissenschaften, Berlin, 1978.

[PR86] H. Porta, L. Recht, Continuous selections of complemented subspaces, in: *Geometry of Normed Linear Spaces (Urbana-Champaign, Ill., 1983)*, Contemp. Math., 52, Amer. Math. Soc., Providence, RI, 1986, pp. 121–125.

[PR87a] H. Porta, L. Recht, Minimality of geodesics in Grassmann manifolds, *Proc. Amer. Math. Soc.* **100**(1987), no. 3, 464–466.

[PR87b] H. Porta, L. Recht, Spaces of projections in a Banach algebra, *Acta Cient. Venezolana* **38**(1987), no. 4, 408–426.

[PR94] H. Porta, L. Recht, Conditional expectations and operator decompositions, *Ann. Global Anal. Geom.* **12**(1994), no. 4, 335–339.

[PR95] H. Porta, L. Recht, Variational and convexity properties of families of involutions, *Integral Equations Operator Theory* **21**(1995), no. 2, 243–253.

[PR96a] H. Porta, L. Recht, Exponential sets and their geometric motions, *J. Geom. Anal.* **6**(1996), no. 2, 277–285.

[PR96b] H. Porta, L. Recht, Geometric embeddings of operator spaces, *Illinois J. Math.* **40**(1996), no. 2, 151–161.

[PS90] A. Pressley, G. Segal, *Loop Groups* (repr. with corrections), Oxford Mathematical Monographs, Clarendon Press, Oxford, 1990.

[Rae77] I. Raeburn, The relationship between a commutative Banach algebra and its maximal ideal space, *J. Functional Analysis* **25** (1977), no. 4, 366–390.

[Re99] L. Recht, Differential geometry in the space of positive operators (Spanish), *Bol. Asoc. Mat. Venez.* **6**(1999), no. 2, 125–139.

[Rt93] C. Reutenauer, *Free Lie Algebras*, London Mathematical Society Monographs. New Series, 7. Oxford Science Publications. The Clarendon Press, Oxford University Press, New York, 1993.

[Sa88] N. Salinas, The Grassmann manifold of a C^*-algebra, and Hermitian holomorphic bundles, in: *Special Classes of Linear Operators and Other Topics (Bucharest, 1986)*, Oper. Theory Adv. Appl., 28, Birkhäuser, Basel, 1988, pp. 267–289.

[Sf66] H.H. Schaefer, *Topological Vector Spaces*, The Macmillan Company, New York-London, 1966.

[Sr84] H. Schröder, On the homotopy type of the regular group of a W^*-algebra, *Math. Ann.* **267**(1984), no. 2, 271–277.

[Sr87] H. Schröder, On the topology of classical groups and homogeneous spaces associated with a W^*-algebra factor, *Integral Equations Operator Theory* **10**(1987), no. 6, 812–818.

[Sch60] J.R. Schue, Hilbert space methods in the theory of Lie algebras, *Trans. Amer. Math. Soc.* **95**(1960), 68–80.

[Sch61] J.R. Schue, Cartan decompositions for L^* algebras, *Trans. Amer. Math. Soc.* **98**(1961), 334–349.

[Scw64] L. Schwartz, Sous espaces Hilbertiens d'espaces vectoriels topologiques et noyaux associés (noyeux reproduisants), *J. Analyse Math.* **13**(1964), 115–256.

[SW85] G. Segal, G. Wilson, Loop groups and equations of KdV type, *Inst. Hautes Études Sci. Publ. Math.* **61**(1985), 5–65.

[SP54] M.E. Shanks, L.E. Pursell, The Lie algebra of a smooth manifold, *Proc. Amer. Math. Soc.* **5**(1954), 468–472.

[So66] J.-M. Souriau, Quantification géométrique, *Commun. Math. Phys.* **1**(1966), 374–398.

[So67] J.-M. Souriau, Quantification géométrique. Applications, *Ann. Inst. H. Poincaré* **6**(1967), 311–341.

[Swa77] R.G. Swan, Topological examples of projective modules, *Trans. Amer. Math. Soc.* **230**(1977), 201–234.

[Sw65] S. Świerczkowski, Embedding theorems for local analytic groups, *Acta Math.* **114**(1965), 207–235.

[Ta03] M. Takesaki, *Theory of Operator Algebras III*, Encyclopaedia of Mathematical Sciences 127, Operator Algebras and Noncommutative Geometry 8, Springer-Verlag, Berlin, 2003.

[Th82] R.C. Thompson, Cyclic relations and the Goldberg coefficients in the Campbell-Baker-Hausdorff formula, *Proc. Amer. Math. Soc.* **86**(1982), 12–14.

[Th89] R.C. Thompson, Convergence proof for Goldberg's exponential series, *Linear Algebra Appl.* **121**(1989), 3–7.

[Tr67] F. Trèves, *Topological Vector Spaces, Distributions and Kernels*, Academic Press, New York-London, 1967.

[Tu05] B. Tumpach, *Variétés Kählériennes et Hyperkählériennes de Dimension Infinie*, Thèse de doctorat, École Polytechnique—CMLS, 2005.

[Up84] H. Upmeier, A holomorphic characterization of C^*-algebras, in: *Functional Analysis, Holomorphy and Approximation Theory, II (Rio de Janeiro, 1981)*, North-Holland Math. Stud., 86, North-Holland, Amsterdam, 1984, 427–467.

[Up85] H. Upmeier, *Symmetric Banach Manifolds and Jordan C^*-algebras*, North-Holland Mathematics Studies, 104. Notas de Matemàtica, 96. North-Holland Publishing Co., Amsterdam, 1985.

[Up86] H. Upmeier, Some applications of infinite-dimensional holomorphy to mathematical physics, in: *Aspects of Mathematics and Its Applications*, North-Holland Math. Library, 34, North-Holland, Amsterdam, 1986, pp. 817–832.

[Va72] F.-H. Vasilescu, On Lie's theorem in operator algebras, *Trans. Amer. Math. Soc.* **172**(1972), 365–372.

[Va82] F.-H. Vasilescu, *Analytic Functional Calculus and Spectral Decompositions*, Mathematics and its Applications (East European Series), vol. 1. D. Reidel Publishing Co.; Editura Academiei, Dordrecht-Bucharest, 1982.

[Wa71] F.W. Warner, *Foundations of Differentiable Manifolds and Lie Groups*, Scott, Foresman and Company, Glenview-London, 1971.

[Wei69] A. Weinstein, Symplectic structures on Banach manifolds, *Bull. Amer. Math. Soc.* **75**(1969), 1040–1041.

[We80] R.O. Wells, Jr., *Differential Analysis on Complex Manifolds* (second edition), Graduate Texts in Mathematics, 65. Springer-Verlag, New York-Berlin, 1980.

[Wi90] D.R. Wilkins, The Grassmann manifold of a C^*-algebra, *Proc. Roy. Irish Acad. Sect. A* **90**(1990), no. 1, 99–116.

[Wi93] D.R. Wilkins, Homogeneous vector bundles and Cowen-Douglas operators, *Internat. J. Math.* **4**(1993), no. 3, 503–520.

[Wi94] D.R. Wilkins, Infinite dimensional homogeneous manifolds, *Proc. Roy. Irish Acad. Sect. A* **94**(1994), no. 1, 105–118.

[Wo77] W. Wojtyński, Banach-Lie algebras of compact operators, *Studia Math.* **59**(1977), 263–273.

[Wo04] J.A. Wolf, Principal series representations of direct limit groups (see preprint math. RT/0402283).

[Wu01] T. Wurzbacher, Fermionic second quantization and the geometry of the restricted Grassmannian, in: *Infinite Dimensional Kähler Manifolds (Oberwolfach, 1995)*, DMV Sem., 31, Birkhäuser, Basel, 2001, pp. 287-375.

Index